The

Grandeur

of

Evolution

by

Joseph C. Boone, PhD

January 2019

Contents

Chapter 9: Strange Creatures in Distant Lands

Chapter 10: Genetic Ancestry

Preface

The word evolution in its most general sense simply means change. Mountains slowly erode away while others are created. Entire continents slide over the Earth, sometimes colliding, but breaking apart at other times. We observe stars dying violently while others are being formed from giant clouds of gas and dust. Animals become extinct, but the fossil record documents the appearance of many new forms sporadically distributed in time. If we know where to look, evidence of evolution is all around us. We will investigate those changes associated with the Earth, the Solar System, the distant stars, the Universe, and the life forms on planet Earth. Evolution is the amazing and incredibly fascinating story of these changes.

Our primary goal will be to investigate the evolution of living things. Why should we study evolution? Compared to other great breakthroughs in science, the idea that life forms of today evolved from life forms of the past is perhaps the most exciting and illuminating concept ever formulated. An understanding of the evolution of life is an extremely valuable tool if one is to comprehend many of the mysteries of the living world. It is only as one studies evolution that a comprehensive and unified picture of life on Earth emerges.

As we shall see, the answers to many puzzling questions can suddenly be understood in the context of evolution. In addition, evolution allows

us to make many predictions about what to expect in future observations and experiments. Perhaps there are those who *know in their heart* that evolution is a great falsehood, but even they would be more knowledgeable about the living world if their answers were based on an understanding of evolution.

No one would attempt to teach chemistry without introducing the periodic table of the elements because an understanding of the periodic table allows one to predict the behavior of elements in a chemical reaction. Likewise, trying to make sense of the living world without an understanding of its evolutionary history is utterly hopeless. Even those who refuse to believe that evolution happened must be impressed by the wealth of knowledge that one gains by studying this process. The famous geneticist Theodosius Dobzhansky once remarked, "Nothing in biology makes sense except in the light of evolution." I am sure there must be people who learn pieces of information without understanding the interconnections that exist between them. These people must just memorize chemical reactions without ever seeing the underlying principles provided by the periodic table, and they must tediously learn biological information without the unifying idea that evolution furnishes.

Science is a self-correcting area of study. Discoveries and ideas that are formulated by one scientist are scrutinized by other scientists. In their search for the best description of nature, scientists have a strong incentive for discovering errors in other's work. Einstein became the most well known scientist in history because he was able to show that Newton's *laws* were incorrect. If I could discover a fatal flaw in evolution, I would become an instant celebrity in the scientific community and would probably win a Nobel Prize in Biology. All right, there is no Nobel Prize in Biology, but the instant celebrity part is true.

A strong motivation for writing this book was a desire to counteract some of the misinformation that is spread by organized groups expounding what they like to call *scientific* creationism. Much of their information is in the form of half truths, obsolete ideas, and quotes taken out of context. These quotes are carefully selected to create the impression that there is a great deal of doubt about the validity of evolution.

Because of the dishonest methods often used, it takes a great deal of patience to sit quietly and listen to a strict creationist trying to discredit science. Although evolution is their primary target, all of science must suffer under their attack. The evidence in support of evolution spans the scientific disciplines of biology, geology (especially paleontology, that branch of geology dealing with the study of fossils), chemistry, and physics. If evolution is to be discredited, these other areas of science must be brought into question as well.

In their very nonscientific writings, creationist's generally do not spend much time detailing even the general aspects of their own beliefs. Most of their publications consist of attacks on evolution. Their assumption seems to be that if they can cast doubt on some of the details of evolution then creationism will somehow be proven in the process. However, that is not the way science operates. New ideas must stand on their own merits, but creationists seldom clearly state their positions on various topics such as the age of the Earth or the definition of *kind* as it is used in the Bible. It would be very difficult to prepare a creationism versus evolution book because not much is written about the details of creationism. I know of no book that clearly expounds the creationism position, indeed it seems that there is a wide range of creationist positions.

Even if a creationist were able to find some piece of evidence that invalidates one important aspect of evolution, this evidence would not prove creationism. It is impossible to prove one point of view by invalidating some other point of view. To convince someone of your position, you must clearly show that your position is in better agreement with the observational data. No matter how much negative information one musters against evolution, it will not add one bit of support for creationism. The goal of this book is to present evolution clearly and let the evidence speak for itself, which after all is how science operates. It would be refreshing if creationists used the same honest tactic.

Before I proceed too much further, perhaps we should define an important term. The word *creationist* is an unfortunate choice since when many speak of creationists they do not mean any individual who believes that the Universe was created by God. Science does not dispute the existence of God, and it does not dispute His power or the belief that God created the Universe. If God guides the lives of individual people, He surely has guided the course of evolution. However, the creationists referred to in most circles are certain fundamentalist Christians who have somehow concluded that

Genesis states explicitly that the Universe was created several thousand years ago and that the Sun, Earth, Moon, and all the living organisms were created in just six consecutive 24-hour days. I generally prefer to call these individuals *strict creationists*.

Most strict creationist literature concentrates on trying to create the impression that much of science is done carelessly, and that many scientific findings are suspect. For example, they try to cast doubt on our ability to determine the age of rocks by radiometric dating. Much effort has been spent trying to convince the general public that scientists have overlooked some very simple but critical problems that invalidate their methods. How can creationists claim that scientists do not understand radioactivity well enough to determine the ages of rocks, when scientists are able to use these same principles of radioactivity to build atomic bombs and nuclear power plants?

Strict creationists believe that the Universe is only a few thousand years old, but what *positive* evidence can they produce to convince us of this statement? If the Universe is as young as they say, how can we see a *nearby* galaxy whose light took millions of years to travel to us? The light must have left the object before the Universe was created. Are we to believe that scientists understand light well enough to build lasers, X-ray machines, and make holograms, but do not understand how light travels from one point to another? The Universe is either very old or has been made to appear very old. Surely an honest God would not try to deceive us by making it only appear that the galaxies are at great distances when they are not.

Do strict creationists really believe that all the plants and animals were made during six consecutive 24-hour days? If so, why is there such a distinct ordering to the way the fossils were laid down? The fossil record is particularly rich during the past 600 million years. Even if you do not believe the 600 million years, why were no trees deposited during the first 30 percent of this time? Also, why were there no flowering plants present in the ancient forests that formed the vast coal deposits in the eastern United States? Could it be that flowering plants evolved later, or is God playing another trick on us? Or, as some have suggested, did He create things so they merely appear old. Perhaps everything (including our memory) was created yesterday. We can only hope that God is not that deceptive.

Scientists can use DNA to identify rapists or to determine the father of a child. People believe in this evidence so strongly that it can be used as evidence in a court of law. DNA studies tell us that humans are more closely related to chimpanzees than they are to monkeys or dogs. How can creationists have enough confidence in our scientific knowledge of DNA that they would use it to condemn someone to death, but not believe that scientists know enough about DNA to determine how humans, chimpanzees, and dogs are related?

As we study evolution, I hope that your knowledge of the living world will be expanded, and that you will have an even greater sense of awe about the magnificent creations on this planet. Evolution is like a special pair of glasses that allows us to comprehend many of the subtle patterns present in the living world. Suddenly the living world becomes more understandable and less foreboding. Let us begin with a brief preview of things to come.

How did so many of the plants and animals of Hawaii become so unique? Most are found no where else in the world. Why is there no evidence that reptiles or mammals (other than bats) ever lived on the islands? Is one likely to find large mammals such as tigers on an isolated island?

Suppose you extract any protein (such as thyroxin) from the following: a human, chimpanzee, monkey, porpoise, kangaroo, lizard, and fish. Which animal's protein will be most similar to the human protein? Which animal's protein will be the next most similar to the human protein? The next, etc.? We expect that closely related animals will share a more similar body chemistry than animals that are more remotely related. All life forms have slightly different body chemistry, but why is the body chemistry of a porpoise closer to that of a human than it is to a fish?

Why are we not surprised that birds have scales on their legs that are identical to the scales on a reptile? Most mammals are color blind, but they generally have highly developed hearing and a keen sense of smell. What might these characteristic traits have to do with the dinosaurs?

The bone pattern of all mammals can be understood as modifications of the same basic bone pattern. For example, all mammals have a femur, tibia, humerus, and radius although these bones may vary in shape from mammal to mammal. From the long-necked giraffe to the whales with no apparent neck at all, most mammals have seven vertebrae in their neck. Could it be that they all evolved from a common ancestor with seven vertebrae?

As we explore the intriguing subject of evolution, we will learn the answers to these and many other fascinating questions. First, however, we would like to investigate the nature of science. What is science? How does it operate, and how does it differ from other disciplines? Why is creationism not a science? Creationists (and at least one of our former presidents) have pointed out that evolution is *only a theory*. The implication being that theories like evolution are wild speculations in contrast to scientific *facts* like gravity and electricity. We will address this delusion in the first chapter.

Chapter 1: A Look Into the World of Science

Some Interesting Questions

The greatest value of evolution is that it explains many of the mysteries of the living world. There are many questions that defy understanding until they are examined through the lens of evolution. We will investigate much of the evidence in support of evolution later, but at this time it might be interesting to consider just a sampling of the fascinating questions that can be understood through the lens of evolution.

Male bees and wasps do not have a stinger. How do we explain this observation? In the context of evolution, the answer is quite simple. The stinger evolved from a modified egg-laying device that was used by the ancestors of bees and wasps to insert eggs into plants and other insects. Since males do not lay eggs, they did not have this egg-laying device and, therefore, did not evolve a stinger (Figure 1.1).

Figure 1.1 Wasp and Stinger Although it is not obvious, the male European Paper Wasp (top) is harmless since it does not have a stinger. [Quartl] Below is an unidentified species of wasp that is obviously a female. A drop of venom can be seen on the end of her stinger. [Pollinator]

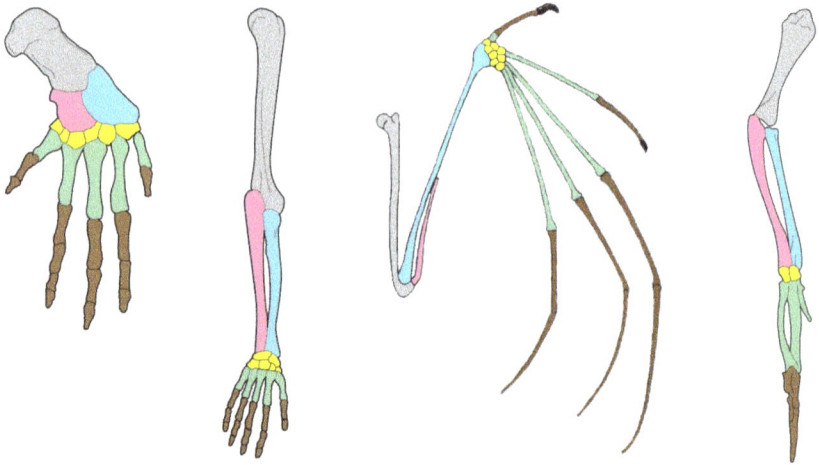

Figure 1.2 Homologous Bones *(from left to right) Right arm of whale, human, bat, bird. The similarity of the bones in a bat wing, whale flipper, and human arm demonstrates that they are more closely related to each other than the bat wing is to a bird wing or the whale flipper is to a fish fin (not shown). The bones are the humerus (gray), radius (blue), ulna (pink), carpals (yellow), metacarpals (green), and phalanges (brown). [Volkov Petrovich]*

The bones of a bat wing and a whale flipper are more similar to the bones in the human hand than they are to the bones of a bird wing and fish fin respectively (Figure 1.2). How did this happen? Although all these structures have a certain degree of similarity, the bat and porpoise are more closely related to humans than they are to either birds or fish. Even though bird wings and bat wings are used for the same function, their bones are less similar than are the bones in a bat wing and a human hand.

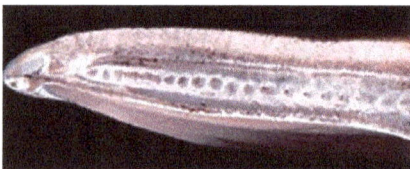

Figure 1.3 Baleen Whale Teeth *The upper jaw of this embryo Fin Whale has been exposed to show the tooth buds. Strong evidence that baleen whales evolved from ancestors that had teeth. [Alex Aguilar]*

Likewise, the whale flipper and the fish fin are both used for swimming, and yet the bones of the flipper more closely resemble those of the human hand. Regardless of its mode of life, any given mammal will share more anatomical similarities with another mammal than it will with a non mammal. These and other similarities tell us that mammals evolved from a common ancestor with these features, which is why we classify them as mammals.

Baleen whales do not have teeth, but they have long plates of flexible material called baleen that hangs down from their upper jaws. Baleen is used to filter food (small plants and animals) from the ocean water. The fetus of a baleen whale develops a full set of teeth that are not covered by enamel and never erupt through the gums, but are completely absorbed before birth (Figure 1.3). How does evolution explain these teeth? The fossil record shows us that the ancestors of all types of whales had teeth, and this toothed ancestry is still displayed in the embryonic development of baleen whales. In addition, baleen whales have teeth forming genes that are non functional.

Dolphins do not have hind legs. So why do dolphin embryos develop hind leg buds that never develop into legs? The early development of a dolphin proceeds much like any other mammal. It even has hind leg buds which develop into hind legs in other mammals (Figure 1.4). However, the leg buds of dolphins are absorbed in later embryonic stages and never develop into legs. This is strong evidence of their four-legged ancestry.

Many of the life forms found on islands are unique. How did these life forms originate? Most islands have been isolated for millions of years. If some type of plant or pair of animals happened to reach the island and was able to become established, its descendants would have evolved into new species as they adapted to their

Figure 1.5 Dodo On the island of Mauritius, the flightless Dodo evolved from a pigeon ancestor about 10 million years ago. It was about one meter tall. Within 60 years of its discovery, it had been driven to extinction by humans. The last sighting was in 1662. DNA evidence suggests that the Nicobar Pigeon, found on several islands off Southeast Asia, may be its closest living relative. [photo, Oxford University Museum of Natural History/Ballista]

new environment. If there are several islands in the group or different environments on one of the larger islands, several closely related species will likely evolve. Examples of animals confined to specific islands include a family of Hawaiian birds called the honeycreepers, the Darwin finches of the Galapagos Islands, the extinct Dodo of Mauritius (Figure 1.5) and its close relative, the extinct Rodrigues Solitaire, the kiwis and the extinct moas of New Zealand, insectivorous mammals called solenodons of Cuba and Hispaniola, the lemurs of Madagascar, and the kangaroos of Australia. In Chapter 9 many of these animals will be discussed.

How do we explain the observation that each chromosome replicates itself before meiosis? This replication appears to be a complication that is not necessary. Meiosis is the process by which eggs and sperms are formed in

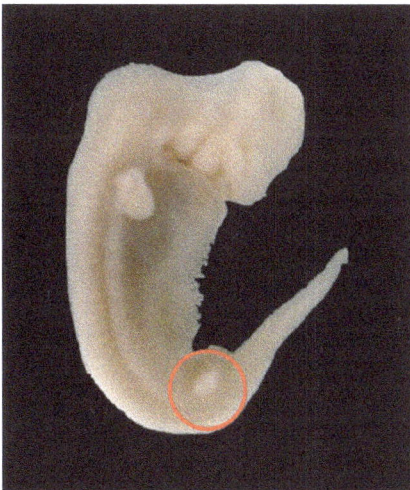

Figure 1.4 Spotted Dolphin Embryo The early development of a dolphin proceeds much like any other mammal. Hind leg buds are formed (circled), but are later absorbed. [J.G.M.Thewissen/NEOMED]

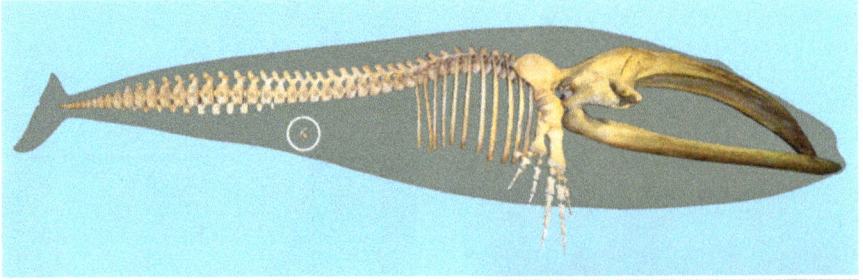

Figure 1.6 Whale Hips Right whales have three bones on either side of their pelvic region (circled above). The photo (right) shows the bones from the right side of a North Atlantic Right Whale. They are the pelvis, femur (f), and tibia (t). [skeleton and contour, H.Zell; bones, Martyn Gorman]

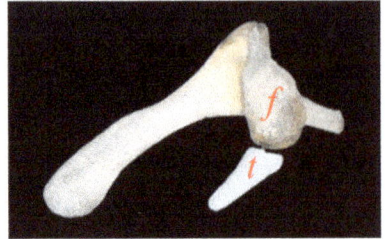

higher forms of life. The complexity of meiosis can be understood if we view meiosis as a slight modification of simple cell division, a process we call mitosis. Meiosis evolved from mitosis. (See Chapter 6)

Whales do not have back legs. Then why do they have pelvic bones that are buried in muscle tissue and are not attached to any other bones? Whales evolved from four legged mammals that were the ancestors of all modern day whales and dolphins. The hind

spurs

Figure 1.7 Snake Legs Snakes evolved from lizards with legs, and this Boelens Python, has remnants of their hind limbs. [Mokele]

limb bones of these mammals have not been entirely lost (Figure 1.6).

Boa constrictors have rudiments of hind legs, and two lungs (Figure 1.7). How are these structures explained? Although all snakes have two lungs, in most snakes one lung is greatly reduced in size to make the body more streamlined. Again, in the context of evolution we know that snakes evolved from walking lizards that possessed two lungs, four legs, and a pelvis. The boa constrictor is a *primitive* snake that is not as highly evolved as other snakes. Their leg bones and pelvic bones have not been completely lost.

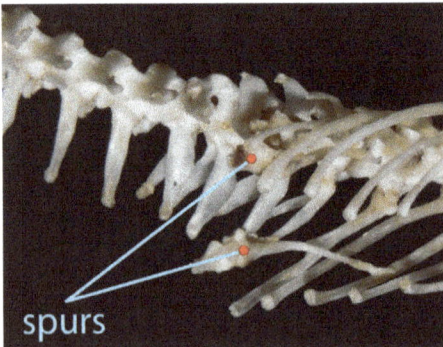

Some salamanders that live in caves have non functional eyes covered by skin (Figure 1.8). How do we explain this? These blind salamanders must have evolved from salamanders that had functional eyes. Since they spend their life in dark caves, eyes are not useful. In a dark environment, eyes might even be a disadvantage to these salamanders since their eyes could easily be injured in the dark and get infected. This type of injury

Figure 1.9 Anal Spurs The anal spurs of an albino Burmese Python. [Dawson]

Figure 1.8 Texas Blind Salamander This cave salamander has lost its eyesight, and its vestigial eyes are covered with skin. [Joe N.Fries/USFWS]

could potential put the salamander's life at risk.

The cave salamander's eyes that no longer function as eyes are called *vestigial* structures. Vestigial structures no longer perform their original function. Many are nonfunctional, or their function is greatly reduced. Some vestigial structures may be used for other simple purposes. For example, the vestigial hind legs of the python protrude from the snakes body, and their ends are covered by spurs that are used in mating (Figure 1.9).

There are many examples of vestigial structures in nature that provide evidence of the evolutionary process. Many birds have lost the ability to fly, as their wings are too small to sustain flight. Rails that managed to reach many isolated islands evolved into flightless species (Figure 1.10). There are also flightless ducks, flightless pigeons like the Dodo, and even a flightless parrot (the Kakapo of New Zealand). Unfortunately, since the arrival of humans, many of these flightless birds have become extinct.

Fossils are found in layers of sedimentary rock, and there is a very definite pattern to the layering. For example, dinosaur bones are never found in rock layers that contain the bones of elephants, horses, humans, or any of the other large mammals whose

bones are so common in younger (higher) layers. How was this pattern created? We conclude that dinosaurs lived before the large mammals evolved. After the dinosaurs became extinct, the larger mammals evolved and when these mammals died, some of their bones were trapped in layers of sediment on top of the dinosaur bones.

Sharks shed their teeth and millions of fossil shark teeth have been found in

Figure 1.10 Flightless Rails The poorly named Tasmanian Nativehen (top) is only found on the island of Tasmania, but a hen is a female bird and they are not all hens. [JCB] The Takahe (below) is only found in New Zealand. [Maungatautari Ecological Island Trust]

certain fossil bearing layers. However, none are found with the trilobite fossils of the earlier time periods we call the Cambrian and Ordovician. This is no mystery if we realize that sharks evolved about a hundred million years after the first trilobites appeared. In Chapter 7 we will study some of the details of the fossil record.

How does evolution explain the Giant Panda's thumb? The so-called *thumb* is a modified wrist bone called the radial sesamoid, but it does serve the purpose of aiding the panda as it feeds on bamboo leaves (Figure 1.11). Evolution does not always lead to the best solution to a problem. Indeed, evolution often does not lead to any solution at all, as evidenced by the great numbers of species that have become extinct throughout history. Many of the uniquely adapted island animals have become extinct since the introduction of foreign species by humans in more recent times. Some of the introduced species are obviously better suited to life on the islands than are many of the native species.

Most people know that there are no penguins in the Arctic and no Walruses in the Antarctic. Kangaroos are only found in Australia, and Giraffes are only found in Africa. You will not see Tigers in Africa and sloths only live in South America. The reason so many plants and animals are only found in specific regions on the Earth can be easily explained by continental drift and evolution. We will investigate this subject in Chapter 9.

The Words We Use

Since we will be dealing with information from the discipline of science, an understanding of the methods, vocabulary, and reasoning used in science is essential. As with any discipline, there are many words used in science that may be unfamiliar to the nonscientist. Unfortunately, even scientists have distorted the meanings of some words through casual use.

When certain words gain a measure of prestige, they may be used to help sway the listener to a particular point of view. For example, calling a survey a *scientific* survey may cause it to carry more weight with the general public. The word scientific has been used so carelessly that many people no longer clearly understand its meaning.

We use words to sway opinions or even mislead our listeners. If an advertisement reads, *Free Pizza*, does it really mean you will get a pizza without paying anything? Not likely! If you read the fine print, the ad probably means that you get the free pizza only after you buy something first, like another pizza (Figure 1.12). In other words, for a certain amount of money, you can buy two pizzas. If that is what was meant, why didn't the ad say two pizzas for the regular price of one? Does the word *free* accurately describe the offer?

Figure 1.11 Panda's Thumb Not a real thumb, but a modified wrist bone used to grasp bamboo, the panda's favorite food. Evolution does not always find the best solution. The right front paw is shown. The "thumb" is the pad on the right. [Kelly Yates]

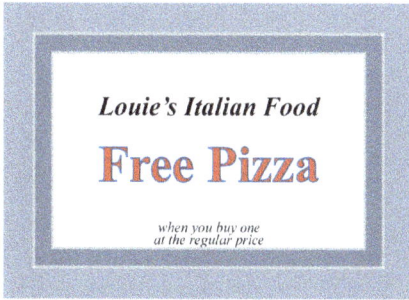

Figure 1.12 Free Pizza *Is a pizza really free when you have to buy something to get the "free" one? [JCB]*

Many statements in advertising are carefully designed to get your attention or mislead the potential customer. It has become an accepted practice to try to create an illusion about an item that may not accurately reflect the true nature of the product. The intention is to mislead the public to sell a product.

Words used improperly to create an impression eventually lose their original meanings and become less useful in communication. What does the word *free* mean? When words are used in a careless manner, it becomes extremely difficult to understand what is being said. Understanding the exact definitions of words is a prerequisite for accurate communication. Many arguments are brought to a quick end or become less intense when the parties involved carefully define the words they are using. Sometimes they even discover that they agree on the issue.

Because some words are not precisely defined, it is often difficult to decide if they apply. To illustrate, let us ask: What is a scientist? Is an engineer a scientist? Is a physician a scientist? They are not generally considered to be scientists, however, some engineers and some physicians are scientists. Is a student who has taken a biology course a scientist, or a student with an undergraduate degree in chemistry? Is

the holder of a doctorate in physics a scientist? These, like many questions, cannot always be answered with a simple yes or no.

In this book, our main objective is to gain an in-depth understanding of evolution. Is evolution a theory or a fact? As we continue, we will discuss the meanings of several words such as fact and theory that are often misunderstood, but are extremely important if we are to understand the nature of science. In many cases, even members of the scientific community may not agree on the usage of the terminology. Our goal is to establish a mutual understanding of how these words will be used in subsequent chapters of this book.

Data and Its Interpretation

An important aspect of science is the collection of information or data while performing experiments and making observations. Data collection is often called the *fact* gathering phase of science. However, the word fact is not precisely defined in science. Most *facts* in science are pieces of information that most scientists would call data.

A certain degree of interpretation by the observer often accompanies the presentation of scientific data. Those not familiar with scientific methods can sometimes confuse the data with the conclusions based on the data. To illustrate the terms we have been discussing, consider the following example.

If someone sees a dead limb fall from a tree during a thunderstorm, the observer might say it is a fact that this particular limb fell from that particular tree. However, would a person who did not see the limb fall consider it a fact? For the following discussion, let us suppose that no one saw the limb fall, but we find the dead limb lying under a tree. Would we then conclude that the limb fell from that particular tree? It is possible that the wind blew the limb from a nearby tree, or perhaps some animal carried it from a more distant tree (Figure 1.13).

Before we conclude anything, we will need to make more observations and perform some experiments. Suppose we observe that the tree is a Sugar Maple, and we determine that the limb came from a Sugar Maple. We also search the surrounding area and are unable to discover any other Sugar Maples. After collecting this data, could we conclude that the limb came from that Sugar Maple? Perhaps, but it is still possible that someone carried the limb from another Sugar Maple quite a distance away, or perhaps the limb came from a nearby tree that we overlooked in our search.

Searching for something without success does not always demonstrate that the thing was not there. If I search my house for my car keys and fail to locate them, I cannot be sure that they are not in the house. On the other hand, if I search my house for a lost elephant and fail to find it, I would be very sure that the elephant was not in the house. When we draw conclusions based on a search, we must consider the nature of

the object and the nature of the search area. Were we looking for a needle in a haystack or a needle in a teacup? Perhaps we need to collect more data.

Suppose we examine the limb and the tree with a microscope and discover that the break on the limb matches a break on the tree. We even decide to analyze samples of the tree and the limb, and we find they are the same chemically. With this additional data, even the most ardent skeptic would probably be convinced that the limb came from the tree under which it was found. However, it is still possible that the limb was broken off an identical tree, and the break just happened to match a break on the tree to which the limb was transported. Remember, no one saw it fall from the tree.

However, the probability that this remarkable set of coincidences took place is so remote that few would doubt that the limb did indeed fall from the Sugar Maple. Nevertheless, a more correct statement would be to say that, based on our interpretation of the observations, we conclude that the limb fell from the Sugar Maple under which it was found. Note also that we did not prove that the limb fell from the tree. We do not prove things in science! We only gather data that supports our interpretation of the observations. In casual discussions, the word *fact* is often used when either *data* or *interpretation of an observation* is a more accurate description.

Most people think it is a fact that the Earth orbits the Sun. Eventually, certain observation were made that did support the idea. We do not usually give the evidence when we say that the Earth orbits the Sun. However, this statement is more accurately described as an interpretation of the observational data, or a conclusion supported by the data.

Figure 1.13 Dead Limb *Which tree did the limb come from? [JCB]*

The Naming of Ideas

There is much more to science than just the collection and interpretation of data. The real power of science comes from our ability to predict the outcome of experiments and observations even before they are performed. To accomplish this feat, we need a statement or formula that will tell us how to make the prediction. These statements are given many names: postulates, theories, principles, or laws.

In some science books you may even find an outline of how the naming of these ideas is supposed to evolve. After some initial investigations, a scientist will formulate an idea or educated guess that is supposed to be called a hypothesis. When the hypothesis is supported by a large number of observations or experiments, the hypothesis is supposed to become a theory. If the theory makes accurate predictions over a *long* period of time and becomes universally accepted by the scientific community, the theory is supposed to become a law or principle.

Although this procedure may sound like a reasonable system, no official scientific group is commissioned to make the necessary decisions about just when these name changes are to be made. Consequently, the word theory is used for a wide range of ideas. Many new ideas with little supporting evidence are called theories, and some of the most widely accepted ideas in science are also called theories. The scientific community does not have a definite vocabulary to differentiate between a highly successful theory and a new idea that has little supportive evidence.

If scientists do not have a clear naming system, we should not be surprised that statements such as the one claiming that evolution is *only a theory* can mislead a nonscientist.

A speculation with little supporting evidence should be called a hypothesis or simply a speculation, but these are such weak words that few people use them. Most new ideas reported in scientific journals are called theories by their author, probably because the word theory carries more weight than the words conjecture and speculation.

Even though science has failed to provide us with a standardized vocabulary in this area, we will discuss some of the words that are commonly encountered in science.

Postulates

Physics is a discipline that strives to fully understand the basic rules under which the various components of the Universe operate. The job of a physicist is to observe, classify, and experiment with objects in nature, and to formulate general statements that explain the behavior of these objects. The most fundamental of these general statements are known as postulates or axioms. Postulates are assumptions that can be used to accurately predict the results of experiments and observations made in the natural Universe. As was noted above, the scientific vocabulary has not been carefully defined, and postulates are often called theories, laws, or just equations. Postulates differ from interpretations of observations in that postulates are comprehensive statements that allow the scientist to make predictions even before observations or experiments are performed. For example, a physicist can use the postulate of gravity to predict where a planet will be located at any time in the future. This postulate is called the law of gravity.

Theories

Sometimes the word theory and postulate are used interchangeably, but sometimes the word theory is used when a more comprehensive concept is implied. When physicists refer to the theory of quantum mechanics, they are not just alluding to the basic postulates of quantum mechanics, but also the predictions and investigations that are associated with this area of study. In nonscientific discussions, theory is often used to imply a wild speculation, but in science, a theory is often a well-verified idea. The nonscientist should be aware of the wide range of meanings that this word encompasses. As with many scientific terms, there is no official definition for the word theory.

As noted, the word theory is sometimes used to describe a comprehensive area of study. The theory includes all the basic postulates of the theory, any predictions that the postulates imply, observational data associated with these postulates, and interpretations of that data. Examples of comprehensive theories include the theory of special relativity and the theory of quantum mechanics. Ironically, these two theories provide our most accurate description of the mechanical workings of the Universe!

Many of the postulates that were formulated in earlier times were called laws. Some postulates that are commonly referred to as laws include, Newton's law of gravity, the laws of thermodynamics, Coulomb's law, Ohm's law, and Newton's laws of motion. These laws are not fundamentally different from the postulates associated with various comprehensive theories such as the theory of quantum mechanics or the theory of general relativity that have been proposed in more recent times. In fact, all the laws listed above have been found to be incorrect under certain circumstances, and newer theories are more accurate descriptions of nature.

The word law is no longer used for new ideas, perhaps because so many laws have been found to be incorrect, or at least limited in their application. Nonscientists often have the mistaken idea that a law is somehow more valid or factual than a *theory*. However, laws are not more valid than theories, and in some cases the reverse is true.

In 1916, Albert Einstein introduced the world to a new theory called the general theory of relativity (Figure 1.14). Today, this theory represents the most accurate description we have of *gravity*. The postulates of Einstein's theory more accurately describe gravity than does Newton's law of gravity. Einstein's theory gives the correct answer to several gravitational problems that cannot be accurately answered using Newton's law of gravity. One of these problems was

Figure 1.14 Albert Einstein A photo taken in 1921 at a lecture in Vienna when he was 42 years old. [Ferdinand Schmutzer]

recognized soon after Newton proposed his law.

As Mercury orbited the Sun, astronomers noticed that it did not follow the exact path predicted by Newton's law of gravity (Figure 1.15). Because of its outstanding success in explaining so many other observations, people were reluctant to believe that Newton's law was incorrect.

Other explanations for Mercury's strange behavior were proposed. One solution to the problem required the existence of another planet near the Sun. However, astronomical searches failed to discover any objects of the size required to explain the observed deviations in Mercury's orbit. Finally, Einstein's general relativity theory explained the peculiar motion.

If Newton's law of gravity is not correct, why do we still teach it? Unless a problem involves extremely strong gravitational forces, or other very specialized conditions, Newton's law of gravity can approximate Einstein's theory quite well. Newton's law is a special case of Einstein's theory. So you see, Newton's law is not *incorrect* in most cases, but it does have certain restrictions, and we cannot apply it to every gravitational problem as originally hoped. In addition, Newton's law has a great advantage over Einstein's theory in that it can be stated in a much simpler form and, therefore, it is easier to understand than Einstein's theory.

Finally, the mathematics used in association with Newton's law is much simpler than the mathematics required by Einstein's theory. The difference in mathematical complexity is similar to the difference between high school math and college calculus. For most problems dealing with gravity, both theories give essentially the same answer. Therefore, we merely determine under what circumstances it

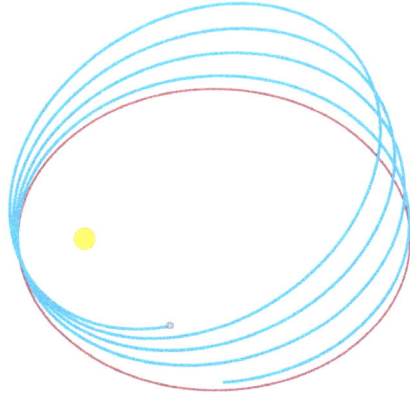

Figure 1.15 Orbit of a Planet *This figure shows the orbit of an isolated planet around a star as predicted by Newton's law of gravity (red) and as predicted by Einstein's general theory of relativity (blue). The blue orbit is what we observe. [I, KSmrq]*

is permissible to apply Newton's approximation and use it only in those cases.

Because scientists have failed to carefully define these terms, the public generally believes that a scientific law is an absolute truth when in actuality, most scientific laws are not completely correct. On the other hand, the general public tends to think of theories as being more speculative, but some theories provide us with our most accurate descriptions of nature. In science, most of the truly great ideas are called theories!

Although we have briefly discussed comprehensive theories, remember that the scientific community is not in uniform agreement as to the accepted usage of the word theory. Often it is used for every idea in science. For example, some scientists speculate that a large asteroid or comet collided with the Earth about 65 million years ago (65 mya), and the resulting explosion caused the extinction of the dinosaurs. This extinction theory is similar to our tree limb example. Given enough data, we might conclude that the asteroid theory is correct. (There is already a great deal of evidence in support of this extinction theory.) An important difference between simple ideas and comprehensive theories is that comprehensive theories can be used to make numerous predictions about the outcomes of future studies. Many of the simple ideas in science that are often called theories lead to few if any predictions.

When the general public reads about a theory that the evolution of intelligence might be due to the complexities faced by apes in their search for a varied diet, and then a scientist tries to tell those people that gravity is also a theory, is it any wonder that there is some confusion? With such casual usage of the word theory, is it any surprise that creationists are able to cast doubt on evolution by simply stating that it is only a theory? The implication being that it is not a *fact* like gravity, but some wild speculation. Ideas of restricted application should not be called theories. Less confusion would arise in both the scientific and nonscientific communities if such ideas were consistently referred to as hypotheses or speculations or conclusions based on observations.

A hypothesis is a simple idea that is proposed to explain certain observations. Most hypotheses are of the type that if enough additional data were gathered to support it, we would consider it to be a conclusion based on the observations. Copernicus speculated that the planets orbit the Sun. His original idea may have been a hypothesis, but we now have enough evidence in support of this hypothesis that most scientists consider it a conclusion based on the observations. Since our current theory of gravity predicts that the Earth should orbit the Sun, our observations support the theory of gravity.

The Postulates of a Theory

The postulate formulating stage in the scientific process involves a type of reasoning called *inductive reasoning*. After sufficient data has been gathered on certain objects or events, inductive reasoning is used to formulate a general postulate (or postulates) that will predict the behavior of these objects. Inductive reasoning is the reasoning we use to formulate a general statement based on specific observations. Simply stated, it is the reasoning of experience. We have all heard that experience is the best teacher, however, like so many old sayings this one is extremely unreliable. Conclusions based on experience are often in error!

We will look at a very simple example to illustrate the process of using inductive reasoning to formulate a postulate. Suppose we observe the dogs in our neighborhood and notice that all of them are covered with brown fur. Because of the data we have collected, we might propose a postulate, which states that all dogs are brown (Figure 1.16). We would very likely be accused of being a little hasty in basing our postulate on such a limited amount of data, but the above example does illustrate the inductive process, and the fallibility of inductive reasoning.

Scientific postulates must explain all of what has been observed or demonstrated in past experiments, but to be widely accepted by the scientific community, a postulate must be able to correctly predict future discoveries, or the results of future experiments. In the above example, our dog postulate would predict that all the dogs we see

Figure 1.16 All Dogs Are Brown *And this one is also very cute.*

in the future should be brown. Suppose we observe more dogs (collect more data), and they all are brown. Although this data does not prove that our postulate is correct, it does give us more confidence in our postulate.

Notice that it is impossible to prove that a postulate is true in all cases. Considering our dog postulate for example, even if we were to search for years, how could we prove that we had seen every dog in the world? How could we prove that all future dogs will be brown, or that all the dogs that lived in the past were brown? A postulate is a statement that cannot be proven. However, as more of its predictions are verified, we gain more confidence in the postulate.

What happens if we observe something that contradicts a postulate? For example, suppose we observe a dog that is not brown. Of course this piece of data would prove that our postulate must be wrong, but there are two options open to us. We might have to throw out the postulate and start over, but it might be possible to amend the postulate to cover the observed exceptions. Perhaps the dogs that are not brown have long hair. Our new restricted postulate might state that all short-haired dogs are brown. Since the new postulate is restricted to short-haired dogs, it cannot be applied to long-haired dogs. (We might also have to carefully define what is meant by short hair.)

The distinction between a scientific postulate and a conclusion based on observational evidence can be fuzzy. A postulate is an assumption we make because it explains many of our observations, while a conclusion based on the evidence is specific statement that we accept as true because the evidence is so overwhelming. In the tree-limb example, one could always argue that there is not enough evidence to say that the limb came from the tree in question. Since no one saw the limb fall, some might say that they do not believe that the limb fell from that tree. However, if no amount of evidence will satisfy us, we will make little progress toward understanding the mysteries of the Universe.

Postulates such as those associated with Einstein's theory of general relativity cannot be proven correct in all circumstances. It would be impossible to accurately investigate all the gravitational problems in the Universe and prove that the postulates give the exact answer in every situation. If postulates (and, therefore, the theories to which they apply) cannot be proven, why do scientists have such a great deal of confidence in many of them? The answer is quite simple. Theories that are widely accepted by the scientific community have been used to make many accurate predictions about future investigations. However, there is always a possibility that some future prediction will invalidate the theory or a part of the theory.

A comprehensive scientific theory creates a unified picture out of many observations, some of which may seem totally unrelated. The reason for the success of science can be traced to the ability of theories to predict the results of future experiments and observations.

An understanding of a few scientific theories will enable a scientist to explain and predict the results of innumerable experiments. Millions of individual pieces of data and hundreds of thousands of problems can be unified into an understandable picture by a few comprehensive theories.

When a scientific theory has an overwhelming amount of supportive evidence, some may be tempted to say that the theory has become *fact*. However, theories never become facts. As we have stated, the word *fact* is best avoided in science, and probably should be avoided in other disciplines as well.

Testing a Theory

If postulates are statements that can never be proven, under what conditions would scientists accept a theory based on these postulates? As stated above, the real test of a theory is how well it explains and predicts the results of experiments and observations.

Scientific theories contain general postulates that are used to make specific predictions. This phase in the scientific process involves the reasoning of logic, called *deductive reasoning*. Deductive reasoning is logical reasoning that uses one or more general statements (called premises) to formulate a specific conclusion or prediction. A premise can be either a postulate or an observation.

To illustrate, let us return to our postulate that all short-haired dogs are brown. If someone tells us that Rover is a short-haired dog, then according to our dog postulate and the observation that Rover is a short-haired dog, we would predict that Rover must be brown. We have not proven that Rover is brown. We have only concluded that Rover must be brown if our premises

are correct. If a conclusion based on deductive reasoning is found to be incorrect, one or more of the premises (either the postulate or the observation, in this example) must be incorrect. As we noted earlier, postulates may turn out to be wrong since inductive reasoning was used to formulate them.

The next stage in the scientific process is to attempt to verify the prediction. In our dog theory, for example, Rover should be examined. If Rover is not brown, of course our postulate is in trouble, and it must be discarded or modified again. However, if we can verify that Rover is brown, we will have an additional piece of data to support our theory, and our confidence in the validity of the basic postulates of the theory will be slightly increased. The most widely accepted scientific theories have made many accurate predictions, increasing our confidence in the theory.

An example of this predictive process occurred in the mid-1800s. Newton's law (postulate) of gravity stated that every object in the Universe should attract a given object. For example, the Sun, Moon, and every other object in the Universe attracts the Earth. However, after calculating the forces on Uranus due to the Sun and all the known planets, it was observed that Uranus did not follow the predicted orbit. Using only Newton's equation (postulate) of gravity, calculations showed that the deviations in the orbit of Uranus could be explained if it were being attracted by a more distant planet. The location of the mystery planet was calculated, and those computations led to the discovery of Neptune in 1846 (Figure 1.17). Though the discovery of Neptune did not prove Newton's law of gravity, it gave scientists more confidence in the theory.

Figure 1.17 Neptune *This photo was taken by the Voyager 2 spacecraft. The Great Dark Spot is a giant hurricane-like storm about the size of the Earth. The white clouds are made of methane ice. [NASA]*

Falsifiable or Predictive Theories

To be of any practical use to scientists, the postulates of a valid scientific theory must be stated in such a way that definite predictions can be made and verified. For example, the theory of gravity says that any two objects in the Universe will attract each other with a certain force, not most of the time, or if they feel like it at the moment, but absolutely always. Every time someone tests the theory of gravity, the results must always fit precisely with the predictions of the theory (assuming the experiment was performed carefully). Just one indisputable experiment that does not fit the predictions of a theory is all it takes to invalidate the theory. We should point out, however, that if data from an experiment does not seem to fit a widely accepted theory, the theory is not immediately canned. Usually, it is the data or its interpretation that is eventually shown to be in error, not the theory.

Since a scientific theory must make definite predictions, the possibility that one of these predictions will turn out to be incorrect always exists. If even one well verified prediction is found to be wrong, we must conclude that the theory is false. Theories that make definite predictions are vulnerable to falsification and are said to be falsifiable.

Being falsifiable means that it must be conceivably possible to prove a theory false. Notice that although it is not possible to prove that a theory is correct, it must be *conceivably* possible to prove that a theory is false. Of course, a theory found to be false would no longer be an accepted theory. We sometimes say that scientific theories must be testable. Testable not only means that a theory must make accurate predictions, but it also implies that certain observations could disprove the theory. In our dog theory, simply locating a short-haired dog that is not brown would disprove the theory. Therefore, our dog theory makes definite predictions and is said to be falsifiable.

Theories with postulates that could never be proven incorrect are of no value in science. As a simple example of such a theory, I might claim that a family of elves lives in my house (Figure 1.18). These elves come out at night and play with various items, but they are very careful to replace everything when they finish. They hide whenever anyone comes near, which is why it is impossible to detect their presence. No one has ever seen them, and no one ever will because they are too clever. If you devise a plan to photograph them, or catch them, or detect their presence in any way, they will find you out and foil your plans. As you can see, the postulates of this theory have been carefully stated so that it is impossible to prove them

Figure 1.18 Elves *A group of elves that might have lived in England during the time of Shakespeare (around 1590). [Arthur Rackham]*

incorrect. I do not wish to enter into a discussion about the validity of the elf theory, but merely wish to state that it is not a scientific theory. A theory such as this one extends beyond the realm of science.

Religions are generally based on non-falsifiable postulates that cannot be studied scientifically. For example, the idea (postulate) that some prayers are answered is not falsifiable. Apparently all prayers are not answered, and there is no way of predicting exactly which prayers will be answered. Since the statement that some prayers are answered does not make definite predictions, this idea is not falsifiable. Although the prayer idea is not falsifiable, it may still be true that some prayers are answered. However, it is not a scientific postulate and cannot be studied by traditional scientific methods.

Historical Narratives

When making investigations in a scientific manner, reproducing the exact event of interest is not always possible. Suppose you are interested in how a particular vase shattered as it fell from a table. When it hit the floor the pieces flew in all directions and scattered around the room. You can mark the position of each piece, weigh it, note its orientation, and make other measurements that might be relevant to the collision. You can study the data from this event, but you cannot repeat the experiment. You might even make some interesting discoveries by studying the data from the collision. For example, the number of pieces as a function of their weight might be interesting, or a plot of the number of pieces as a function of their distance from the impact point might present an interesting pattern. However, you cannot repeat the original collision. Many investigations in science are constrained by the realization that certain events took place that cannot be exactly repeated. The evidence for such events must be measured, analyzed, and conclusions must be drawn without the luxury of repeating the experiments. These non-repeatable events are historical in nature.

Historians tell us that many Europeans became aware of the New World after a man named Christopher Columbus sailed to America in the year 1492 (Figure 1.19). Is this statement a conclusion based on the evidence, a portion of a historical account, or something else? There is a great deal of evidence to support this statement about Columbus, but it is not a conclusion in quite the same sense as the statement that says the Earth orbits the Sun. At any given time, a scientist can make the measurements (collect the data), which can be interpreted as evidence that the Earth orbits the Sun.

The evidence that Columbus sailed to America, however, is of a different nature. One can look up old papers and make inquiries that might lead one to conclude that Columbus did sail to America, but there are no experiments that can be carried out today to

Figure 1.19 Columbus Lands in the Americas A part of the historical narrative. [John Vanderlyn]

demonstrate that Columbus sailed to America in 1492. Perhaps all the papers were forged. The Columbus expedition could have been a well-orchestrated hoax (although someone must have told Europe about America). Assuming the Columbus event did take place, it cannot be repeated. We must be satisfied with studies based on the data that remains.

By their very nature, historical events cannot be verified by the same methods used to verify other predictions. Since historical events have already taken place and cannot be repeated, historical accounts merely attempt to create a consistent picture from the available data. In science, that generic word *theory* is often called to action when referring to historical accounts. We will call these theories historical narratives to differentiate them from non historical scientific theories, such as the theory of general relativity or the theory of electricity and magnetism.

Non historical scientific theories attempt to explain the results of scientific experiments, and historical narratives attempt to construct a self-consistent picture that agrees with all the evidence. In certain areas of science we encounter various historical narratives. These narratives include the Big Bang that describes the formation and early evolution of the Universe, continental drift as it attempts to reconstruct the shapes and locations of the continents in ages past, and organic evolution that attempts to reconstruct the family trees of various organisms. The tree-limb illustration that we used earlier was an example of a historical narrative.

Since these narratives describe events in times past, they often contain many details. To ascertain one detail of the narrative, sufficient evidence must be available to allow one to draw a conclusion. Reaching this conclusion is similar to reasoning used in a court of law. Insufficient evidence requires that no conclusions can be reached about that particular detail. For example, the current data is insufficient to determine who invented the wheel or when it happened.

If enough details are known, historical narratives can be used to make certain predictions, just as is done with non historical scientific theories. For example, if Columbus did open up the Americas to Europe in 1492, we would not expect to find evidence of English firearms in North America before 1492. We would not expect to find evidence of horses in the Americas before the voyage of Columbus, and we would not expect to find evidence that Europeans knew of American turkeys, corn, or tobacco before 1492.

Like the Columbus story, a major portion of evolution is a historical narrative. The general idea of evolution is that the first life forms were very simple, but the descendants of these early forms changed in various ways, and some groups evolved into very different looking organisms. Present day organisms are the modified ancestors of earlier life forms. Darwin called this idea descent with modification. Studies in the field of evolution attempt to discover just what changes have occurred, when they took place, and if these new pieces of information are consistent with the other details of the emerging evolutionary narrative.

Because of its predictive ability, the evolution narrative is one of the most valuable *theories* in all of science. The theories of quantum mechanics and special relativity are probably the most essential of all the physical theories, and evolution performs a similar role in biology. Even fragmented knowledge of the evolution narrative allows one to make numerous predictions and greatly increases one's understanding of the living world.

For example, evolutionary evidence leads us to conclude that humans evolved from an ape-like ancestor a few million years ago. Since other evidence suggests that dinosaurs became extinct about 65 million years ago, we can predict that no evidence of humans associating with dinosaurs will ever be found. In addition, if humans and chimpanzees evolved from a common ancestor only five to eight million years ago, we would predict that their body chemistry should be very similar. It should be much more similar than the body chemistry of a human and a dog for example. Many of these chemical tests have been performed, and as we will see, the results are consistent with our expectations.

If a jury weighs the evidence and finds a defendant guilty, the crime cannot be repeated to check the verdict. Likewise, when scientists weigh the evidence and conclude that mammals evolved from reptiles, we cannot repeat history to check the conclusion. The events have already happened, and we cannot repeat the experiment. We can only study the evidence and try to assemble the pieces into a consistent picture.

There are, however, important differences between a legal trial and the scientific investigation of evolution. The scientific evidence is always available to scientists for examination, and the search for new evidence continues in a never-ending process. Although conclusions are drawn along the way, the jury of scientists is always deliberating. Some additional piece of evidence may show that a conclusion is incorrect. A scientific trial is always in session.

The details of evolutionary changes are quite numerous and involved, and many questions have not been answered at this time. For example, did mollusks evolve from segmented worm-like ancestors, from a type of flatworm, or from some other group of organisms? Many relationships are not known at present, but many have been painstakingly pieced together from the fossil record and molecular studies. Evolution, like any viable branch of science, is an ongoing study.

On our scientific journey, we will explore some of the evidence that has been used to reconstruct the evolutionary history of life. As with any historical narrative, the story is not complete. Many of its gaps must await new discoveries before we are able to fill in the details, and many gaps may never be filled since some evidence may have been destroyed.

Ockham's Razor

Given any body of evidence, it is sometimes possible to formulate more than one postulate to explain the data. For example, in our illustration of the tree limb, it is possible that the tree limb was carried a great distance and laid under the Sugar Maple. Perhaps life forms from another world brought the limb from their home planet and left it under the tree. When two or more explanations describe the observations, we find that nature seems to prefer the simpler one. The principle of picking the simplest explanation or theory is known as Ockham's Razor (or Occam's Razor).

Ockham's Razor explains why most scientists remain skeptical that some UFO's are space ships from other worlds (Figure 1.20). Even if there are life forms on other planets scattered throughout the Universe, the energy and time required to travel even from a nearby star would make the trip extremely difficult and, therefore, highly improbable. Because a visit by beings from another world would be such an extremely incredible accomplishment, (certainly one of the most remarkable events in recorded history), the evidence for such a visit would have to be absolutely beyond reproach to gain wide acceptance.

A few ambiguous pieces of evidence are not sufficient to be interpreted as a visit by creatures from another world. Invoking Ockham's Razor, we interpret most UFO sightings as lights, meteors, airplanes, or other common phenomenon. Some sightings must remain just what the name says, unidentified flying objects. Even though we may not be able to explain all UFO sightings, there is not enough evidence to jump to the remarkable conclusion that they are the space ships of visitors from another world.

Newton's law of gravity carried with it the implication that the Earth must orbit the Sun. However, if you choose to believe that the Earth is fixed in

Figure 1.20 UFO One of many UFO photos that have been circulated. [George Stock]

space and does not move, the theory could be modified to accommodate this belief. All that is really needed is a footnote, which says that the Earth is special (for some unknown reason) and acts differently than all the other objects in the Universe. The fixed Earth theory would be more complicated than Newton's law of gravity, and Ockham's Razor requires us to pick the simpler theory.

Einstein's formulation of his general theory of relativity also illustrates the use of Ockham's Razor. When Einstein first proposed his theory, it was used to calculate the large-scale structure of the Universe. To the surprise of Einstein, the only possible solutions to his equations required that the Universe be either expanding or contracting. A stable solution was not possible. Instead of making the bold prediction that one day observations would show

20

that the Universe is either expanding or contracting (we now know that it is expanding), Einstein added another term (called the cosmological constant) to his equations that permitted a static solution. Einstein later called the addition of this term his greatest blunder. With the discovery that the Universe is expanding, Einstein's extra term was no longer necessary. Although keeping a small cosmological constant could have satisfied the observations, applying Ockham's Razor we pick the simplest equation by removing the cosmological constant.

Fixing the Rules

Science encompasses many widely varying areas of study. From physics to biology, the theories of science must be consistent with each other. An observation in biology cannot contradict one of the theories of physics. Likewise, one theory in physics cannot contradict another. In the past, great strides in science have sometimes taken place because of the discovery of an inconsistency. James Clerk Maxwell, one of the truly great minds in science, discovered that the four laws of electricity and magnetism were not consistent with each other (Figure 1.21). Maxwell was able to correct this inconsistency, by modifying a law known as Ampere's law (another law that was wrong). He was then able to write four consistent equations that contained all the information needed to describe the electric and magnetic properties of matter. These four equations (known as Maxwell's equations) are the basic postulates of the theory of electricity and magnetism.

However, Maxwell's equations were not consistent with the equations that Newton used to describe the motion of

Figure 1.21 James Clerk Maxwell (1831-1879) One of the world's great physicists.

objects under the influence of various forces.

When an observer watches the motion of an object (an airplane for example), Newton's equations of motion are used to describe the position, speed, and acceleration of the airplane. If a second observer, who is moving relative to the first observer, watches the same airplane, the second observer uses a similar set of equations to describe the motion he sees. Notice that among other things, the two observers will not measure the same speed for the airplane, since the observers are moving relative to each other. For example, an observer moving toward the airplane will measure a faster speed for the plane than an observer who is not moving toward the airplane (Figure 1.22).

The equations used by the two moving observers are related to each other by another set of equations that we call transformation equations. They are called transformation equations because they transform the equations of one observer into the equations of the other observer.

Scientists noticed that the equations that transformed Newton's equations of motion were not the same as the transformation equations that were used for Maxwell's equations. This inconsistency guided Einstein in the formulation of his special theory of relativity. Einstein discovered that Maxwell's equations were correct, and that Newton's *laws* of motion were wrong. However, Newton's laws of motion are still being taught, because Einstein's equations reduce to those of Newton when the speeds involved are small compared to the speed of light.

Therefore, Newton's equations give an answer that is accurate enough for most of the problems encountered in everyday life. This example is another case in which the theory of Einstein is more accurate than the laws of Newton.

Evolution has been attacked, because some claim that it violates the second law of thermodynamics. If there were a real conflict, one of the two would have to be altered or discarded. The origin of this claimed violation seems to be a failure by some to fully comprehend how to apply the second law. Fortunately, neither has to be discarded because evolution does not violate the second law.

In one of its popularized forms, the second law of thermodynamics states that natural processes tend to move an isolated system toward a state of greater disorder. However, the disorder that the second law refers to is at the *microscopic* level. Furthermore, disorder is a very poor word to use when discussing the second law.

It is easy to confuse someone by using examples like tidying up your room (Figure 1.23) or watching a car in a junkyard fall apart (both are *macroscopic* examples). Is the

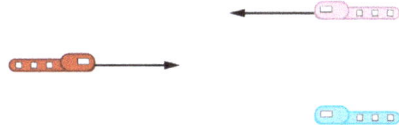

Figure 1.22 Relative Speeds Suppose we are sitting in the blue train. Also suppose we see Mary on the pink train passing us at 50 km/h, and we see the red train approaching us at 50 km/h. If we ask Mary how fast the red train is moving, what will she say? Her answer should be 100 km/h. Einstein discovered that at speeds near the speed of light, you cannot just add the two speeds. If we saw the two trains moving at half the speed of light, Mary would see the red train approaching her at only 0.8 times the speed of light. [JCB]

thermodynamic disorder increasing or decreasing in these examples? We will see in a moment.

The physical quantity that measures the disorder of a system is called the *entropy*. The second law says the entropy of a *closed* system must either increase or stay the same, but it cannot decrease. So, if one thing loses entropy, something else in the system must gain entropy. We often talk about the change in entropy when something happens.

For solids and liquids, the amount of energy they gain or lose is related to their change in entropy. For example, consider a hot cup of coffee sitting on a table. As the coffee cools, it gives energy to the surrounding air. The loss in entropy of the coffee is proportional to the energy it disperses into the environment. As energy is dispersed by the coffee, the entropy of the coffee decreases. On the microscopic level, the molecules are less chaotic, but the entropy of the surrounding air has increased.

Obviously, if it is cold enough outside, a glass of water set outside will eventually freeze as it disperses energy into the surrounding environment. The entropy of the ice

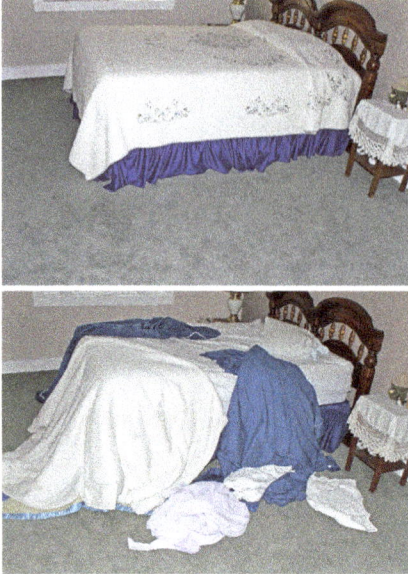

Figure 1.23 Which Room Has More Thermodynamic Order? Which room configuration has less entropy (which is the more ordered system)? Hint: It is not the neatest looking one. [JCB]

will then be *less* than the entropy of the original glass of water.

Similarly, if you drop a rock, it gains energy of motion as it falls. Ultimately, after the rock hits the ground, the net result will be a slight decrease in the entropy of the rock since the rock has dispersed some of its gravitational energy into the environment. In a hydroelectric plant, the gravitational energy given up by falling water is harnessed to generate electricity.

So, a messy room with many things thrown on the *floor* will have a lower entropy than a room with the clothes hanging neatly in a closet. Likewise, an engine lying in pieces on the ground is in a lower entropy state than an engine suspended in a car. We now see how order and disorder can be very misleading terms.

You can calculate the entropy of a glass of water by doing a thought experiment. Imagine removing energy from a glass of water. As the energy is removed, the temperature drops until it reaches the freezing point. As more energy is removed the water freezes, but ice still contains much energy. Water ice is much warmer than carbon dioxide ice (commonly called dry ice). If we keep removing energy, the temperature of the ice continues to drop.

Eventually you will reach a point where no more energy can be removed. The temperature at this point is called *absolute zero* or zero *Kelvin* (K). It is the lowest temperature any object can possibly have. Its entropy is zero!

Now that we have the ice at 0 K, let us begin to add a little energy (the unit of energy is called a Joule). Suppose we add enough energy to raise the temperature to 1 K. We divide the energy added by the temperature (1 K) and write it down. We continue adding small amounts of energy, dividing by the temperature, and recording the numbers. These numbers are the changes in entropy of the ice and then of the water. When we reach room temperature, adding up the numbers will give the entropy of the glass of water at room temperature. Room temperature is about 293 K or 20 °C or 68 °F.

Obviously if you start with twice as much water you will have to add twice as much energy to bring it up to room temperature. Therefore, twice as much water will have twice as much entropy.

As plants and animals grow, their entropy increases, (they become more *disordered*). This is because the entropy of water (the main ingredient in life forms), is proportional to the mass of the object.

Therefore, the entropy of a 100 kilogram adult is more than the entropy of a 10 kilogram child. It is about 10 times more. What does it mean to say the adult is ten times more disordered than the child? When this adult dies and becomes a pile of dust, the dust will have a much lower entropy (have more order) than either the child or the adult. These examples illustrate why order and disorder are such poor word choices for entropy.

The Sun is dispersing energy to the rest of the Universe, so its entropy must be decreasing (Figure 1.24). If the entropy of a star continually decreases, how could a star exist? Doesn't a star have to obey the second law of thermodynamics that says the entropy of a *closed* or *isolated* system must increase? The question is easily answered if we consider the entire Universe.

The Sun disperses a great deal of radiation (light) energy out into the Universe. If we calculate the change in entropy of the entire Universe, which has been changed by the radiation from the Sun, we find that the entropy of the entire Universe is indeed increasing with time, but there are many smaller pockets within the Universe where the entropy is decreasing. Some of these smaller pockets are places where stars shine, and water freezes.

Therefore, the entropy of the Sun or any star can decrease with time, as long as the entropy of the entire Universe increases. This example illustrates the kinds of errors that can easily result when we try to summarize a rather sophisticated theory, like the second law of thermodynamics, with a short descriptive sentence.

The argument against evolution is that evolution is a process that seems to produce more complex organisms as time passes. However, we have just seen that a large complex organism has *more* entropy than a small simple organism. Complex and simple are also very poor words to use when talking about the second law. As we will see,

Figure 1.24 Entropy of the Sun *The Sun's entropy is decreasing since it is dispersing energy to the rest of the Universe. The entropy of the entire Universe, however, is increasing. The photo shows the Sun during one of its more active times. A large prominence (cloud of hot gas) can be seen rising from its surface (upper left). [STEREO Project/NASA]*

evolution involves changes in the DNA of organisms. This is a random process and has nothing to do with the second law of thermodynamics.

Neither the Sun nor the Earth is an isolated system, and evolution does not take place on an isolated Earth. The energy supplied by the Sun obviously plays an important role in many events (like growth and evolution) that take place on the Earth.

Not only must all scientific theories and ideas be consistent with one another, they are also assumed to be universal in their application. Unless evidence is presented to the contrary, we assume that the rules do not change with time or from place to place within the Universe. As we look at distant stars and galaxies, we find that their behavior can be explained using the same physical theories that operate here on Earth. We have no reason to believe that gravity or electric charges behave differently at other locations in the Universe. Indeed, we see stars orbiting other stars, in obvious submission to the law of gravity.

The light we see today coming from the most distant galaxies, left these objects billions of years ago, and yet that light appears to have been produced by the same kinds of atoms that we find here on Earth. We, therefore, assume that the rules by which nature operates do not change with time or location. This assumption does not mean that the Universe is static. Objects change continually, but we have no evidence to suggest that there have been changes in the laws of physics that determine the behavior of these objects.

We have gone out of our way to describe how theories could be proven wrong and how they may have to be modified as we discover new information. However, we know the major theories of science are substantially correct since they are able to describe nature amazingly well. Even Einstein's ideas, as revolutionary as they were, did not cause us to throw out Newton's work. As we stated earlier, Newton's laws are still used in most everyday situations.

Evolution

The idea of evolution was originally introduced as an interpretation of certain observations from the fields of paleontology (the study of fossils) and biology, especially the geographical distribution of different plant and animal species. Today, evolution incorporates information from many other disciplines as well.

Although scientists discussed evolution before Darwin's time, Charles Darwin and Alfred Wallace formulated the first reasonable explanation for the evolutionary process. Their idea was that natural selection caused evolution. Darwin's book, *On the Origin of Species by Means of Natural Selection*), was originally published in 1859.

Years before, Jean Baptiste de Lamarck had proposed that evolution occurred because organisms have some inner urge to become more complex. Lamarck apparently believed that the *Supreme Creator of all life* instilled this inner power in the organism. Lamarck is often credited with the idea that acquired characteristics are inherited, although this idea was widely held by many others of his time.

The inheritance of acquired characteristics states that changes caused by the environment can be passed on to the offspring. For example, if an individual uses a given set of muscles until the muscles become enlarged, the offspring of this individual may inherit a slightly enlarged set of muscles. Since there seems to be little evidence to support it, Lamarck's idea is not held in high esteem by the scientific community.

Darwin formulated the idea of natural selection by observing organisms in nature. He noticed that the individuals in a given population of organisms displayed a great deal of variation. For example, variations in size, color, and physical abilities such as speed and strength. Interestingly, when Darwin proposed the process of natural selection, he did not know that genes controlled the variations of a trait. Darwin also realized that all organisms produce more offspring than could possibly survive to reproduce.

Although the number of organisms in a population fluctuates somewhat, the average number must be stable over relatively long periods of time. If a population was not stable, either their numbers would decrease and the population would likely become extinct or their numbers would increase and eventually exhaust their food supply.

Populations are stable if each mating pair produces an average of two offspring that survive to reproduce (Figure 1.25). Since a single female can produce large numbers of offspring in her lifetime, most offspring must die before they are able to reproduce. In recent times the number of humans has grown explosively because the average couple has produced more than two surviving offspring.

Darwin reasoned that those individuals who are best suited to the environment would have a better chance of surviving to the reproductive age. Those individuals will be the ones that produce young and, therefore, pass their traits on to the next generation.

For example, animals that eat the leaves of certain tall trees might have an advantage if they were taller and could more easily reach the highest leaves. This advantage might be especially important during times when food is scarce, and other animals had eaten most of the lower leaves. The taller animals of the group would have a better chance of surviving and passing their genes for height to the next generation. Over the years, we might expect the population to evolve into very tall animals. A scenario similar to this simplified one could explain the evolution of the giraffe.

Although our scenario may sound reasonable, such stories are speculative in that they attempt to explain *why* things happen, whereas science primarily attempts to describe *how* things happen.

As we have explained, there is an important difference between scientific data and a scientific theory. If you drop a rock and watch it fall to the ground, you have collected a piece of data for your scientific journal. If you carefully observe the path taken by a rifle bullet after it leaves the barrel of a gun, you have more data for your journal.

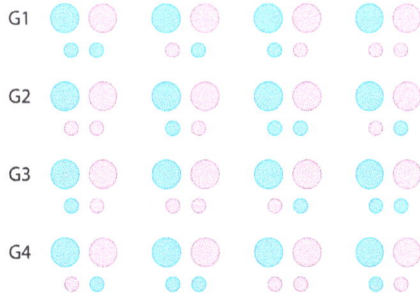

Figure 1.25 A Stable Population In the *first generation (G1) of a population of 4 breeding pairs, each pair has 2 offspring (small circles). The grown offspring pair up and form the next generation (G2), etc. In a real situation there would be more fluctuations, but the diagram illustrates the principle of a stable population. [JCB]*

However, if we wish to explain how the rock and the bullet behave, we use the theory of gravity. Natural selection is a theory that explains the evolution of a population.

When we discuss evolution, or what some scientists carelessly refer to as the *fact* of evolution, we are referring to the changes in life forms that have occurred through time. Did some type of evolutionary process produce the life forms we see today? In other words, do the life forms living today share a common ancestor from the past? For example, do birds and reptiles share a common ancestor from the distant past? Since we did not directly observe a population of reptiles evolve into a population of birds, some might say that we cannot answer this question. However, the data in support of evolution is so overwhelming that the vast majority of scientists have little doubt that evolution did take place. (Remember the tree-limb example above.)

The fossil record provides ample evidence to support the conclusion that life forms have changed substantially through time. This conclusion based on the observational data is the general theme of the historical narrative of evolution.

The most widely accepted theory for the observed changes is natural selection as proposed by Charles Darwin. Natural selection is not a broad general statement that can be used to make definite predictions, but simply a set of conclusions based on observations. However, just as Copernicus suggested that the Earth orbited the Sun before the relevant observations were made, Darwin suggested the idea of natural selection before much data had been collected to support the idea. That foresight is what separates Newton, Einstein, and Darwin from the rest of us who are much better at hindsight.

We have learned a great deal about biology since Darwin's time, and our knowledge of natural selection has increased as well. With our current knowledge, natural selection can be summarized as follows:

1. Individual members of a given population display a great deal of variation, and genes control most of these differences.

2. All organisms produce more offspring than could possibly survive to the age of reproduction.

3. Those individuals that are best suited to the environment will have a better chance of producing offspring and passing their advantageous traits (genes) on to the next generation.

Statements 1 and 2 are simply conclusions based on observations that were made even before the time of Darwin. However, an analysis of statement 3 has led some to believe there is a problem with natural selection. It has been pointed out that *advantageous traits* are defined to be traits that are more likely to be passed on to the next generation. Statement 3, therefore, appears to be a repetitious statement or a tautology. (An example

of a tautology is the statement that says all bachelors are single.)

The important part of statement 3, however, is that it says some traits are more likely than certain other traits to be passed on to the next generation. In other words, Darwin saw evolution as a directed process rather than a purely random process. If evolution were purely random, then all traits would have the same probability of being passed to the next generation.

Darwin decided to place the label *advantageous* on those traits that were more likely to be passed to the next generation. For example, in the arctic, where the ground is often covered with snow, the trait of white hair might be passed to the next generation preferentially over the trait of black hair. No matter what name you give to these traits, if white hair genes are more often passed to the next generation, the population will change over time. The population will evolve! Eventually, all the individuals in the population will have white hair, and the genes for black hair will disappear from the population.

Since Darwin proposed his idea, many examples supporting this type of differential selection have been documented. For example, a species of moth known as the Peppered Moth lives in the woods of England. This moth generally has white wings flecked with darker spots, but a few are dark gray to black in color (Figure 1.26). The lighter varieties blend in well with the tree trunks on which the moths are often found, making them difficult to spot by their natural predators. Therefore, the lighter moths are more likely to survive and pass their *light colored genes* to the next generation.

During the industrial revolution, however, coal soot from the factories darkened the tree trunks around the industrial cities. The lighter moths became easier to see than the darker varieties, and the birds that preyed on these moths ate more of the lighter moths. Soon the lighter colored moths decreased sharply in numbers, and the darker varieties became more common. The population of moths changed in response to the environment. On the sooty tree trunks, the darker coloration was apparently an *advantageous* characteristic, and soon the darker moths were the dominant form. If the factories had continued polluting the air, the *light colored genes* might have eventually been eliminated from the population.

House sparrows were introduced into North America in 1844. Unfortunately, they have now spread over the entire continent and have become a nuisance. However, differences in widely separated populations are already apparent. In the Southwest, the House Sparrows are paler in color and smaller than the sparrows living in more northern locations. We are observing the first stages of evolution in these localized populations.

Groups of finches on the Galapagos Islands have been studied extensively, and it has been documented that certain

Figure 1.26 Peppered Moths A dark variety and a light variety moth on a tree trunk. [Martinowksy from nl]

bill sizes have a higher probability of being passed on to the next generation. During several drought years it was observed that the relative numbers of seed producing plants changed. As the size of the available seeds changed, the survival rates of the birds with different size bills changed also. When most of the available seeds are large, those birds with the larger bills have more reproductive success. The population evolved as the environmental conditions changed.

When most of the available seeds are small, those birds with smaller bills are more successful. A permanent change in the weather would eventually produce a permanent change in the bill size of finches.

Evolution of many other groups of organisms has been well documented. In just a few decades, codling moths, that once attacked only apples, have evolved races that feed on plums and walnuts. Many insects have evolved varieties that are resistant to certain insecticides, and strains of bacteria have become resistant to certain drugs.

It should be noted that natural selection only explains how populations evolve, but we are often tempted to explain why the changes occurred. In other words, we often attempt to figure out why a certain trait is advantageous. It seems fairly obvious that the dark peppered moths had an advantage because they were harder to see on the soot-covered trees. However, trying to figure out why a certain group of reptiles evolved feathers is more speculative. We may be able to determine how it happened, but trying to explain why is much more difficult and uncertain. Many lively discussions involve the why questions.

The above examples of evolution are so well documented that even hard-core creationists accept them as examples of evolution. However, they do not believe that many small changes similar to the ones we have observed can accumulate to the point where one population evolves into a different species.

The word microevolution has been coined to describe those small changes that have not yet produced new species, while evolution above the species level is often called macroevolution. Microevolution has been observed, but macroevolution above the species level has not been directly observed. (Actually, a condition called polyploidy can produce a new species of plant in just one generation. This condition will be discussed in Chapter 10.)

Many people have little trouble visualizing how a population of moths can change color over the years in response to industrial pollution, but they have a hard time visualizing how many of these small changes can produce a new species or a new family of organisms. They find it difficult to imagine how thousands of little changes could eventually transform one group of mammals into dogs and an identical group of mammals into cows. Indeed, it would be very hard to visualize how these changes could have occurred if new genes were not introduced into the population from time to time.

These new genes are produced by mutations, and we will discuss some of the ways mutations arise in Chapter 5. Using a little math it can easily be shown that if just two or three mutations were injected into each of two populations every generation, the two populations would be as different as dogs and cows in less than 100 million years. This number of mutations is consistent with the numbers observed. We will address this topic in Chapter 10.

Evolution says that all the organisms we see today evolved from life forms of the past. A group of today's organisms is different from its ancestor group. We would expect a population of zebras living today to be significantly different from its ancestral population of say 500,000 years ago. Today's population could be so different that we might classify them as a different species.

It should be noted that we are using the word population, not individual. The word evolution is used to describe the changes that occur over time to a given population of organisms. As it matures, a single organism may change significantly during its own lifetime, but these are not the changes of evolution. Populations evolve, individuals mature.

Evolution and Probabilities

The *gene pool* of a population of organisms is the collection of all the genes possessed by the breeding members of the population. As the members of the population mate, have offspring, and die, the gene pool changes. A population will evolve if its gene pool changes over time. The evolution of a population of organisms is really just a problem based on statistical probabilities. Is it possible for you to flip a normal penny and have it come up heads every time? Although theoretically possible, it is highly unlikely that the penny would never come up tails, just as it is highly unlikely that a population could remain unchanged through time.

Mutations introduce new genetic variations into a population, and natural selection is the driving force that causes a population to evolve in a certain direction. However, even if natural selection were not operating, a population that reproduces sexually would have to evolve due to the random nature of sexual reproduction. If new genetic variations were not continually created by mutations, every population of organisms would eventually become homogeneous.

Consider some population of organisms (for example an isolated population of dogs). When a pair of these dogs has babies, each offspring will receive half of his or her chromosomes from each parent. (Many genes are linked on structures called chromosomes.) The genetic make up of each offspring is a purely random process. Since it is unlikely that both parents would pass on all of their chromosomes to their offspring, some of their chromosomes (and the genes on them) will be lost when the parents die (Figure 1.27).

In each generation, chromosomes can be lost by this random process until every chromosome pair is identical and the population becomes homogeneous. At that point, all the individuals would have identical genes except for those differences on the chromosomes that determine the sex of the organism. This process of homogenization must eventually occur because of the random nature of egg and sperm formation, the random nature of fertilization, and the fact that all the chromosomes (and their genes) in one generation are not always passed to the next generation.

The only thing that keeps this homogenization from occurring is the introduction of new genetic material through mutations. These mutations introduce new variations into the population. Therefore, the genetic makeup of the population (its gene pool) changes with time. In other words, the population evolves. The observation that populations of organisms are not homogeneous is strong evidence that new mutations are being continually introduced into the populations and that the populations

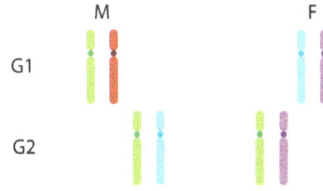

Figure 1.27 Lost Chromosomes Consider *a pair of chromosomes carried by the male (top left), and a pair carried by the female (top right). Each passes one member of the pair to each of their offspring (G2). Since the process is random, the father could pass his green chromosome to each child, and the red chromosome would be lost. [JCB]*

are evolving. Notice that a population would evolve even without the introduction of mutations, but the evolution would be towards a more homogeneous population.

Evolution is like a dice game where each organism brings its own dice. (The dice are like the genes of the organisms.) If some individual can run faster, or is more camouflaged, or can see better, or has some other trait that gives them an advantage over the other members of the population, it is like that individual is playing with loaded dice (or has advantageous genes). It is more likely to be a winner, and being a winner means it is more likely to survive and reproduce, thus passing the loaded dice on to its offspring and the next generation. An organism that does not have loaded dice (no advantageous genes) is more likely to lose the game (it will not have offspring that survive to reproduce), and the unloaded dice will be lost. This is how the gene pool changes and the population evolves.

The only difference between microevolution and macroevolution is one of degree. In a few years or a few hundred years a population evolves a little bit, but in a few thousand years or a few tens of thousands of years, a population evolves so much that we might call it a different species.

Macroevolution is not easily observed simply because it takes a longer time to happen than the time over which we have been making careful observations.

Many organisms living today look and behave very much like their ancestors, others have changed a great deal in appearance. The rate at which a group of organisms changes in appearance is certainly not the same for every group of organisms.

Life forms that look very similar to their ancestors are often referred to as *primitive*, but this word is in many ways an unfortunate choice. Less specialized or less evolved would be better descriptions. As used here, the word primitive is not meant to imply that the organism is in any way less suited for its mode of life than we are for ours. Many populations of organisms become well adapted to their own little niche and subsequently undergo few obvious changes.

The word *advanced* is often used for organisms or parts of organisms that have changed a great deal from their ancestors. Again, this word should not be interpreted as meaning better, but just different. Terms like greatly modified or more specialized would be more accurate descriptions than advanced. Horses have more advanced feet (more highly evolved) than do primates such as humans. Our hands and feet have the same number of bones as did the earliest mammals, but the bones in the foot of the horse have been greatly modified from the primitive condition of early mammals. The dinosaurs were more advanced than the primitive bacteria. However, the dinosaurs are extinct while many bacteria are doing quite well.

How, Why, and Who

As we said earlier, often we are unable to determine why things happen, but asking the why questions and discussing the various possibilities generates excitement. For example, why are most mammals color blind? Why did certain reptiles evolve into birds? As one might imagine, a great deal of disagreement as to the causes of the observed changes exists. Upon hearing these arguments, the general public may be left with the impression that scientists are disagreeing on whether evolution took place, when they are merely arguing about which mechanisms were more important, or just how quickly these changes were brought about.

As we discuss topics related to evolution in the subsequent chapters of this book, we will present evidence that demonstrates evolution has happened, and we will also describe the mechanisms believed to be responsible for the observed changes. Remember, while the evidence that evolution happened is overwhelming, the reasons *why* certain changes occurred are often very speculative.

Because of its rather complex nature, we will incorporate the historical narrative of evolution, the important evolutionary principles (such as natural selection), and the ideas about why the changes occurred into an integrated summary and call it the evolutionary theory. Most scientists undoubtedly have this type of comprehensive picture in mind when they speak of the *theory* of evolution.

In terms of contributing to our understanding of what we observe in the living world, all other ideas become insignificant when compared to evolution. Simply stated, evolution makes sense out of our observations in the fields of biology, biochemistry, and paleontology. Studying biology without a through understanding of evolution would be like studying chemistry without an understanding of the periodic table of the elements, or like studying the motions of the planets without comprehending Newton's *law* of gravity and the *laws* of motion.

As we observe a leaf floating to the ground, we can describe its path using gravity and air resistance. The blowing breeze can be described by the air masses set into motion by the Sun's energy and the rotation of the Earth. We can explain how the leaf is no longer needed by the tree and is shed as the tree becomes dormant during the winter months. However, if someone asks why the leaf falls, we cannot really answer that question. A physicist might reply that it is due to gravity. But why was gravity made to work the way it does? The physicist might go on to explain about the gravitational interaction and mumble something about the Earth and the leaf exchanging gravitons, but he is still only describing how gravity behaves, not *why* the rules of gravity exist. We do not know why there is gravity or why atoms seem to be composed of positive nuclei and negative electrons. We can only describe how they work.

As stated earlier, science tries to answer how questions, but not why questions. However, if we try to classify all questions into the *how* group and the *why* group, we will soon encounter difficulties.

For example, what do we know about mammal vision? We know that there are light receptors in the eye that respond to the light energy entering the eye. These receptors send signals to the brain, and the brain interprets the signals into a visual picture. From experimentation we know that the black-and-white receptors in the eye (called rods) are more sensitive to light than the color receptors (called cones).

In addition, we observe that most mammals are active in the morning, night, and evening hours when the light is less intense and colors are more subdued. It seems reasonable to conclude, therefore, that rod vision (black-and-white vision) would be more useful than color vision under these circumstances. Are we only able to determine how mammals see, or do we have some idea why their vision is dominated by rods? In other words, do we know why most mammals are color blind?

Although this book is not intended to be a creation versus evolution book, another important aspect of science should be pointed out. Science can explain how things operate, and sometimes perhaps it can explain why things happen the way they do, but it certainly does not presume to know if someone created the rules that all objects are forced to obey, or who that someone is.

Who made the Universe, the Sun, the Earth, the plants, and the animals (which includes humans)? As scientists we cannot answer that who question. Most religious people believe that there is a God who created the life of this Earth, and that He also formulated the law of gravity and all the other physical principles that we and all other objects in the Universe are compelled to obey. No scientist can argue scientifically that they are wrong. We cannot scientifically disprove the existence of a God. As we have seen, these beliefs lie outside the realm of scientific investigation. Remember, religions are based on non falsifiable ideas and cannot be investigated by the methods of science.

Does evolution conflict with the idea that a God created everything we observe? The answer to this question depends on your exact view of creation (Figure 1.28). If God made the living creatures of the world, then as scientists we can only determine how He did it. The preponderance of the evidence points to evolution as the method used to create the vast array of creatures that now inhabit the Earth. Contrary to popular belief, evolution does not dispute the existence of God as the creator of all things. Evolution merely attempts to describe how the creation was accomplished.

In some religions, there have been interpretations of dogma or scriptures that conflict with evolution. For example, evolution may conflict with some interpretations of the book of Genesis in the Bible, namely that the Universe, the Earth, and all the living organisms were created in just six consecutive 24-hour days. Others, however, claim that the Hebrew word, which is translated as *day*, can also mean a long time. No matter how one interprets these passages, an overwhelming amount of scientific evidence points to a very old Earth and Universe. If the Earth and Universe are not billions of years old, then God has added apparent age to the Universe to deceive us. Why would God create

humans with a curious mind and then deliberately try to deceive those who dare inquire into the secrets of the Universe?

As we will see, the evidence indicates that the various life forms were not all created at the same time. The earliest organisms that we have discovered were one-celled organisms, and increasingly complex forms appeared over hundreds of millions of years. If the Earth and all its living organisms were created in just six consecutive 24-hour days, then why does all the evidence indicate that these spectacular changes occurred over millions of years?

Surely we do not live in a Universe created by a dishonest God. In science, we assume that the evidence was created by events that happened in the past and that some supreme being did not create the evidence to appear that something happened when it did not. A devious God could have created us a few seconds ago with a programmed memory of things that never took place.

In addition to the six consecutive 24-hour days, there are many passages in the Bible that apparently cannot be taken literally. For example Psalm 93:1

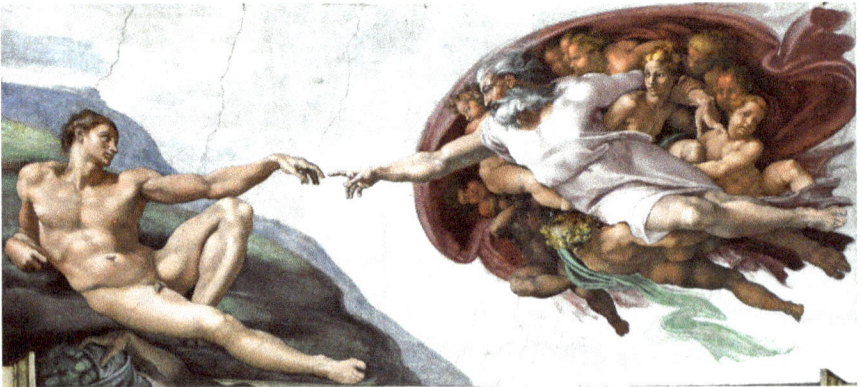

Figure 1.28 Creation of Adam One of the frescos painted by Michelangelo Buonarroti on the ceiling of the Sistine Chapel sometime between 1508 and 1512.

says, "... Indeed the world is firmly established, it will not be moved." This phrase was once believed to be proof that the Earth did not orbit the Sun.

Ecclesiastes 3:14 says, "I know that everything that God does will remain forever, there is nothing to add to it and there is nothing to take from it, for God has so worked that men should fear him." This statement was once used to support the belief that animals could not become extinct since God made them. We now know that many creatures have become extinct. Sadly, humans are now devastating vast areas of the world and causing the extinction of many of the unique and wonderful creatures that inhabit this planet we call Earth. Humans are the only animals capable of comprehending the idea of God, and yet many seem to have little reverence for the creation that some attribute to Him.

Some have claimed that evolution makes God unnecessary. However, one could just as easily support the claim that discovering the details of evolution elevates the image of God, and increases our appreciation and reverence for His work. Instead of picturing God in a playroom making various animals out of clay, He becomes the great designer who formulated the rules, knowing what would be produced and how it would happen. In this context, evolution is the wonderful story of the creation of God's creatures.

Recently, creationists have been advocating an idea called intelligent design. The idea is that when one studies nature, many complicated ideas and intricate patterns are revealed. Indeed, many of the details are so complicated that we do not understand them at this time. Many diseases have not been cured, many other problems have not been solved, and our understanding of some scientific topics is very rudimentary. Intelligent design proponents believe that the complexity of nature speaks of an intelligent designer who is God.

It is possible that an intelligent designer formulated the principles that rule the Universe, but the introduction of an intelligent designer does not replace or even augment our understanding of gravity, electricity and magnetism, chemical bonding, evolution, or any other scientific idea. It does not make it easier for us to find cures for diseases such as AIDS, cancer, and Alzheimer's disease, or guide us in our search for the basic laws that govern the Universe. In addition, the introduction of an intelligent designer is a not a falsifiable theory, so the idea cannot be disproved. It is, therefore, outside the realm of science. If an intelligent designer designed the Universe, then the mission of the scientific community is to discover the details of that design and to study the creation process used to implement that design.

Just as it appears that the laws of physics were used to form the stars and planets, a preponderance of the evidence points to evolution as the process that was used to shape the living world. The idea that life forms were created by evolution may seem hard to believe. However, many ideas in biology are even more unbelievable. It is certainly an even more amazing idea to think that a person who is made of hundreds of trillions of cells could have been created from a single cell formed by the union of an egg and sperm. Yet biologists tell us that is exactly how many life forms are created. The amazing process of reproduction perpetuates life, and the evidence tells us that evolution is the process by which new diverse life forms are created.

Evolution is a process of change. Change is observed in many parts of the Universe. The Earth changes, the stars change, life forms change, and even the Universe as a whole has evolved over time. Before life could form and prosper on Earth, the proper conditions had to be prepared. Many important events contributed to the establishment of these conditions, some of which occurred long before the first organisms appeared in Earth's early seas. A few important events that we would like to investigate occurred in the first few minutes after the formation of the Universe itself. Our investigation will begin there.

Chapter 2: The First Ten Billion Years

Our Place in the Universe

As we gaze up at the sky on a clear night, it may seem as though the stars are distributed randomly throughout the sky. Closer examination, however, reveals that most stars are concentrated in a milky band forming a great circle around the entire sky. This band, known as the Milky Way, is composed of billions of stars, most being so far away that their light blends into a cloudy ribbon. Its true nature is revealed only with the aid of a telescope. This immense assemblage of approximately 200 billion stars, including our own Sun, along with many gigantic clouds of gas mixed with small solid particles called dust, is named the Milky Way Galaxy (Figure 2.1).

Even one billion (1000 million) is such a huge number that if you began counting one number each second, it would take more than 30 years of continual counting to reach one billion. No one has ever counted to a billion! Probably no one has even counted to one million, a task that would require more than two weeks of continual counting.

The average distance between stars in the Milky Way is unimaginably large. Alpha Centauri, the nearest star to the Solar System, is about 300,000 times as far away as the Earth is from the Sun. This distance is about the average distance between the stars in our neighborhood of the galaxy.

Large distances in astronomy are popularly measured in light years. A light year is the distance traveled by a pulse of light in one year. In one-second, light will travel 300,000 kilometers (186,000 miles), which is a little over 7 times around the Earth in one-second! Most people think of the Sun as being relatively close, and yet it takes almost eight and a half minutes for sunlight to make the journey to Earth. More incredible still, it takes over 4 years for light to reach us from our nearest neighbor, the star called Alpha Centauri.

The distribution of its stars gives the galaxy a shape that somewhat

Figure 2.1 Milky Way Galaxy *The bright central region is in the constellation of Scorpio. The Large Magellanic Cloud is the bright pink fuzzy patch below the center and almost half way to the right edge of the photo. The Small Magellanic Cloud is below and to the left of the LMC. The short fuzzy line below the center and three-fourths of the way to the left edge is the Andromeda Galaxy. [Axel Mellinger/Univ. of Potsdam, Germany]*

resembles a fried egg. Most of the stars are concentrated near the center or nucleus of the galaxy, and the stars that surround the nucleus are concentrated in a plane (called the galactic plane), giving the galaxy a highly flattened appearance. When we look at the Milky Way, we are looking into the plane of our galaxy.

The galaxy is held together by the gravitational attraction of its members. Stars orbit the center of the galaxy much like the planets of the Solar System orbit the Sun. The diameter of the galaxy is about 100,000 light years, and our insignificantly small Solar System is located about 30,000 light years from the galactic center. The Solar System is moving at the incredibly fast speed of about 300 kilometers (180 miles) each second! Still, it takes approximately 200 million years to complete just one trip around the galactic center. One galactic *year* ago, the dinosaurs were just establishing their long reign on Earth.

To get a feel for the distances we have discussed, imagine making a model of the galaxy. If we made the diameter of our model galaxy two kilometers across, the distance between the Sun and the nearest star would be about 5 cm (2 inches), and a microscope would be required to see the circles representing the orbits of Jupiter and Saturn. The orbit of Saturn, which is ten times larger than the Earth's orbit, would be about the size of a single bacterium. (Bacteria are so small that 500 lined up in a row would be about one millimeter long.) The Earth's orbit around the Sun would be too small to see with a regular microscope in this model!

As we look out beyond the stars in our own galaxy, we see millions of other galaxies, some of which look similar to the Milky Way. One of the most famous of these galaxies, and one of our closest neighbors, is the Andromeda Galaxy (Figure 2.2). Andromeda is about 2 million light years away. The light from Andromeda that enters our eyes tonight began its journey 2 million years ago, about the time man's ancestors were discovering

Figure 2.2 Andromeda About 2 million light years away, this nearby galaxy covers more than two degrees of sky (four Moon diameters). On a clear dark night, the bright nucleus can be seen with the naked eye in the constellation of Andromeda. [NOAO/ AURA/NSF/B.Schoening, V.Harvey/REU program]

how to make chipped stone tools, and Andromeda is one of our nearest neighboring galaxies.

It is important for us to remember that when we see an object that is very far away, we are not seeing it as it is today, but as it looked in the past. We see many galaxies that are billions of light years away. Therefore, we are seeing these galaxies, as they looked billions of years ago.

The Milky Way and Andromeda galaxies along with a couple dozen other galaxies belong to a rather small cluster of galaxies known as the Local Group. The Large and Small Magellanic Clouds (Figure 2.1) are small irregular shaped galaxies belonging to the Local Group that are very close satellite galaxies of the Milky Way. They can be seen with the naked eye, but only in the Southern Hemisphere. They were seen by Magellan when he sailed around the world during the years 1519 to 1522 (Magellan, of course, died before the voyage was completed).

Most galaxies we see are clustered in groups that can include as many as a thousand galaxies (Figure 2.3). These clusters of galaxies are moving away from us, and the more distant clusters are moving away at greater speeds than the nearby clusters (Figure 2.4). In fact,

Figure 2.3 Cluster of Galaxies About 500 million light years away, this cluster of about 200 galaxies is called the Hercules Cluster because it is located in the constellation of Hercules. Most of the objects in the photograph are galaxies. The angular width of the photo is about 24 arc minutes. Compare this to the angular measure of the full Moon which is about 30 arc minutes or one-half a degree. [ESO/ INAF-VST/OmegaCAM]

Figure 2.4 Expanding Universe Clusters of galaxies are moving away from the Local Group at speeds proportional to their distance. This "motion" is produced by the expansion of space. Each cluster would see the same picture. [JCB]

the distance to a cluster is proportional to its speed of recession. This relationship means that the distance between any two clusters of galaxies in the Universe is getting greater with time.

Einstein told us that in empty space it is impossible to say that one object is moving and another is not. All we can say is that one object is moving relative to another. There is no way to determine if the clusters of galaxies are moving away from us, or if we are moving away from them.

There is no fabric of space to judge motion. If space were filled with some type of material, it would be possible to determine if an object was moving though it. For example, you can easily tell if you are moving through water because you can see the water rushing by you. However, since space is empty, there are no reference points to tell if you are moving or not.

One of the great ideas of Einstein was that there is no way to determine absolute motion through space. Therefore, in agreement with Einstein, the relative motion of the clusters of galaxies is interpreted to mean that space itself is expanding. Since the Universe is composed of all the galaxies and the space between them, this relative motion means that the entire Universe is expanding. It is like the clusters are glued on a piece of invisible rubber, and the rubber is being stretched, or perhaps like the clusters are glued on an invisible balloon, which is getting larger.

A supernova is an exploding star. Studies of a particular type of supernovae in clusters of galaxies have recently indicated that the rate of expansion of the Universe is increasing with time. To explain this observation, it has been suggested that the Universe is filled with some sort of energy that produces a repulsive force, opposing the attractive force of gravity. The source of this energy is unknown, but it has been given the name *dark energy*.

Some sort of dark energy may explain the increased expansion rate, but another possibility is that our theory of gravity is not complete and needs to be amended. We know that Einstein's theory is not exactly correct because it is not consistent with quantum mechanics, but a more complete theory has yet to be formulated.

Although the clusters of galaxies are moving apart, the galaxies within a given cluster are attracted to each other gravitationally. An interesting puzzle is that even though the galaxies within a cluster are held together gravitationally, there is not enough visible material in the cluster to account for the attraction.

Another surprise is that even single galaxies (like Andromeda) do not have enough visible matter to hold on to their own stars. To measure the mass of a galaxy, we observe the individual stars as they orbit the center of the galaxy. Using Newton's law of gravity, we can calculate the mass of the galaxy, just as we can calculate the mass of the Sun by observing the speed of a planet and the size of its orbit around the Sun. To explain the missing mass detected by these observations, astronomers have postulated that some sort of matter is present that holds the individual galaxies together and also holds the clusters of galaxies together. Since this matter has not been seen or detected by usual methods, it must be some exotic form of matter that has not been discovered yet. It is referred to as *dark matter*.

The observational data seems to be telling us that most of the Universe is hidden from our view. This shadow Universe is filled with a form of energy that we do not understand, and mysterious matter that we have only detected by its gravitational influence on the ordinary matter we do see. Current observations and theories suggest that only 4% of the matter in the Universe is ordinary matter and the other 96% is dark matter and dark energy (23% dark matter and 73% dark energy). The exact nature of the dark energy and dark matter are two of the greatest unanswered questions in physics.

If space is expanding and the clusters of galaxies are getting farther apart, they must have been closer together in the past. If you go back in time far enough, the Universe was very small and all the matter must have been greatly compressed. Using the expansion rate of space and our model of the Universe we calculate that the Universe began its expansion about 13.7 billion years ago (13,700 mya). The initiation of this expansion of space is called the Big Bang. The name Big Bang is a little misleading since it was not an explosion of matter into empty space, but an expansion of space itself, that happens to contain some matter.

Distance Measurements

So far we have given some information about the Milky Way Galaxy and the Universe, but have not included any statements about the methods used to determine this information. Scientists are just as interested in how information is determined as they are in what the information is. If the procedures used are suspect, how valuable can the information be? Therefore, when possible, we will include a discussion of some of the methods used to determine the information.

Astronomical distances can be measured by one of two methods. The distances to the nearest stars can be measured directly by triangulation, a method known as trigonometric parallax (Figure 2.5). This method is the one surveyors use to measure distances on the Earth.

Once we have measured the distances to several nearby objects by triangulation, the distance to a more remote object can be obtained by comparing its brightness with the

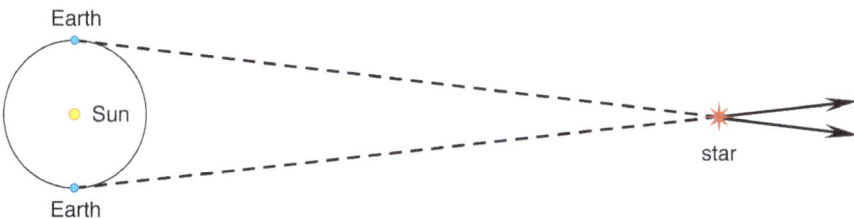

Figure 2.5 Stellar Parallax *From Earth, we see nearby stars in different directions during the year. Knowing the diameter of the Earth's orbit, the distance to a nearby star can be calculated from the Earth-star-Earth angle (the angle between the two arrows). [JCB]*

brightness of a similar object of known distance. The fainter the object looks, the further away it must be. We might call this second procedure the method of similar objects.

By the method of trigonometric parallax, the distance to the nearest few thousand stars can be measured. Once the distance to a star is obtained, a very important property can be calculated, namely the amount of light the star gives off each second. This quantity is the star's true or intrinsic brightness.

Partly for historical reasons, astronomers calculate a quantity related to this brightness that is called the absolute magnitude. To point out the obvious, we realize that how bright a star appears has no bearing on how bright it really is. As seen by us, the brightest appearing star in the sky is the Sun, not because it gives off more light than any other star, but simply because it is so much closer to us than any other star. If we could place all the stars at the same distance, the Sun would be a star of average brightness.

The intrinsic brightness or absolute magnitude of a star depends on its surface temperature and size. We can understand why if we consider a familiar example. The hotter you make the filament in a light bulb (usually by sending more current through it), the brighter will be the light. This is why a flashlight gets fainter as the batteries die. Old batteries cannot send as much current through the bulb as new batteries. Therefore, the filament does not get as hot.

Another way of increasing the light output is to get a larger filament area. The easiest way to accomplish this is by getting another flashlight. Two hot filaments (two flashlights) produce twice as much light as one because they have twice the emitting area.

The surface temperature of a star can be determined approximately by its color (Figure 2.6). The coolest stars are red. Although red stars emit light of all colors, most of the visible light is red, so we see them as red. Stars that are a little hotter will emit mostly orange light, and stars like the Sun emit mostly yellow light. Because of the sensitivity of our eyes, objects that emit mostly in the yellow-green region of the spectrum look white to us. The hottest stars appear bluish-white to our eyes.

Since we know how the apparent brightness of a light depends on the distance, (if the distance is doubled, the object will appear one-fourth as bright) we could calculate the distance to a star if we somehow knew how bright the star really was (its intrinsic brightness). Fortunately, by studying the nearby stars and stars in clusters, it was discovered that the intrinsic brightness could be determined from the spectrum of the star. A stellar spectrum is created by passing starlight through a prism,

Figure 2.6 Color of Stars A star's color is related to its surface temperature. The coolest stars are about 3,000 K. They give off light of all colors, but most of the light is in the red region of the spectrum so they look red. Stars of 4,000 K emit mostly in the orange. Our yellow Sun is about 6,000 K. At 8,000 K stars emit mostly yellow-green light, which is where our eyes are most sensitive, so they look white. Stars hotter than about 10,000 K look bluish-white. *[JCB]*

which separates the light into its component colors.

The dense gasses inside a star produce light of all colors. However, as this light passes out through the more rarefied gases near the surface of the star, certain colors are removed, creating dark places in the spectrum (Figure 2.7). These dark places are called dark lines or absorption lines. At a given temperature, every different kind of gas produces its own unique pattern of absorption lines.

By carefully examining these absorption lines, astronomers can determine the surface temperature of the star, the star's size, its chemical composition, and many other interesting properties of the star. The intrinsic brightness can also be determined since it depends only on the star's surface temperature and its size. Finally, we can calculate how far away the star must be to appear as faint as we see it.

Even our nearest neighbor galaxies are so far away that most ordinary stars are too faint to be seen individually. Only the brighter stars can be seen as individual objects. Some of these brighter stars are a type known as Cepheid variables, which are important tools for measuring the distances to some of the nearer galaxies. The brightness of a Cepheid variable pulsates in a regular way. That is, they get larger, then smaller, and then larger,

somewhat like a beating heart. As the size of the star varies, so does its temperature and brightness.

By studying Cepheids in the Small Magellanic Cloud, it was discovered that the average intrinsic brightness of these stars is related to the time it takes for the star to vary in brightness. This relationship is not too surprising since we expect larger stars to be brighter, and it should take longer for these larger stars to pulsate. Therefore, by measuring the period of variation, the average intrinsic brightness can be found.

From its average intrinsic brightness and its average apparent brightness, the distance to the star can be calculated as it was in the above example. Since it is not necessary to take a spectrum of the star (a process that requires more light than just a photograph), this method can be used for greater distances than the spectrum method described above.

Several methods have been devised for measuring the distances to the more remote galaxies. For example, the distance to an unknown galaxy can be approximated by comparing its brightness to a similar galaxy whose distance is known. Although this method is not accurate enough for most purposes, it does give us an idea of the immense distances involved when we look at distant galaxies.

The method commonly used to find the distances to clusters of galaxies is

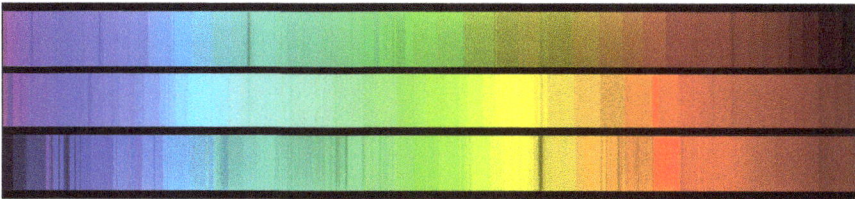

Figure 2.7 Stellar Spectra The top spectrum, from a very hot star, shows four dark absorption lines of hydrogen, in the violet, blue-violet, blue-green, and red regions. The middle spectrum is from a yellow star like the Sun, and the lower spectrum is from a cooler red star. The very dark line in the yellow is produced by sodium gas. [NOAO/AURA/NSF]

the brightness of certain supernovae. As we will see later in the chapter, near the end of a star's life, the star may become unstable and a large portion of the star may be blown off in a giant explosion. Such an exploding star is called a supernova. For a few weeks, a supernova can be as bright as an entire galaxy. By studying supernovae in nearby galaxies, astronomers have discovered that the absolute peak brightness of a certain type of supernovae is very constant. Knowing the real brightness of the supernova and how bright it appears to us, the distance to the galaxy in which it is located can be determined.

For more detailed information on methods of determining distances, the reader is referred to any of the numerous elementary astronomy texts. Before we leave the topic of distance determinations, however, it should be pointed out that a simple calculation can show that the galaxies must be at great distances. First, we know the brightness of the Sun. Sirius, one of the closer stars to the Sun, is about 100 times brighter than the Sun. The brightest stars are about 10,000 times brighter than Sirius.

With our present telescopes, we should be able to see these brightest stars to a distance of about 100 million light years. Yet, other than an occasional supernova, we do not see individual stars in the vast majority of galaxies. Most are so far away that the starlight blurs into a diffuse glow. Galaxies that are too distant to distinguish a few bright individual stars must be more than 100 million light years away. There are only a few thousand galaxies closer than 100 million light years, but billions of galaxies lie beyond this distance.

Size of the Milky Way

How do we determine the size of the Milky Way Galaxy and the Solar System's location in it? This question becomes even more perplexing when one is told that the numerous gas clouds in the galaxy contain dust that makes it impossible to see very far in those directions in the galaxy. The solution to this problem begins by observing other galaxies.

Distributed spherically about the centers of other galaxies, we see large clusters of stars called globular clusters (Figure 2.8). The distribution of these clusters extends out near the edges of their respective galaxies and well above the plane of the galaxy. Therefore, it seems logical to conclude that the globular clusters in the Milky Way Galaxy are also distributed spherically about the center of our galaxy. By measuring the distances to the globular clusters in our galaxy and making a plot of their distribution (Figure 2.9), we can determine both the size of the galaxy and location of its center.

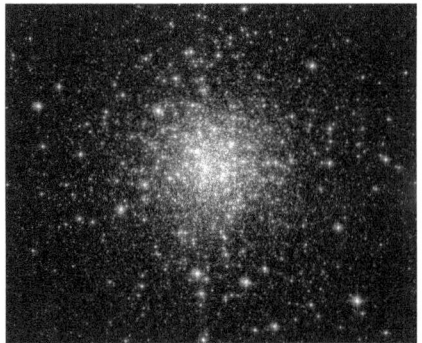

Figure 2.8 Globular Cluster Known as NGC 6934, this cluster of over 100,000 stars is one of over 150 globular clusters in the Milky Way Galaxy. Located in the constellation of Delphinus (the dolphin), it is about 50,000 light years away and about 50 light years across. [NASA/ESA/HST]

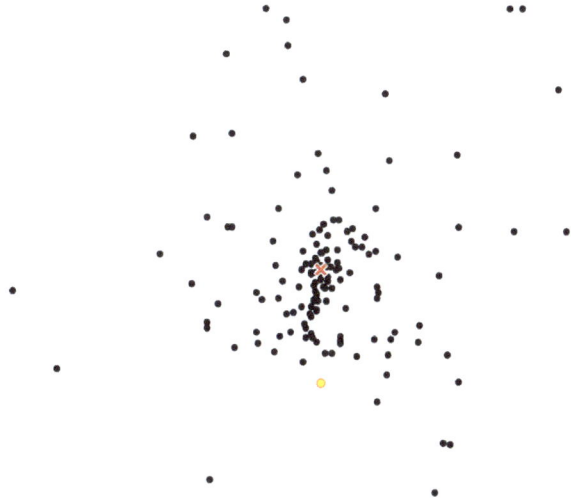

Figure 2.9 Galactic Center *This plot is of the globular clusters within 100,000 light years of the galactic center, projected onto the galactic plane. The red* x *marks the center of the galaxy and the Sun is located at the center of the yellow circle (below the red* x*). Knowing the distances to the clusters, we can determine the size of the galaxy and our location. Note: The orbit of Jupiter around the Sun would be about the size of an atom in this figure! [JCB; data: W.E.Harris, see credits page 414]*

The center of the Milky Way lies in the direction of the constellations of Sagittarius and Scorpius. When visible from the United States and Europe, this area of the sky is very near the southern horizon. Since stars near the horizon are dimmed by the Earth's atmosphere, this region does not usually appear overly impressive. However, when seen overhead, from an area near the Equator, the spectacular brightness of this portion of the Milky Way leaves little doubt that one is looking toward the central nucleus of our galaxy. It is truly a spectacular sight!

Age of the Universe

The expansion of space has a measurable effect on the light emitted by distant galaxies. As the light waves travel through space, the expanding space stretches the light waves. Therefore, the wavelength of the light is stretched and becomes longer (redder) as it travels to us. This effect is referred to as a *cosmological redshift*. The more distant the galaxy, the longer it takes for its light to get to us, and the more the light waves are stretched or shifted to longer wavelengths by the expansion of space (Figure 2.10). Therefore, the more distant a galaxy is (or cluster of galaxies), the greater will be its cosmological redshift.

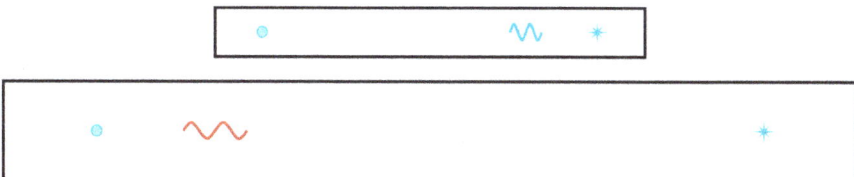

Figure 2.10 Cosmological Redshift *A star located in a distant galaxy emits a blue light wave (top) that travels toward us (the blue circle to the left of the pictures). As the wave travels (bottom), space expands and stretches the wave, causing it to shift to a longer (redder) wavelength. [JCB]*

Since galaxies are composed of billions of stars, it should come as no surprise to learn that their spectra are similar to that of an average star. Remember that stellar spectra have patterns of absorption lines. Since we know the wavelength (color) of the absorption lines when produced in the laboratory, we can easily measure the cosmological redshift of the galaxy, which is the factional change of a wavelength due to the expansion of space. For example, if the wavelength is stretched so it is 75% longer, that would be a 0.75 fractional change.

The cosmological redshift of a certain cluster of galaxies gives us the average rate at which space has expanded since the light we see left that cluster. If we knew the distance to the cluster, we could use the rate of expansion to calculate the time it took for the space between the cluster and us to expand that much. (The distance is equal to the rate multiplied by the time.) This time would be the time that has elapsed since the Big Bang, a time we commonly refer to as the age of the Universe.

There is a slight problem with this explanation. If the rate of expansion has changed with time, using the average rate of expansion for a cluster would give a slightly inaccurate time. To correct for this, we measure the rate of expansion for clusters at various distances, which corresponds to different expansion times. Remember that as we look to greater and greater distances, we are looking back in time. Therefore when we measure the expansion rate at different distances we are getting the expansion rate for different times in the past. We can then determine how the expansion rate has changed with time and correct our age calculations accordingly.

When we determine how the expansion rate has changed with time,

we find that the expansion slowed for the first 8 billion years or so after the Big Bang, but the expansion has been increasing during the last 5 billion years.

We would expect the expansion rate to slow because of the gravitational attraction of the various clusters of galaxies on each other. It was quite a surprise, however, to discover that about 5 billion years ago (5,000 mya), the expansion rate began to increase. What could cause the clusters to begin repelling each other? More accurately, what could cause space to expand at an increasing rate? Just as a car moving down the freeway needs energy to accelerate, there must be some energy source driving this increased expansion of space. The source of this energy is unknown at this time, but as we mentioned earlier, it has been given the name *dark energy*.

If we look at the age of the Universe calculations that have been made during the past 10 years or so, we find that the values have ranged from about 10 to 20 billion years. These ages were calculated using distances measured by more than a dozen different methods. The quoted value of 13.7 billion years is consistent with the average of these values and is the number obtained by analyzing the cosmic background radiation with the current Big Bang model. We will discuss the cosmic background radiation later in the chapter.

The Nature of Matter

Since we know of no way that matter could have been created in the middle of an expansion such as we see today, we assume that initially all the matter was squashed together and was, therefore, very dense and hot. Although it is not essential, we also assume that this matter filled all of space. This assumption is based on the fact that even though we can see out to distances of billions of light years, we do not see an edge to the matter in the Universe. That is, we observe galaxies to the limit of our telescopes in every direction. One should not imagine either an edge to the matter in the Universe or a center of the Universe.

Assuming the Universe has no edge requires that either the Universe is infinite, or that space is curved in such a way that it has no center or edges. Just as we can travel around on the two-dimensional surface of Earth and never find an edge or a center on its surface, it is possible that the three space-dimensions of the Universe are curved in such a way that no center or edge is present.

At the extremely high temperatures and densities believed to have occurred in the first fraction of a second after the Big Bang, the matter would have been in the form of exotic particles, which today are produced only in high-energy accelerators.

The temperature of a substance is a measure of the average energy of motion of the particles in the substance. This energy of motion is called kinetic energy. The kinetic energy depends on the mass and speed of the particle. Therefore, we study the behavior of high temperature particles by means of high-energy accelerators. Particles are smashed into each other at great speeds (high temperatures) to monitor their interactions.

Particle physicists have formulated and tested many of our ideas about the physical processes that must have occurred in the initial stages of the Big Bang. (A particle physicist is a physicist who studies the fundamental particles out of which matter is made.) Since the nature of matter is so important to our discussion, we shall

digress briefly to describe some interesting aspects of particle physics. This discussion will also allow us to understand how a star produces energy, and it will help us to understand the methods used for dating rocks. We will discuss these topics later.

Our theory of the atom is that it is composed of a very heavy, compact, positively charged nucleus surrounded by much lighter negatively charged *electrons*. Except for the common form of hydrogen, all atomic nuclei contain two types of particles, *protons* that have a positive charge and *neutrons* that have no electrical charge.

We call this picture of the atom a theory since no one has ever seen a proton or electron, but by postulating the existence of these particles, we are able to create a consistent picture that explains the various forms of matter and accurately describes the results of every atomic experiment that has ever been performed.

Most people have learned that like charges repel each other and unlike charges will attract each other. This force is commonly known as the electrical force, but physicists often refer to it as the *electromagnetic interaction*. Since moving electrical charges produce magnetic forces, electric forces and magnetic forces must be different aspects of a single force, which we prefer to call the electromagnetic force. Interaction is often preferred to the word force since it is an *interaction* between two charged particles that produces the force they feel.

The chemical properties of an atom are determined by the interactions of the electrons. However, it is the positive charge on the nucleus that attracts the electrons and determines how they cluster around the nucleus. Since the protons carry the positive charge, it is the number of protons that

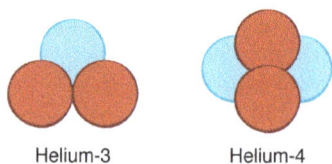

Figure 2.11 Nuclei of Helium Isotopes *Helium nuclei have 2 protons (represented as red spheres). Isotopes of helium differ by their number of neutrons (blue). Helium-4 has 2 neutrons, and helium-3 has only 1 neutron. [JCB]*

ultimately determines the chemical properties of an atom.

Atoms with the same number of protons and, therefore, the same chemical properties, are given the same *element* name. For example, any atom with two protons in its nucleus has the chemical properties we associate with the element helium. The most common form of helium also has two neutrons, but a nucleus with two protons and one neutron is still helium, although it is a different form that we call helium-3. Different forms of the same element are referred to as *isotopes* of that element (Figure 2.11). There can be many isotopes of a given element, differing only by the number of neutrons they contain.

If all protons have a positive charge, they must repel each other in the nucleus of an atom, and yet the particles in the nucleus are held tightly together. Therefore, an even stronger force than the electromagnetic force must be responsible for holding these

particles together. We call this force the nuclear force or the *strong interaction* since it is the strongest interaction known. Neutrons and protons, referred to collectively as *nucleons*, participate in the strong interaction, but electrons do not.

The strong interaction is a short-range interaction. That is, to feel this strong attractive force, nucleons must be very close to each other (Figure 2.12). Since the repulsive electrical force of the protons tries to blow the nucleus apart and the attractive nuclear force tries to hold it together, you might think that the nucleus is a delicately balanced object. Generally, however, this is not the case. Since most nuclei are very small, the attractive nuclear force of the nucleons completely overwhelms the weaker electromagnetic repulsion of the protons.

Protons are positively charged, but in many other ways they are very similar to neutrons. Outside the nucleus, a neutron is unstable and will change into a proton. An electron and a particle called a *neutrino* (actually an *antineutrino*) are also produced in the decay.

If enough energy is available, a proton can be transformed into a neutron. This can happen by the emission of a positively charged particle (along with a neutrino). Except for its electrical charge, this positively charged particle has physical properties

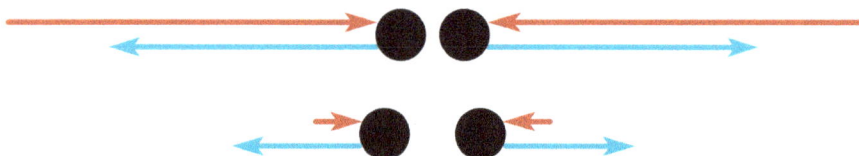

Figure 2.12 Proton Attraction *The nuclear force (red arrows) and electrical force (blue arrows) get weaker as the distance increases, but the nuclear force drops off more quickly. Therefore, the attractive nuclear force dominates when the protons are very close (top), but the repulsive electrical force dominates when they are farther apart (bottom). [JCB]*

very much like an electron. It is a form of *antimatter*, which is the antiparticle of the electron. This *antielectron* is usually called a *positron*.

Although every type of particle has a corresponding antiparticle, the matter in the Universe is only composed of ordinary matter. (We will return to this interesting fact later in the chapter.) Antimatter has only been produced and detected in high-energy accelerator experiments. When a particle of matter and its antimatter particle come together, they completely annihilate each other, releasing energy in the process.

The interaction that allows neutrons to be transformed into protons or vice versa is called the *weak interaction*. It is noteworthy that the weak interaction along with the strong or nuclear, the electromagnetic, and the gravitational interaction are the only four interactions that are known to exist between the numerous particles in the Universe.

Today, along with the familiar nucleons (protons and neutrons), more than a hundred other short-lived particles have been discovered in accelerator experiments.

In one accelerator experiment, very high energy electrons were fired at neutrons and the results were very interesting. Some electrons passed right through the neutrons, but others were deflected as if they had come close to a charged particle. The interpretation of this experiment was that neutrons must be composed of even smaller particles that are electrically charged. Similar experiments showed that protons are also made of smaller charged particles.

In 1963 a highly successful theory was proposed that explained all the known particles and predicted some that have since been discovered. This theory postulates that all the particles

in the grouping that includes protons and neutrons are made of smaller particles called *quarks*. The particles in the group to which electrons and neutrinos belong are not made of other particles, but are believed to be truly fundamental.

There are six different kinds of quarks named: down, up, strange, charmed, top, and bottom. Besides electrons, ordinary matter contains only protons and neutrons, which are only made of down and up quarks. Particles made of the other four kinds of quarks exist only at higher energies such as in high-energy accelerators.

A neutron contains one up and two down quarks. Up quarks are positively charged and down quarks are negatively charged. The positive charge on an up quark is two-thirds as strong as the charge on an electron and the negative charge on a down quark is one-third as strong as the charge on an electron. Therefore, a neutron made of one up quark and two down quarks has no net charge $(+2/3 -1/3 -1/3 = 0)$. A proton is made of two up quarks and one down quark. Therefore, it has a positive charge that is the same strength as the negative charge on the electron $(+2/3 +2/3 -1/3 = +3/3 = +1)$.

We believe that quarks are the only fundamental particles that interact with each other by way of the strong interaction. The strong interaction of the three quarks in a nucleon, produces the attraction that holds the nucleon together. In the nucleus of an atom where the nucleons are very close to each other, the quarks in one nucleon will interact with the quarks in a neighboring nucleon, thus explaining the attraction between nucleons (the nuclear force). The nuclear force or strong interaction is an interaction that only involves quarks, or particles composed of quarks (such as protons and neutrons). Electrons and neutrinos

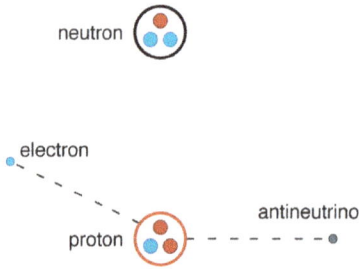

Figure 2.13 Neutron Decay A neutron (top) is composed of two down quarks (blue circles with a charge of –1/3 each) and one up quark (red circle with a charge of +2/3), making it electrically neutral. A neutron decays into a proton (bottom), when a down quark changes into an up quark by emitting an electron and an electron-antineutrino. The proton has a net charge of +1. [JCB]

are not composed of quarks so they do not interact by way of the strong interaction.

When a neutron changes into a proton as was described above, one of the down quarks in the neutron changes into an up quark. In this process, an electron and an antineutrino are emitted (Figure 2.13). Similarly, under conditions of high energy (high temperature) one of the up quarks of a proton can change into a down quark, and the proton will become a neutron. A positron and neutrino will be emitted in this process.

As we have seen, there are only a few particles that can be considered as truly fundamental, meaning they are not made of other particles. If we group these fundamental particles according to mass, they seem to fall into two groups, the heavier particles which are the quarks, and the lighter particles that are the different kinds of neutrinos and electrons.

From accelerator experiments, we have discovered that there are three different electron-like particles, the electron, and two other negatively charged particles with similar properties known as the muon and the

tau. There are also three types of neutrinos, one associated with each of the three electron-like particles (the electron-neutrino, the muon-neutrino, and the tau-neutrino). As their names imply, neutrinos (which means *little neutral*) do not have an electric charge. Neutrinos are very strange particles. They have almost no mass and they travel at almost the speed of light.

It is possible to group fundamental particles into repeating patterns called families or *generations* (Figure 2.14). The generation of particles out of which ordinary matter is formed includes the up quark, the down quark, the electron, and the electron-neutrino.

In high-energy accelerators we can produce the second generation of elementary particles whose members are the charmed quark, the strange quark, the muon, and the muon-neutrino.

At still higher energies, a third generation has been discovered which includes the top quark, the bottom quark, the tau, and the tau-neutrino. You might wonder if there is a fourth

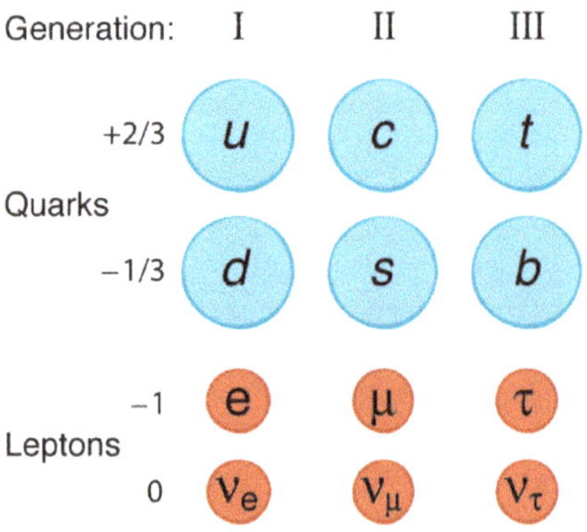

Figure 2.14 Elementary Particles There are three generations of elementary particles. Ordinary matter is made of only first generation particles. Although they are represented here as spheres, as far as we know, they are point objects. The electric charge of the particles in a given row is shown at the left. [JCB]

generation that is still waiting to be discovered, but experiments suggest that this is not the case.

Corresponding to each particle listed above is an associated antiparticle. For example, there is an up antiquark, down antiquark, antielectron (called a positron), electron-antineutrino, etc.

As we have seen, quarks or particles made of quarks are the only particles that participate in the strong interaction. Quarks and electron-like particles are charged so they are the only fundamental particles that participate in the electromagnetic interaction (neutrinos do not). All particles or objects made of particles (like animals, planets, and stars) participate in the gravitational interaction. All fundamental particles also participate in the weak interaction.

The weak interaction can perhaps be most easily visualized by considering the following interaction. At very high energies (very high temperatures), an electron can collide and interact with a proton and produce a neutron and electron-neutrino. What actually happens in the collision is that the electron interacts (by way of the weak interaction) with one of the up quarks in the proton and changes it into a down quark and an electron-neutrino (Figure 2.15). In this example, the weak interaction is between the electron and the up quark. We will see later that this reaction is very important in supernovae. This process also illustrates why we prefer to use the term interaction instead of force.

Our theory of the basic interactions (called quantum field theory) is that particles interact by exchanging *field particles*. This interaction can be thought of like the interaction between two people playing catch with a baseball. The interaction between the two people is produced or mediated by the baseball. In a similar fashion, the

electromagnetic interaction (attraction or repulsion) experienced by two charged particles is mediated by a field particle that is exchanged by the charged particles. This field particle is called the photon, which is a particle of light. Two charged particles feel each others presence by the exchange of photons.

When two charged particles interact, they exchange photons that are called *virtual photons*, because the photons involved in an interaction cannot be detected during the interaction.

Field particles are so named because we often talk about a charged particle having an electric field. Whenever we talk about a field, there is always a particle that is associated with the field, or sometimes several field particles. As we have just seen, the photon is the field particle that mediates the electromagnetic interaction.

There are eight field particles called *gluons* that mediate the strong interaction between quarks, and there are three field particles that mediate the weak interaction. Two of the weak field particles (known as W bosons) have opposite electric charges (Figure 2.15), and the third is a neutral particle called the Z boson. The field particle for the

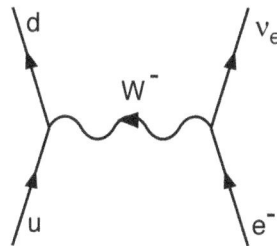

Figure 2.15 Weak Interaction Schematic diagram showing an up quark (moving up from lower left) interacting with an electron (moving up from lower right) to produce an electron-neutrino (upper right) and a down quark (upper left). The field particle that mediates the interaction is a negative W boson (middle of the figure). [JCB]

gravitational interaction is called the graviton, but it has not yet been detected.

One question that puzzled particle physicists about the fundamental particles is why they have mass. One idea is that there must be another field the fills all of space and gives particles their mass. This field, which is called the Higgs field, produces a drag on the various particles. The more the drag, the harder it is for the particle to move and, therefore, the more massive it appears to be. This is just how we define mass. A truck is harder to move than a bicycle (even in outer space), so we say the truck is more massive. With the introduction of the Higgs field, the idea of mass is made obsolete, and it is replaced by the strength of the Higgs interaction.

The particle associated with the Higgs field that mediates the Higgs interaction is called the *Higgs boson* (named after Peter Higgs, one of the men who postulated its existence). These Higgs bosons interact strongly with the up quark and therefore the up quark appears massive. On the other hand, Higgs bosons interact weakly with the electron and, therefore, the electron is not very massive. Without the Higgs bosons, our theory of elementary particles predicts that all particles would be massless.

The European Organization for Nuclear Research (also known by the French acronym CERN), built the largest particle collider in the world, and in 2013, the Higgs boson was discovered at that facility.

A proposal in 1989 to improve communications between the many collaborators associated with CERN led to the formation of the World Wide Web in 1990.

Except for gravity, the theory that specifies how the fundamental particles interact with each other is called the Standard Model of particle physics. The Standard Model is the most comprehensive theory we have for explaining how nature works.

However, it does not explain several important observations such as why the Universe is filled with matter, but not antimatter, and it cannot explain the nature of dark matter or what causes the Universe to expand at an increasing rate (the dark energy).

The Standard Model also cannot explain why there are three generations of fundamental particles, even though ordinary matter is made only of particles belonging to the first generation. Although the Standard Model has been a spectacular achievement, extending it to address these questions is one of the major goals in physics today.

Earlier, we said that the Universe is expanding, but it is only the space that is expanding, not the matter. The clusters of galaxies themselves are not expanding and neither are other objects in the Universe like galaxies, stars, planets, animals, or atoms. Now we can understand why this is so. For example, consider an atom. The size of the atom is governed by the strength of the electromagnetic interaction between the positively charged nucleus and the negatively charged electrons. The only way the size of the atom could be changed is by changing the strength of the electrical interaction. There is no evidence that this is happening or ever has happened. Similarly, the size of an atom's nucleus is governed by the strong interaction, and the only way to change the size of a nucleus would be to change the strength of the strong interaction.

The size of the Sun, stars, and planets is governed by the strength of the gravitational interaction, which also appears to be constant. Since the strengths of the basic interactions have not changed, the sizes of objects made of the fundamental particles will not change as space expands.

The Big Bang

Let us now continue to describe some of the important astronomical events that we believe occurred during the past 13.7 billion years and see what evidence there is to support these statements. We assume that the fundamental theories of physics that accurately describe the observations today have not changed since the Big Bang began. As we look out into space, we are looking back in time. The light which took millions and even billions of years to reach us can be explained by the same theories we use today. We do not invent new theories in science unless the data demands it.

During the first fraction of a second after space began to expand, the temperature was unbelievably high, and the matter must have been in the form of the fundamental particles, quarks, electrons, neutrinos and their corresponding antiparticles. Along with the matter and antimatter, there were large quantities of light energy.

When we say light, most think of visible light, but there are many forms of *electromagnetic radiation* as it is more correctly called. Electromagnetic radiation comes in bundles or *quanta* containing a definite quantity of

energy. A quantum of light is called a *photon*, and a violet photon contains more energy than a red photon.

Listed in order of increasing energy contained in a quantum, the forms of electromagnetic radiation include radio waves, microwaves, infrared light, visible light (red, orange, yellow, green, blue, violet), ultraviolet light, X-rays, and gamma rays (Figure 2.16).

Although some are called waves, some are called light, and some are called rays, they are all similar. They differ only in their wavelength and the amount of energy contained in one quantum. At the high temperatures that existed in the early part of the Big Bang, most of the light energy was in the form of gamma rays (the most energetic quanta of electromagnetic radiation). We will discuss the role of this radiation later.

Earlier, we learned that a particle and its antiparticle can annihilate each other, releasing energy in the process. This energy is emitted as gamma rays. Also, a particle and its antiparticle can be created from a gamma ray. In the very early Universe, various particles and their antiparticles were being continually created and annihilated in this manner.

When the Universe began to expand, the temperature and density dropped rapidly. (The density is the mass divided by the volume.) As the temperature dropped below a certain value, there were not enough energetic

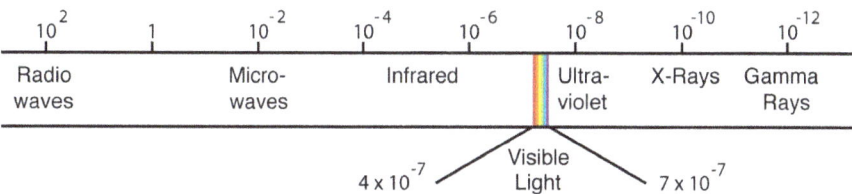

Figure 2.16 Electromagnetic Spectrum Arranged from lowest energy quanta (left) to highest energy quanta (right). The numbers are the wavelengths in meters. The wavelength of a wave is the distance from one crest of the wave to the next crest. [JCB]

gamma rays to create particle and antiparticle pairs, and the matter quickly annihilated the antimatter.

For some unknown reason, more matter had been created than antimatter and when the antimatter was annihilated, the early Universe was left with an excess of ordinary matter. Why the Universe contains only matter and no antimatter is a problem that hopefully will be explained by future high-energy accelerator experiments and by a modification of the Standard Model of particle physics.

As the Universe expanded during the first few seconds, the temperature and density continued to drop. Although all types of quarks were initially present, higher energy quarks soon decayed leaving only up and down quarks. These quarks began to stick together in triplets to form protons and neutrons. Since up quarks are less massive than down quarks, up quarks are more easily produced and they outnumbered the down quarks at the time the triplets formed. Consequently, as the quarks began to stick together more protons (two up quarks and one down quark) were made than neutrons (one up and two downs).

This sea of protons and neutrons expanded and cooled further. After about four minutes, when the temperature dropped below one billion Kelvin, an important process began. (Temperature on the Kelvin scale is essentially the same as Celsius at this high temperature.) Some of the protons and neutrons stuck together to form larger nuclei. Since the ratio of protons to neutrons at this time was approximately 7 to 1, about 25% of the mass formed helium while about 75% remained as hydrogen (single protons). These percentages were calculated from well-known physics principles,

and they agree with what we observe in the Universe today.

We might imagine that all the elements were made at this time. However, studies of nuclear reactions tell us that few nuclei heavier than helium were created. Accelerator experiments have demonstrated that elements containing 5 or 8 nucleons are unstable and cannot be formed under these conditions. If 8 nucleons will not stick together, how can you add particles to get a nucleus containing 9 or more nucleons? We believe, therefore, that it is extremely unlikely that significant quantities of the heavier elements were made under the conditions found in the Big Bang.

Reactions such as three helium nuclei combining to form a carbon nucleus do occur, but only at temperatures above 100 million Kelvin and at extremely high densities. In the Big Bang, by the time sufficient quantities of helium had been produced, the temperature and density would have already dropped to such low values that very few of these reactions could have taken place. Therefore, at this point in the expansion, just a few minutes after the start of the Big Bang, the nuclear reactions had ceased.

Although few nuclei heavier than helium were produced in the Big Bang, small amounts of the less common isotopes of hydrogen (deuterium) and helium (helium-3) were formed. Scientists believe that the Big Bang provides the only mechanism for making deuterium and helium-3 in the quantities we find present today. From accelerator experiments performed at different energies (temperatures), and by measuring the abundance of these isotopes in the Earth, Sun, and other stars, we can calculate what the temperature and density must have been when these isotopes were being formed in the Big Bang.

This information is of interest because it has a bearing on whether the Universe will expand forever or whether the gravitational attraction of the galaxies will eventually halt the expansion, in which case the matter will begin to collapse. If the temperature was extremely high and the density relatively low when deuterium was formed (a very violent expansion), the Universe will expand forever. On the other hand, if the temperature was lower and the density higher (a less violent expansion), we expect that gravity will eventually stop the expansion and cause the Universe to collapse. The current data seems to favor the ever-expanding model. However, the data is surprisingly close to that critical value that divides these two cases, so there is still some room for doubt.

An interesting question that has been asked is why the value is so close to the critical value. The critical quantity that is generally talked about is the average density of the Universe. If the average density of the present Universe is less than some critical value, the Universe will expand forever, but if the average density is greater than this critical value, the density of the Universe will be large enough to stop the expansion. The possible range in values is enormous, so why should the actual value be close to the critical value?

Since the average density of the Universe it is so close to the critical value, many scientists believe that the value must be exactly the critical value for some physical reason that has not yet been determined. Einstein's theory of general relativity states that the density of matter in the Universe determines the curvature of space. If the average density of the Universe is

near the critical value, that means that space is very nearly flat.

One proposal that answers the critical density problem (and a few other problems) states that there might have been a period of very rapid expansion early in the Big Bang, which made the space essentially flat. The idea is analogous to making the surface of a sphere almost flat by expanding the sphere. A small area on the surface of a huge sphere appears very flat. This early expansion is called a period of *inflation*, but the cause of this hypothetical inflation is not known.

Just after the Big Bang, along with the matter, space was still filled with large quantities of electromagnetic radiation that was being absorbed and reemitted by the matter. At that time, the matter was still so hot that the electrons were not orbiting nuclei, but the electrons and nuclei were moving about independently. In other words, the atoms were completely ionized. A hot ionized gas with matter and radiation interacting in this way is called a *plasma*.

Thick plasmas are opaque because everywhere you look you see electrons that are emitting the light that they just absorbed, or the electrons are scattering photons. Light is coming from everywhere in the plasma. If we had been present then, we would have seen a hot glowing gas filling all of space (before we were vaporized by the intense radiation). The illusion would be similar to what one would see inside the Sun or inside the flames of a forest fire. Our vision would have been severely limited because hot gas is not transparent, but resembles a dense fog.

The next event of interest did not occur until a few hundred thousand years after the initial Big Bang (about 380,000 years after the Big Bang based on recent data). As the plasma cooled below about 3000 Kelvin, the nuclei began to capture electrons and neutral atoms were formed. (For comparison, the surface temperature of the Sun is about 6000 Kelvin.) Once the atoms became neutral, the radiation could no longer interact with the matter. That is, the electrons would no longer absorb and reemit the light that was present. The bright opaque cloud of hot gas that had filled all of space became transparent.

Before this time of clearing, the plasma would have been like the opaque gas in a flame, but after the atoms became neutral, the gas would have been more like the transparent gas in the Earth's atmosphere (light passed right through it). Had we been there, we would have seen the space all around us begin to clear. We would not have seen all of space clear in an instant, because it takes light longer to get to us from more distant regions.

A sphere of transparent gas centered on us would, therefore, grow larger (at the speed of light), and we would be able to see to the wall of hot opaque plasma at greater and greater distances as time went on. We would see light coming from the plasma at the edge of our transparent sphere, and as our transparent sphere expanded, we would see light coming from even more distant plasma.

Because of the expansion of space, the light reaching us from the matter at the edge of our transparent sphere would experience a cosmological redshift as it traveled to us. As time went by, we would continue to see light emitted by the matter at the edge of our ever-expanding transparent sphere, but this light would be even more redshifted as it traveled to us (Figure 2.17).

Today, about 13.7 billion years later, we should still see light coming from the edge of our expanding transparent sphere, but cosmologically redshifted

by a very large amount. Indeed, we do see radiation coming from every direction in space! Although we sometimes say it is radiation from the Big Bang, the radiation is not really from the Big Bang, but from the time several hundred thousand years after the Big Bang when the light and matter stopped interacting and space became transparent.

This radiation, known as the cosmic background radiation or *cosmic microwave radiation*, is coming from the edge of that ever-expanding transparent sphere. But, it is greatly redshifted to the microwave region of the electromagnetic spectrum due to the expansion of space that occurred as the radiation traveled to us. The light is from the 3,000 K (Kelvin) hot plasma, but redshifted by a factor of 1000 so that it emulates the kind of radiation that comes from a 3 Kelvin object. (The most recent measurement gives a value of 2.735 K.)

Cooler objects emit more of the longer wavelengths of light, so as the light is redshifted it appears as if it is coming from a cooler object. For this reason, the cosmic microwave radiation is sometimes called the 3 degree black body radiation. The microwaves even pass through the gas and dust clouds in the Milky Way. The red-shifted light from the time of clearing described above is the only idea that has satisfactorily explained the cosmic microwave radiation that we detect.

Another way of understanding the cosmic microwave radiation is to remember that as we look out to greater and greater distances, we are also looking further back in time. If we look out to a distance of 13.7 billion light years in any direction, we see what happened 13.7 billion years ago. What we see 13.7 billion light years away is space still filled with the hot (3000 K) plasma just before the time of clearing.

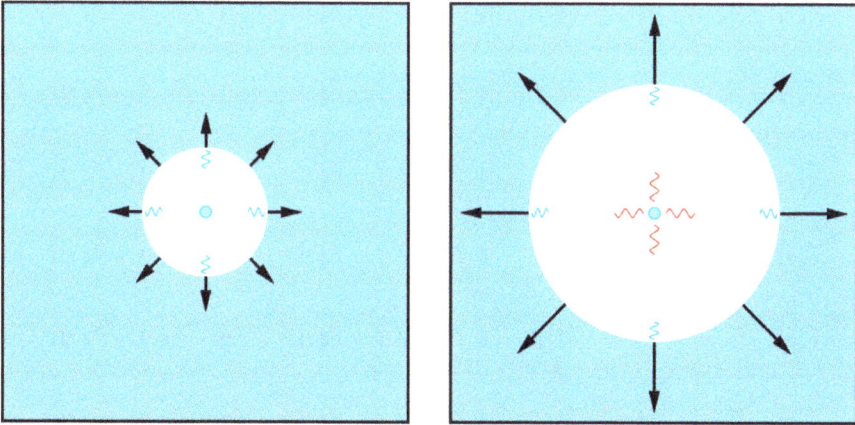

Figure 2.17 Expanding Transparent Sphere *When space became transparent, it took a while for the light from various places to reach us. We would have seen a sphere of transparent gas (centered on us), growing at the speed of light with hot plasma beyond the edge of the sphere (blue region). Because the Universe is expanding, the plasma at the edge of our transparent sphere (left) is moving away (arrows). The light emitted by the plasma (blue waves) will be stretched by the expansion of space as it travels towards us. Later (right) the redshifted light from the first picture is just reaching us so we finally see the edge of the plasma in the first picture. In the mean time, the transparent sphere has grown. The plasma is moving away faster (longer arrows) because it is further away. Since the light emitted by this plasma (blue waves) will take longer to reach us, it will be redshifted (stretched) even more by the expansion of space before it reaches us. [JCB]*

In the time since the radiation was emitted, space has expanded by a factor of about 1000 times. Therefore, the radiation has been stretched or redshifted 1000 times and appears as a cold (3 K) cloud. (The hot plasma is much further away than 13.7 billion light years because the space between the emitting gas and us has expanded in the 13.7 billion years since the light was emitted.)

An amazing thing about the cosmic microwave radiation is that when we look in opposite direction in space we measure the same temperature for the radiation (2.735 K). However, the matter emitting this radiation from opposite sides of the visible Universe is separated by tens of billions of light years. Since the matter on opposite sides of the Universe was never close together, why should their temperatures be so close to the same value?

This question could be answered if the Universe experienced a rapid expansion early in the Big Bang (the inflation mentioned above). For example, a baseball sized volume of space could have expanded to a volume the size of the visible Universe. This small initial volume would have been nearly the same temperature and what we are seeing today is this small volume filling all the visible Universe. The physical process that could have caused this hypothetical expansion or inflation is unknown.

Formation of the First Stars

The time of clearing was an important event for the future of the Universe. From that time forward, the matter and radiation would, for the most part, evolve independently. Stars would eventually form and produce light, but not the abundance of light that was present before the time of clearing. As the matter and radiation stopped interacting, the matter began cooling and gravity, the weakest of all the forces in nature, could at last begin to impose its important influence on the matter. Observations of the cosmic background radiation show that the first stars formed about 200 million years after space became transparent.

Although space was expanding, the matter in the Universe must have contained some random motions. Just as the wind or flowing water contains little whirlpools or eddy currents, so also did the gases filling space at this early time. Regions of higher than average density collapsed because of their gravity, and stars began to form. Stars formed in galaxies and galaxies formed into clusters of galaxies.

These first stars contained only hydrogen and helium since few heavy elements were formed in the Big Bang. The Sun and planets contain many elements heavier than helium, so the Solar System could not have been formed at this early time. The heavier elements had to be created before the Sun and Earth could form. Therefore, we would expect the Solar System to be much younger than the age of the Universe. Current measurements indicate that the Solar System is about 4.6 billion years old.

Since the giant gas clouds from which the galaxies formed must have already started to fragment by the time of clearing, astronomers felt that the cosmic microwave radiation should not be completely uniform. We should observe cooler and hotter regions corresponding to the denser and less dense regions. Indeed, observations by three different satellites have measured the intensity of the radiation. The most recent probe, the Planck space telescope, gives a detailed picture of the fluctuations in the cosmic microwave radiation (Figure 2.18).

Some of the details of the formation of galaxies and why we have clusters of galaxies have not been fully worked out, so there are still some unanswered questions. Clustering, however, seems to be the rule in nature rather than the

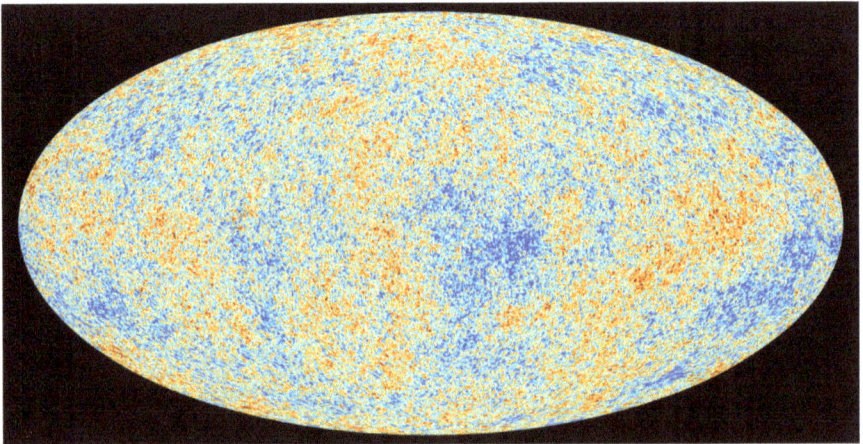

Figure 2.18 Cosmic Microwave Radiation This map shows that after the Big Bang, the gas became clumpy as it cooled. Gravity caused the denser cooler regions (dark blue) to collapse and form stars and galaxies. [ESA/Planck Mission]

exception. Galaxies are formed in clusters containing hundreds or even thousands of galaxies. Within the galaxies, stars are formed in clusters rather than singly. Instead of remaining uniformly spread out in space, much of the gas tends to collect into large clouds, which subsequently form star clusters (Figure 2.19).

The reason galaxies were able to form is because the expansion rate of the Universe is very close to the critical value that separates the ever-expanding model of the Universe from the collapsing model. Had the expansion been much more violent, matter would have expanded too fast for the gravitational attraction of local condensations to form galaxies. If the expansion had been much less violent, matter would have collapsed back on itself before galaxies had a chance to form.

As we stated above, this *lucky coincidence* may be due to a more fundamental theory which, when discovered, will explain this coincidence of nature. The inflation idea can explain why the density is near the critical value, but the cause of the inflation during the early Big Bang is not known. What might have happened during this early period depends on the behavior of elementary particles, but the details are not known at this time.

Before we leave the topic of galaxy formation, we might ask if there is any clear evidence that the Universe has evolved. Do we see galaxies change substantially over time? To answer this question we need to again remind ourselves that as we look out further and further into space, we are also looking further and further back in time. Since it takes light about four years to travel to us from the nearest star, when we look at that star, we are seeing it, as it looked four years in the

Figure 2.19 Star Formation *About 1500 light years away in the Orion Nebula, this cluster of young bright stars imbedded in a cloud of gas and dust is called the Trapezium Cluster. We know the stars are young because bright stars do not last very long. The photograph was taken in infrared light, which allows us to look deep inside the dust clouds. [ESO/M.McCaughrean (AI Potsdam) et al.]*

past. As we look at a galaxy 10 million light years away, we are seeing the galaxy as it looked 10 million years ago (10 mya).

As astronomers look to greater distances (and, therefore, further back in time) they see that things were quite different in the past. If we look back in time beyond 10 billion years ago (10,000 mya), (by looking out to distances beyond 10 billion light years), we see that high luminosity galaxies called *quasars* were about a thousand times more numerous than they are now. Quasars are very young galaxies with very energetic central regions that get dimmer with time. We see these galaxies are evolving.

Nuclear Reactions in Stars

A star is a mass of gas that has been compressed by its own gravity until its center becomes hot enough for nuclear reactions to begin. Remember that as a gas is compressed, it gets hotter. When nuclear reactions begin, the energy produced by these reactions keeps the pressure high enough to prevent the star from collapsing further.

Even crude calculations based on the well-known gas laws (theories) show us that the center of the Sun must have a temperature of several million Kelvin. At these high temperatures, neutral atoms do not exist. The collisions are so violent that the electrons cannot remain attached to the nuclei. The gas is completely ionized so the electrons and nuclei (mostly protons) are traveling and interacting independently. In other words, the gas is a plasma.

When hydrogen gets hotter than a few million Kelvin, nuclear reactions begin. These nuclear reactions have been well studied in accelerator experiments. Even at temperatures of several thousand Kelvin, when two hydrogen nuclei (protons) collide they never really get very close to each other. This is because the repulsive electrical force stops the protons before they can approach each other very closely. As the temperature increases, the faster moving protons get closer before they are stopped by the electrical repulsion.

At temperatures of a few million Kelvin, however, some of the protons get extremely close during a collision. If the protons get close enough, they will feel the attractive nuclear force, and may stick together. When they stick, one of the protons changes into a neutron, and a deuterium nucleus is formed. Remember, deuterium is the isotope of hydrogen that contains one neutron. This process is called *nuclear fusion* because two protons fuse into one deuterium nucleus. During this fusion reaction, a large amount of energy is released as gamma rays. Nuclear fusion is the primary energy source of the Sun and most other stars.

You might remember that when a proton changes into a neutron, a positron is also produced (and an electron-antineutrino). The neutrino interacts only weakly with the matter in the star so it easily escapes into space, but the positron soon finds an electron and they are annihilated, giving off energy as gamma rays. The deuterium that is produced soon combines with another proton and helium-3 is formed, releasing more energy. Eventually, by means of several possible reactions, helium-4 (the common isotope of helium) is formed.

The energy from this chain of nuclear fusion reactions is released as gamma rays near the center of the star. As this energy journeys to the surface of the star, the electromagnetic quanta get absorbed, reemitted, and scattered numerous times. The matter in a star is a plasma, which is why it interacts strongly with the electromagnetic waves. The plasma can absorb quanta of one wavelength (energy) and emit the energy as two or more quanta of a lower energy (longer wavelength). Before reaching the star's surface, the energy from a single high-energy gamma ray (quanta) is transformed into thousands of quanta of lower energy, many of which are visible light quanta, called photons.

The nuclear fusion reactions described above are similar to the reaction that occurs in a hydrogen bomb (which is why they have been so well studied). A hydrogen bomb is fueled by deuterium and tritium (tritium is an isotope of hydrogen with 2 neutrons). Tritium is very unstable and is not found in nature so it must be artificially produced. The D-T fusion reaction as it is called is used in hydrogen bombs because it will react more easily than the proton-proton reaction that occurs in the Sun and most other stars.

Stars that are converting hydrogen into helium have central temperatures from about 5 to 35 million Kelvin. When helium is produced, it is in the same predicament as the helium produced in the Big Bang. There are no stable elements with 5 or 8 nucleons in their nuclei. Therefore, a star continues to make helium until the hydrogen runs out. Since there is not much mixing between the central part of a star (the core) and the outer layers (the envelope), only the hydrogen in the central portion of the star is converted into helium.

As hydrogen is used up in the core, the energy production drops and the core pressure is no longer able to counteract the inward pull of gravity. Therefore, the core of the star begins to collapse. As the core collapses, it gets hotter and denser, and when the temperature reaches about 100 million degrees, triplets of helium can begin to form carbon. (A helium nucleus is called an alpha particle so this process is known as the triple alpha process.)

As the core gets smaller and hotter, the energy produced pushes out the envelope of the star and the surface of the star gets larger and cooler. At this stage in the star's life, the star has become a huge star with a relatively cool surface. Cooler stars are redder in color and we call these bloated stars *red giants* (Figure 2.20). Red giants can be larger than the Earth's orbit!

A relatively small amount of energy is released in the triple alpha process, but this energy stabilizes the collapsing core. The reaction requires a high pressure and temperature because two helium nuclei will not stick together. (Remember that there are no stable nuclei with 8 nucleons.) In order for helium to react, three helium nuclei must come together at the same time, an unlikely happening unless the

Figure 2.20 Red Giant This 20 million year old cluster of stars in the constellation of Cassiopeia is known as M103. As seen by its color and brightness, one of the massive stars has evolved to the red giant stage (just below and to the left of center). [NOAO/ AURA/NSF/H.Mathis, N.A.Sharp]

helium is at a high temperature and density.

Notice that this process could not happen in the Big Bang because the matter was getting cooler and less dense with time. In the interior of a star, however, the matter gets hotter and more compressed after the helium is produced. If the star is massive enough, the temperature will continue to rise in the core and many other nuclear reactions are possible. We believe most of the elements up to and including iron and nickel are made in the interiors of various stars. If the star is not massive enough, the nuclear reactions may stop before iron is made. For example, the star could end up with a carbon-oxygen core.

Elements much heavier than iron are not made in significant numbers in the interiors of normal stars because it requires energy to produce these very heavy nuclei. The star would have to supply the needed energy, and this drain in energy would cause the core to collapse quickly. Remember, it is the continual outward flow of energy produced by the nuclear reactions that maintains the high internal pressure of the star. This high internal pressure creates an outward push that balances the inward pull of gravity and keeps the star from collapsing.

Life Expectancy of Stars

The most important property of a star is its mass. Since even today, most of the matter in the Universe is still the hydrogen and helium produced in the Big Bang, there are few variables that can influence the final form of a star, other than its mass. Since a massive star has a stronger gravity than a smaller mass star, the gas in the massive star gets compressed more and gets hotter.

A massive star will be hotter on the surface, hotter at the center, and larger in diameter than a star of less mass. Since a more massive star is hotter in its core, the conversion of hydrogen into helium proceeds at a faster rate than in lower mass stars. Therefore, the more massive the star, the brighter it will be.

The brightest star in the sky is Sirius. Sirius appears bright for two reasons, it is relatively close to us and it is intrinsically a fairly bright star. If Sirius and the Sun were placed at the same distance, Sirius would be about 100 times brighter than the Sun. That is, it is giving off 100 times more energy each second than the Sun does. Therefore, to keep this energy flowing it must be converting hydrogen into helium 100 times faster than the Sun does. If it had the same amount of fuel (hydrogen) to begin with, Sirius would only live one-hundredth as long as the Sun. Sirius is about 2 times the mass of the Sun, so if we consider its mass, we would expect the lifetime of a star like Sirius to be about 1/50 that of a star like the Sun.

We define the lifetime of a star to be the time from its formation (when nuclear reactions begin) until it runs out of hydrogen in its core. The lifetime of a star like the Sun has been calculated, and found to be about 10 billion years. A crude measure of the lifetime of the Sun can be gotten from the rate at which hydrogen is being converted to helium. From accelerator experiments, we know how much energy each reaction produces, and we know how much energy the Sun is giving off. Therefore, we can calculate the rate at which hydrogen is being converted into helium. Since we also know how much available hydrogen the Sun had to begin with, we can calculate how long it will take to use up all its hydrogen. We will see later that the Sun is about 5 billion years old, so it has lived about half of its life already.

If a star like the Sun has a lifetime of about 10 billion years, a star like Sirius must have a lifetime of approximately 1/50 of this time or about 200 million years. Recall that 200 million years is the time for the Sun to orbit the galaxy. By similar reasoning we can calculate that a star 100 times brighter than Sirius, say one of the bright stars in the trapezium region of Orion, must have a life span of only about 10 million years (Figure 2.19). This is extremely short compared to the age of the Universe, and some stars are even brighter and, therefore, more short lived than this.

Very massive stars only live a few million years! We also observe that hot bright stars, that must have been just recently formed, are invariably found associated with the gas clouds out of which they were made (Figure 2.19). Stars with longer lifetimes have had enough time to drift away from their birthing clouds before their hydrogen is used up. The deaths of very large, short-lived stars were extremely important to the formation of the Earth and its ability to support life.

Death of a Star

We have seen that many of the elements necessary for life were formed by nuclear reactions in stars. How did these elements become part of the Earth, and where were the elements heavier than iron produced? The answers to these questions can be found by studying the deaths of stars. Without the heavier elements that are released when a star dies, a planet like the Earth could not have formed, and life would not have been possible.

As long as nuclear reactions are occurring in the central regions of a star, the supply of energy keeps the interior gases hot enough to counteract the inward pull of gravity, and the star remains relatively stable. When the nuclear fuel begins to run out, however, the high pressure in the core cannot be maintained, and the core of the star begins to collapse because if its own gravity. Can the collapse be stopped? The answer to this question depends on the mass of the collapsing core.

If the mass of the core is not much greater than the mass of the Sun, the electrons in the core will be able to halt the collapse, and a stable object called a *white dwarf* will be formed (we will describe white dwarfs a little later).

A full understanding of how electrons stop the collapse of a white dwarf requires some knowledge of the theory of quantum mechanics. Strange as it may seem, quantum mechanics tells us that for particles moving at a given speed, the lighter the particle, the more room the particle must occupy. This statement (known as the *uncertainty principle*) means that at a given temperature, light electrons will need to occupy much more space than the heavier nucleons. Equivalently, the uncertainty principle says that if an electron is to be squeezed into a smaller volume, the electron must be given *more* energy. Notice that the uncertainty principle has nothing to do with the charge of the particle. It acts the same on electrically neutral particles.

Now we can understand why atoms are so very large. A hydrogen atom is a proton and an electron. The volume of space occupied by the electron determines the size of the atom. If the negative electron is attracted to the positive proton, why doesn't the electron stick to the proton and take up much less space? Why is the electron always well separated from the proton? If we made a model of a hydrogen atom and let a tennis ball represent the proton, the electron would be about two kilometers away on average.

The answer to why the atom is so large can be found in the uncertainty principle. If we wish to confine the electron to a volume much smaller than the size of a hydrogen atom, we would have to give the electron *more* energy because of the uncertainty principle. On the other hand, it would also require energy to pull the electron away from the nucleus because of the electrical attraction. The size of the atom is where the electrical attraction on the electron by the nucleus is balanced by the resistance the electron feels when it is confined to a smaller volume due to the uncertainty principle.

The size of an atom is such that the energy of the electron is a minimum. To make the atom larger, you must pull the negative electrons away from the positive nucleus, which requires energy. On the other hand, squeezing an atom would require that the electrons be confined to a smaller volume and this would also require more energy because of the uncertainty principle.

Now we can understand why it is so very hard to compress solids or liquids. In solids and liquids the atoms are touching so you must make the atoms smaller to compress the material. Again, it is the volume of space occupied by the electrons that determines the size of an atom. It is extremely difficult to compress a solid or liquid because of the resistance of the electrons caused by the uncertainty principle. It is not easy to reduce the size of an electron's living space, although it can be forced into a smaller volume if it is given more energy.

When gravity collapses a star's core after the nuclear reactions have ceased, the electrons try to resist the collapse, but if the star is too massive, the electrons will be unable to stop the collapse. The gravity of a very massive star will squeeze the electrons so much

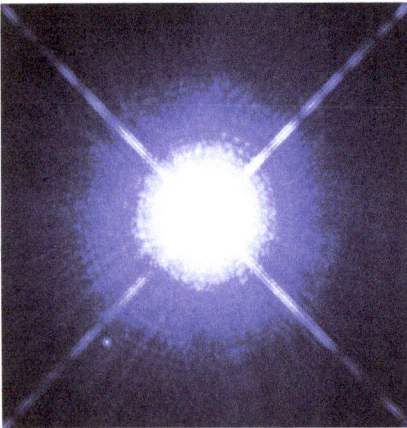

Figure 2.21 White Dwarf Sirius, the brightest star in the night sky, has a white dwarf companion named Sirius B (the faint dot to the lower left). Its mass is about 0.98 times the Sun's mass, but it has a radius of 0.92 that of the Earth. It orbits Sirius once every 50 years and is one of the hotter white dwarfs, with a surface temperature of 25,000 K. [NASA/ESA/H.Bond (STScI), M.Barstow (Univ. of Leicester)]

(and consequently give them so much energy) that they will have enough energy to combine with the protons to form neutrons.

If the electrons are to stop the collapse, the star cannot be too massive. Calculations have shown that for the electrons to stop the collapse, the final mass of the star's core cannot be much more than about 1.4 times the mass of the Sun. If the electrons are able to stop the collapse, a *white dwarf* will be formed. The matter in a white dwarf is squeezed as much as it is possible to squash ordinary matter (matter made of atomic nuclei and electrons).

If, however, the star's core is too massive, the electrons will combine with the protons to form neutrons. (Remember, the uncertainty principle says that at a given temperature, neutrons can live in less space because they are more massive.) The star will collapse to form either a *neutron star* (a solid mass of neutrons) or a *black hole*. (We will discuss these two objects a little later.)

White dwarfs (Figure 2.21) are much more common than neutron stars or black holes. We have measured the sizes and masses of several white dwarfs, and found that they are about the size of Earth but contain about as much mass as the Sun. Most of us think of Earth as being fairly large, but for a star this is quite small, hence the name dwarf. The diameter of the Sun is more than hundred times the diameter of the Earth. Over a million planets the size of the Earth would fit inside the Sun.

The word white comes from the fact that their surfaces are white-hot. White hot objects like the star Sirius have a surface temperature of about 10,000 K, almost twice as hot as our Sun, which is only yellow hot.

Because so much matter is compressed into such a small volume, white dwarfs are very dense. One penny weighs 2.5 grams. If a penny were made of white dwarf material, it would weigh about 400 kilograms, about one third the weight of an average car! When white dwarfs were first discovered, many people found it hard to believe that matter could be so dense. Since then, we have discovered that the Universe contains more strange objects than anyone could have imagined.

Even though we refer to white dwarfs as stars, they are technically not stars, but the end products of stars. No nuclear reactions occur in white dwarfs. They are just hot dense objects that are slowly cooling off. Eventually, when they cool off sufficiently, they will no longer give off visible light.

The fact that white dwarfs are very common indicates that in the past, many stars ran out of nuclear fuel and formed these white dwarfs. Stars like the Sun live about 10 billion years, so many of these white dwarfs must have come from stars that were originally more massive than the Sun. (Remember that more massive stars use up their hydrogen faster, therefore, live a shorter time.)

If this is true, how did such massive stars become white dwarfs? (We know that a white dwarf cannot be much more massive than the Sun.) Obviously, these stars must have ejected part of their mass before becoming white dwarfs. Indeed, we see many stars doing just that! The clouds of gas ejected by these stars were given the misleading name of *planetary nebulae* (Figure 2.22). They were so named because when seen through a telescope the gas clouds of many look round and greenish, superficially resembling the planets Uranus and Neptune. (Nebula means *cloud*.)

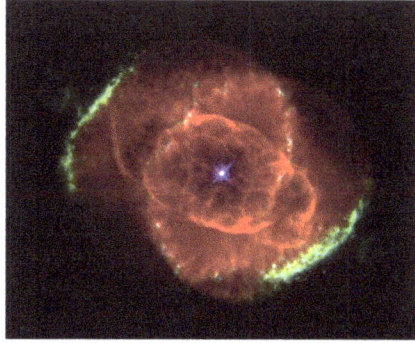

Figure 2.22 Planetary Nebula The Cat's Eye Nebula is a spectacular planetary nebula about 3,000 light years away. Having ejected the complex envelope of gas about 1000 years ago, the central blue star is an evolving (cooling) white dwarf. It is located in the constellation of Draco (the dragon). [NASA/ESA/J.P.Harrington, K.J.Borkowski (Univ. of Maryland)]

After the outer gasses of the star are ejected, the remaining core cools off and becomes a white dwarf. The ejected gas cloud contains some of the heavier elements produced by the star, and as it spreads out, these new elements mix with the other gases in the galaxy (mostly hydrogen and helium). Therefore, the abundance of the heavier elements in the galaxy increases. The next stars that form out of this recycled gas will have more heavy elements than their predecessors did. Indeed, when we study stars that we know are very old (like the stars in globular clusters), we find that they contain very few heavy elements.

The formation of a planetary nebula is a relatively nonviolent way by which a star can eject matter, however, some massive stars blow off a large fraction of their outer envelope in very violent explosions. These explosions are extremely bright and when they occur, we call them *supernova* explosions.

In 1054 AD, the Chinese observed the most famous supernova in recorded history. It was bright enough to be seen

Figure 2.23 Supernova Remnant *The Crab Nebula in the constellation of Taurus is the remains of an exploding star that was observed by the Chinese in 1054 AD. At its center is a neutron star, which formed from the core of the exploding star. The diffuse blue light is produced by electrons whirling around in the magnetic field of the nebula. It is about 6,500 light years away. [NASA/ESA/J.Hester (ASU)]*

in the daytime for several weeks. If we look in that portion of the sky where the Chinese saw this bright star, we see the remnants of the explosion as an expanding cloud of gas.

Today we call this supernova remnant the Crab Nebula (Figure 2.23). We know it is the remnant of the 1054 supernova because we have observed it for many decades and have watched it expand. Using its rate of expansion and the size of the nebula (the rate and the distance traveled), we can calculate when the explosion occurred, and we find that it agrees with the date given by the Chinese.

Supernovae explosions occur when a massive star runs out of nuclear fuel and the electrons are unable to stop the collapse of the core. As the core collapses beyond the white dwarf stage, the electrons combine with the protons and a neutron star is formed. If the mass is not too great (less than about 3 times the mass of the Sun), the neutrons may be able to stop the core

from collapsing further (because of the uncertainty principle), and a stable neutron star will be formed.

However, if the core is too massive, it will continue to collapse into a black hole. In either case, the falling material in the core gains a great deal of energy as it collapses, and this energy must be released quickly. The gravitational energy of the star's collapsing core is the source of the tremendous amount of energy released by the supernova. Some of the energy blows off the outer layers of the original star, sending the gasses hurling out into space at speeds of thousands of kilometers per second.

Supernovae eject heavy elements back into space as do planetary nebulae, but they also produce the heavy elements beyond iron and nickel. During the explosion, many neutrons are ejected at high energies, and in the outer layers of the star these neutrons can be captured by heavy nuclei such as iron to make even heavier nuclei.

The elements heavier than iron and nickel were not made in stable stars because it requires energy to produce them. In supernovae, however, there is plenty of energy available to make these very heavy elements. The abundances of these elements in the Universe agree well with calculations that assume they were made by neutron capture in supernovae. Since the elements beyond iron and nickel are made only in supernovae, they are much less abundant than the elements lighter than iron (Figure 2.24).

So what is a neutron star? A neutron star behaves physically somewhat like a white dwarf. In a white dwarf, the electrons prevent the collapse, but in a neutron star the neutrons prevent further collapse. However, since neutrons are more massive than electrons, they can be squashed into smaller spaces. Remember that the uncertainty principle says that more

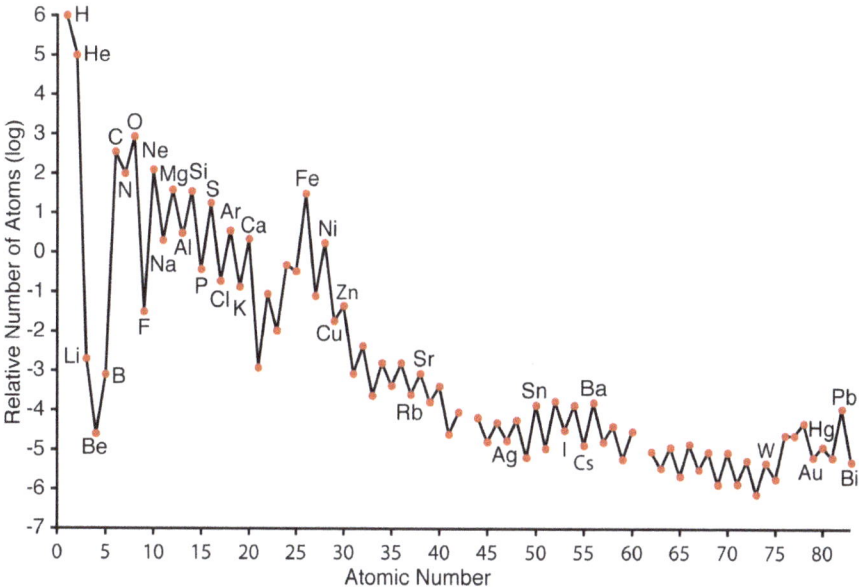

Figure 2.24 Cosmic Abundances of the Stable Elements *Abundances are relative to 1,000,000 hydrogen atoms (log 1,000,000 =6). For example, if you obtained a sample of cosmic matter that contained one million hydrogen (H) atoms, the sample would also contain 100,000 helium (He) atoms (log 100,000 =5), about 1000 oxygen (O) atoms (log 1000 =3), and 0.1 of a potassium (K) atom (log 0.1 =-1). The gaps at 43 and 61 are where the unstable (radioactive) elements of technetium and promethium are found. [JCB; data: E.Anders, N.Grevesse, credits p. 414]*

massive particles can be confined to *smaller* spaces than less massive particles.

So, because of the uncertainty principle, the neutrons resist being compressed by the pull of gravity and a stable star can be formed. A neutron star is a solid mass of neutrons with a mass of about 1.5 to 3 times that of the Sun and a diameter of about 30 kilometers (18 miles). This much mass squeezed into such a small volume means that the density of a neutron star is about 100 billion times that of a white dwarf, a number that is too large to comprehend. A house fly made of neutron star material would weigh about 200 billion kilograms! On the other hand, a house fly made of white dwarf material would only weigh about

20 kilograms (the average weight of a five year old child)!

Do neutron stars really exist? We believe they do. As a star's core collapses to form a neutron star, any rotation the star had will be amplified. Just as skaters spin faster as they pull in their arms, the collapsing star will spin faster as it gets smaller. We expect a spinning neutron star might rotate several times in one-second.

In addition, any magnetic field the star might have had will get stronger as the star collapses. Any matter falling onto the neutron star will spiral in along the magnetic poles of the star and give off radio waves. The radio waves are sent out as a beam of radio waves (like a flashlight) and as the star rotates, the beam sweeps around the sky. (The magnetic field is most likely tilted with respect to the rotation axis as is the Earth's magnetic field.)

If the beam happens to sweep in our direction, we would see a pulse of radio waves every time the beam comes around (like a lighthouse beacon). We see just such objects as these, and before they were identified as neutron stars, they were called *pulsars*, which means pulsating stars. A pulsar is located in the middle of the Crab Nebula. It is a neutron star that was produced by the supernova explosion of 1054 AD.

If the core of a star is greater than about three times the mass of the Sun, the neutrons will not be able to keep gravity from crushing the core even more and it will become a black hole. We believe that when the most massive stars run out of nuclear fuel, they collapse into black holes. A black hole will be formed if a star's core is compressed beyond a certain radius known as the Schwarzschild radius.

Beyond this point, gravity will be so strong that nothing, not even light, can escape the object. The Schwarzschild radius of a star three times the mass of the Sun is about 10 km (6 miles). Notice that this is very close to the size of a neutron star.

Why can't light escape a black hole? If you throw a rock straight up, it slows down and stops momentarily before falling back to Earth. At the top of its flight, the rock lost all of its energy of motion. It was not launched with enough energy of motion (kinetic energy) to escape the Earth's gravity. A way of thinking about what happens to light emitted by a black hole is that the light looses all its energy trying to escape the black hole. Since light is pure electromagnetic energy, there is nothing left; the light has lost all its energy and has disappeared. If light, which is pure energy, cannot escape a black hole, then matter, which is not pure energy, cannot escape either.

Figure 2.25 Gravitational Lens *Light from distant galaxies is distorted into arcs by the gravity of the nearer cluster, especially the large elliptical galaxy to the right of center. The lensing cluster, called Abell 2218, is about 2 billion light years away. [NASA/ESA/Johan Richard(Caltech),Davide de Martin & James Long/ESA/HST]*

Light is pure electromagnetic energy, but according to Einstein's theory of general relativity, this energy should be attracted by gravity. Einstein's famous equation $E=mc^2$ says that energy (E) and mass (m) are different forms of the same thing. In the equation, c^2 is a number that allows you to convert between units of energy and units of mass (c is the speed of light).

Light passing near a black hole will move on a curved path because of the gravitational attraction of the black hole. (Even ordinary objects will bend the path of light rays a little bit.) This bending of light by very massive objects can be seen in certain clusters of galaxies and is called *gravitational lensing*. A spectacular example of gravitational lensing is seen in Figure 2.25. It shows the light from several distant galaxies being bent as it passes by a closer very massive galaxy. The images of the distant galaxies are distorted into arcs by the gravity of the closer galaxy. This produces an image similar to what you see when looking at a light through a goblet of water.

Since black holes do not give off light, you might think they are theoretical objects that cannot be detected. Although we cannot see a black hole directly, they do make their presence known by the effect they have on other objects. As matter falls toward a black hole, it will emit X-rays (before it reaches the Schwarzschild radius). We see X-rays coming from many objects, and some could be black holes. Neutron stars also emit X-rays by a similar mechanism, and the only way to tell the difference is by the mass of the object. Neutron stars cannot be more massive than about three times the mass of the Sun.

We believe a super massive black hole is located near the center of our own Milky Way Galaxy, and several

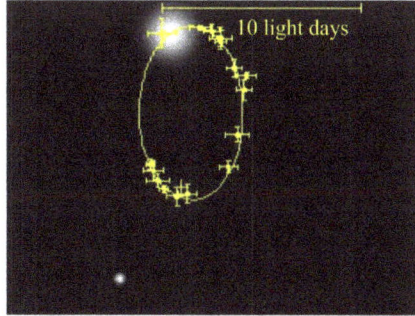

Figure 2.26 Orbit Around a Black Hole
The orbital period of star S2 was found to be only 15 years, but its orbit is 25 times larger than Pluto's orbit (11 light hours across). Therefore, the unseen mass that S2 orbits must be about 4 million times the mass of the Sun! [ESO]

other galaxies appear to have massive black holes at their centers as well. We think they have black holes because the centers of these galaxies are strong sources of X-rays. We can see several of the stars near the center of our own galaxy, and they are orbiting some unseen object that is an X-ray source.

By observing the orbit of one of these stars (named S2), we are able to calculate the mass of the unseen object (Figure 2.26). Even though the orbit S2 is about 25 times larger than the orbit of Pluto, S2 orbits the unseen object in just 15 years (It takes Pluto 248 years to orbit the Sun). The mass of the unseen object was calculated to be about four million times the mass of the Sun! It is much too massive to be anything but a black hole.

Forming the Solar System

Finally, after the passing of many generations of stars, the ingredients needed to create the Solar System were present. The observed abundance of all the elements can be explained by the various nuclear reactions that occurred in the earlier stars (Figure 2.24). Obviously, the Solar System could not have been produced shortly after the Big Bang. Planets similar to the Earth are made of the heavier elements that were formed by earlier stars. We would not expect planets like the Earth to be present around any of the earliest stars that formed in the galaxy. Planets like Jupiter and Saturn are composed mostly of liquid hydrogen and helium and could have been formed any time after the Big Bang.

Jupiter's composition is similar to that of a star, but it did not get massive enough to become a star. Jupiter's gravity compressed the gasses and the central temperature increased. The interior of Jupiter is still so hot that the planet gives off about 3 times as much energy as it receives from the Sun. However, since it did not contain enough mass, the central temperature did not get hot enough for nuclear reactions to begin. Jupiter became a very hot planet instead of a star.

The Solar System is arranged in a very regular way. All the planets orbit the Sun in the same direction, in nearly circular orbits, and in approximately the same plane. All but two (Venus and Uranus), also rotate in the same direction. Because of these regularities it is believed that the Sun, planets, and other members of the Solar System were formed out of the same cloud of gas and dust.

Astronomers use the word dust to describe the small solid particles of matter formed when many atoms or molecules stick together. Dust is composed of ices such as water ice, carbon dioxide ice (commonly called dry ice), methane, and ammonia, and rock-forming substances like iron, silicon, and other materials that normally form solids.

The presence of dust in a cloud of gas is easily recognized because the dust is opaque and blocks our view of more distant objects (Figure 2.27). There is so much dust in the gas clouds situated around the plane of the galaxy that we have difficulty seeing very far into the plane (Figure 2.1). The external galaxies that we observe are located either above or below the band of the Milky Way. Even though the dust is a very obvious component, about 98 per cent of the matter in these clouds is still hydrogen and helium. The dust, however, is extremely important since the rocky or terrestrial planets like Earth are primarily made of the rock-forming substances found in dust.

Figure 2.27 Dark Nebula *The Horsehead Nebula in the constellation of Orion is a cool cloud of gas and dust located in front of a hot glowing cloud of gas. The dust in the dark nebula blocks the light coming from the more distant glowing gas cloud. It is about 1500 light years away. [Canada-France-Hawaii Telescope/Coelum]*

Some problems are so complex that they can only be attempted with the aid of modern high-speed computers. Carrying out the orbital calculations necessary to send a space probe to the Moon or Mars would be very difficult without computers. Working out the details of such a complex process as the formation of the Solar System is a much more difficult problem, and to be sure, many of the details are unknown at this time. However, with the aid of computers, some of the details are being successfully attacked.

The Solar System formed out of a cloud of gas and dust, similar to the ones we see scattered throughout the galaxy. This cloud had a small amount of rotational motion associated with it, and as the mutual gravity of its various parts pulled the cloud together, it rotated faster and faster, and the material formed a flattened disc. As the matter was pulled in towards the center, the faster moving gas and dust was left behind to form the disc. We have discovered several of these discs around young stars (Figure 2.28). As the disc became more compacted, the dust particles collided and stuck together forming larger and larger clumps of matter. Many of these growing clumps formed fairly large objects called *planetesimals* that could accrete matter more efficiently, eventually growing into planets. The craters that we see on the Moon and many planets and their satellites attest to the violent collisions that occurred as these growing objects swept up some of the smaller planetesimals.

As the mass near the center of the cloud collapsed to form the early Sun, it became denser and its temperature began to rise. Radiation from this protosun drove off the lighter material such as hydrogen, helium, and the more easily vaporized ices. This process left the inner portion of the Solar System enriched in the heavier less-volatile materials that formed the terrestrial or rocklike planets of Mercury, Venus, Earth, and Mars.

Further from the forming Sun, however, hydrogen, helium, and other gasses could collect to form the larger Jovian type planets of Jupiter, Saturn, Uranus, and Neptune. Remember that the original gas and dust cloud was composed mostly of hydrogen and helium. As the planets formed, so too

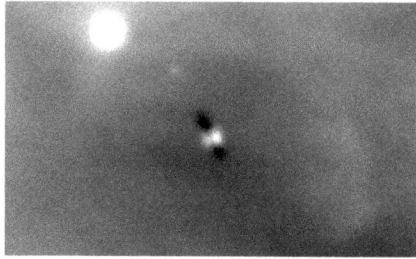

Figure 2.28 Protoplanetary System About 1500 light years away in the Orion Nebula, planets may be forming in this disc of gas and dust around a young star. The disc is 30 times larger than Pluto's orbit. [ESO/ M.McCaughrean (AI Potsdam) et al.]

did their satellites. Jupiter and its larger satellites may have formed like a miniature Solar System, but most of the smaller satellites of Jupiter and Saturn were planetesimals that were captured as they passed close to one of the giant planets.

The Sun and planets began to form about 4.6 billion years ago (4,600 mya). The determination of this date and the dates of other events in the history of the Earth is so extremely important that the next chapter will be devoted to the methods we use for determining the ages of rocks.

Chapter 3: Aging Rocks

Radioactivity

As we saw in the last chapter, the protons and neutrons in the nucleus of an atom attract each other with a force called the nuclear force or the strong interaction. To feel this strong attractive force, the nucleons have to be very close to each other. The positively charged protons also repel each other because of the longer-range electromagnetic force, but since the nuclear force is stronger at short distances, most nuclei are held together very tightly.

However, if a nucleus is too large, it may not be completely stable. Because of the short-range nature of the nuclear force, nucleons on opposite sides of a large nucleus may be so far apart that their nuclear attraction is greatly reduced. Since some of these nucleons are protons, their electrical repulsion may be almost as strong as their nuclear attraction. Under these conditions the nucleus will be unstable and will eventually change into a different nucleus. This size constraint is

why we find no stable nuclei larger than bismuth, which has 209 nucleons (83 protons and 126 neutrons).

An unstable nucleus will generally emit various kinds of particles until it becomes a stable nucleus. Unstable nuclei are said to be radioactive, and the particles they emit are referred to as radiation. The general use of the word radiation is somewhat unfortunate here, because this kind of radiation is composed of small bits of matter and is very different from the electromagnetic radiation we discussed in the last chapter. Electromagnetic radiation, of which light is only one example, is not composed of material particles, but consists of bundles of pure energy in the form of electromagnetic waves.

Some radioactive nuclei, especially very large nuclei like uranium and plutonium, emit helium nuclei as a way of becoming smaller. Remember that a helium nucleus contains two protons and two neutrons. A helium nucleus is emitted instead of a single proton or neutron because the helium nucleus is a very tightly bound object. A fast moving helium nucleus is called an *alpha particle*, and nuclei that emit alpha particles are said to be emitting alpha radiation.

Because of the high speed with which they are emitted, alpha particles have a great deal of energy and can damage biological tissues. Because they are so large, alpha particles usually cannot penetrate the dead outer layer of the skin. However, alpha emitters can cause cancer if swallowed or inhaled. Waste from mining minerals like uranium and phosphate ore contain alpha emitters that can become airborne or get into the water system. Radon gas is an alpha emitter that can collect in basements and easily dissolves in water.

In stable nuclei, the number of neutrons is about as large or larger than the number of protons. Generally, the bigger the nucleus, the greater is the percentage of neutrons. The reason for this excess of neutrons in larger nuclei can be understood if we remember that for a given separation, the nuclear force between any pair of nucleons has the same strength whether it is a pair of protons, a pair of neutrons, or one of each.

However, since two protons also experience a repulsive electric force that works against the nuclear force, the net attractive force between two protons is not as strong as the force between two neutrons or between a neutron and a proton. The effect becomes more important in the larger nuclei because as the protons get further apart, the nuclear attraction drops off in strength faster than the electrical repulsion.

Therefore, for larger nuclei, a more tightly bound nucleus results if there are more neutrons than protons. The uncharged neutrons add strong attraction to the nucleus without adding any electrical repulsion. This extra attraction adds stability to larger nuclei containing many positively charged protons. If a nucleus does not have the right ratio of protons to neutrons, it will be unstable and will emit a particle to help correct the situation.

These unstable nuclei may emit an alpha particle, or may convert a neutron into proton or vice versa depending on which is needed. A fast moving electron or positron is also emitted along with the appropriate neutrino.

We commonly refer to the emitted electrons or positrons as beta particles or beta radiation. Because the beta particles are emitted at such a fast speed, beta radiation can be harmful to the human body and other biological tissue. Beta emitters are found in the waste products from atomic power plants and in certain substances used in medical procedures.

Beta particles are more penetrating than alpha particles. They can pass through the outer skin layers and harm the living cells. They are more harmful if the beta emitters are swallowed or inhaled.

Often when alpha or beta particles are emitted, the nucleus is left in an excited state. The nucleus will then emit a gamma ray (a high energy electromagnetic wave) to get rid of its excess energy. Gamma rays, like X-rays, can travel through the body and cause damage to internal organs.

Along with alpha and beta emission, another type of radioactivity is important to radiometric dating. Although the orbiting electrons generally remain well separated from the protons in the nucleus, when an electron does get close to a nucleus that has an excess of protons, there is a chance that it will combine with one of the protons to produce a neutron. This process occurs in some nuclei and is called *electron capture*. In electron capture, the nucleus is generally left with an overabundance of energy, and this excess energy is given off as a gamma ray.

Radioactivity can be understood within the context of the theory of quantum mechanics, the most successful and most widely verified theory in the field of physics, or all of science for that matter. Quantum mechanics was developed largely in the 1920s, and along with radioactivity, quantum mechanics explains such things as superconductivity, the production of X-rays, the behavior of lasers, the behavior of the electrons associated with atoms (which constitutes most of chemistry), the behavior of the transistor and other semiconductor devices which are largely responsible for the recent computer and electronics revolution. Quantum mechanics explains the characteristics of the light emitted by atoms including the complicated array of wavelengths emitted by various gases, the sensitivity of film to the different colors of light, the operation of light meters in modern cameras, and many more.

To be sure, much of the progress enjoyed by the field of physics during the past 100 years can be directly attributed to the success of the theory of quantum mechanics. What a theory predicts is just as important as what it explains, and indeed, many of the important new inventions and discoveries of this age were predicted by quantum mechanics before they were developed.

As predicted by quantum theory, and as confirmed by experiments, the decay of a radioactive nucleus is governed purely by chance. Each radioactive nucleus has a certain probability of decaying in a given amount of time. Using these probabilities, we are able to accurately calculate how a large number of radioactive nuclei will behave.

The number of nuclei that decay in a given period of time is proportional to the time interval and the number of nuclei present. Mathematically this means that the number of nuclei, which have not decayed, will decrease in time exponentially. The time interval during which one half of the nuclei are likely to decay is called the *half-life* of that particular nucleus. An equivalent definition of half-life is the duration of time in which a nucleus has a fifty-fifty chance of decaying. Since radioactive nuclei decay at a known rate, the percentage of decays that have occurred can be used as a means of measuring time.

The fractional number of radioactive nuclei that decay per second is a quantity we call the *decay constant*. Although we talk about the decay constant, one might ask if the rate can indeed be considered a constant. In the case of alpha and beta decay, experiments have shown that the decay rate is indeed a constant. In the case of electron capture, experiments on some atoms have shown that even when subjected to pressures greater that 100,000 atmospheres, the decay constant is only changed by a few tenths of one percent. This pressure corresponds to a depth of about 350 kilometers below the surface of the Earth. At this depth, the material is not solid, but is able to flow like thick syrup.

Since a measured age is proportional to the decay constant, this change is too small to be of any consequence as far as radiometric dating is concerned. In addition, as we will see later, rocks that are dated could not have been subjected to such high pressures. For all practical purposes, therefore, we find that the decay rates are indeed constant.

The reason for the constancy of the decay rates can be understood by considering the structure of an atom. If we made a model of an atom and used tennis balls for the protons and neutrons, a typical nucleus might be a bundle of tennis balls about one foot in diameter. The electrons for this model would be the size of pin points (as far as we have been able to determine, the electron has no measurable diameter) and would be orbiting the nucleus at an average distance of two kilometers. The size of an atom is completely determined by the volume of space occupied by the orbiting electrons.

Although the nucleus contains most of the mass of the atom, it occupies an insignificant amount of the atom's volume. When involved in chemical reactions, or subjected to high pressures, or temperatures high enough to vaporize the material, the nucleus is not involved. These processes only involve the electrons. Even at the center of the Earth, where the pressure is unbelievably high, the electrons are only about 20 percent closer to the nucleus. In our tennis ball model, the electron would be about 1.6 kilometers from the nucleus instead of two kilometers, still a great distance compared to the size of the nucleus.

In addition, most of the reactions that occur in the nucleus are governed by either the nuclear force or the weak interaction, which are not influenced by the presence of the surrounding electrons. The one exception is electron capture. If the electrons are forced a little closer to the nucleus, there will be a slightly greater chance that a proton will capture an electron. This is why electron capture is slightly more probable under conditions of extremely high pressure.

Let me emphasize again, however, that the rocks which are dated have not been subjected to pressures even close to those values that are required to produce even a fraction of a percent change in the decay constants of nuclei undergoing electron capture. Only one element used in dating (potassium-40) decays by electron capture, and since the calculated age is proportional to the decay constant, a change of a few tenths of one percent in the decay constant would be of no consequence to the ages obtained by this method.

Rocks

Since most of the objects we are able to date are rocks, I shall digress to describe some of the properties of rocks that are of interest. Based on the method by which they are formed, rocks can be divided into three groups.

Igneous rocks are produced by the solidification of molten material. If the cooling is not too rapid, the minerals of a particular kind will tend to collect together and form crystals. A mineral is a substance that is made of molecules that have a specific chemical composition such as common table salt, which contains equal amounts of sodium and chlorine. The chemical formula for salt is NaCl, where Cl stands for chlorine and Na stands for sodium (Na really stands for natrium, an old word for sodium). Salt forms crystals of a very definite shape (a cube). When a rock cools, the mineral crystals are almost never pure, but contain small amounts of impurities. For example, if a salt crystal formed, it would be composed mostly of sodium chloride, but small amounts of potassium chloride might also be present.

A familiar igneous rock is *granite*, which is the gray speckled rock frequently used for gravestones (Figure 3.1). Although all granites are not identical, a typical granite would contain crystals of quartz, also called silicon dioxide [SiO_2], crystals of potassium feldspar or orthoclase [$KAlSi_3O_8$], and crystals of biotite mica [$K(Mg,Fe)_3AlSi_3O_{10}(OH)_2$] which are the black crystals. These crystals are what gives granite its mottled appearance. Igneous rocks are very important in radiometric dating. The age obtained by the various techniques is the time since the rock solidified from a molten state.

Sedimentary rocks are commonly formed by the accumulation of pieces of other rocks and organic material. As weathering breaks down rocks, the small particles that are washed down to lower elevations often accumulate in valleys, oceans, and lakes, along with material from plant and animal life. When this sediment is cemented together by pressure or chemical processes, sedimentary rocks are formed. Since the composition of sediment generally changes with time, sedimentary rocks usually form in layers or strata (Figure 3.1).

Sandstone is a sedimentary rock that is produced when sand is cemented together with silicon dioxide or some other chemical substance. Limestone is another important type of sedimentary rock that is produced when calcium carbonate [$CaCO_3$] precipitates out of ocean water. This precipitation occurs more rapidly in warm water and often contains the shells of various animals. Limestone is an important source of marine fossils. The most common type of sedimentary rock is shale, which is produced by the hardening or lithification of clay and mud.

Sedimentary rocks are unsuitable for radiometric dating because they are composed of small pieces of broken down igneous rocks. Instead of yielding the date when the sedimentary rock was deposited, the dating process will yield the average age of the pieces of igneous rocks that formed the sedimentary rock. However, certain sedimentary rocks that are formed by chemical precipitation can be used for radiometric dating. An example of this type of sediment is a silicate of potassium, aluminum, and iron named glauconite:

[$K_2(MgFe)_2Al_6(Si_4O_{10})_3(OH)_{12}$]

Glauconite forms as green sand-sized pellets on the ocean bottom. Under certain conditions this mineral can be used for radiometric dating.

Metamorphic rocks are rocks that have undergone a change. This change is commonly caused by heat and pressure. Rocks can be subjected to

Figure 3.1 Rock Types Granite (left) is an igneous rock. Gneiss (middle) is a metamorphic rock formed when granite is subjected to heat and pressure. The dark lines show how the crystals have been flattened by the pressure. Sandstone (right) is a sedimentary rock. This specimen has unusually colorful layers. [JCB]

heat and pressure when they are buried beneath other rocks. Marble is recrystallized limestone, and it is produced when limestone is subjected to heat and pressure. Slate is shale that has been metamorphosed.

Either sedimentary or igneous rocks can be metamorphosed. For example, if granite is subjected to heat and pressure, the crystals tend to become flattened, forming a type of rock known as gneiss (Figure 3.1). When igneous rocks are subjected to heat, it still may be possible to date the rock since the atoms are merely rearranged, but not lost. By using several different dating methods, it is sometimes possible to determine when the rock first solidified and when the metamorphism occurred.

Potassium-Argon Dating

Perhaps the easiest dating technique to understand is potassium-argon dating. Potassium is a rather common element that occurs in three-forms or isotopes. The least common isotope is potassium-40, which happens to be radioactive. Potassium-40 can decay to argon-40 by electron capture or to calcium-40 by beta (electron) emission.

Potassium-40 decays to argon 11.2% of the time and to calcium 88.8% of the time. Potassium-40 is called the *parent*, and the argon and calcium produced

are referred to as *daughter products*. The half-life of potassium-40 is 1250 million years.

If one can measure the amount of potassium-40 in a rock and the amount of either daughter product, the age of the rock can be calculated. Since calcium-40 is a common element in rocks, there is no easy way of differentiating the calcium-40 that came from potassium-40 (called radiogenic calcium) from the naturally occurring calcium-40 that was present when the rock solidified.

Argon on the other hand is an inert gas (it does not react with other elements), and since it is a gas, it is not normally found in rocks just after they solidify. Any argon found in a rock, therefore, is the daughter product of potassium-40. To get the age of the rock, the argon trapped in the crystals must be measured along with the potassium-40.

To see how the age of a rock is determined, consider the following simple example. Suppose a rock is formed containing 200 grams of potassium-40. After 1250 million years (one half-life), when we analyze the rock, we will find 100 grams of potassium-40 and 11.2 grams of argon. These 11.2 grams of argon came from the decay of 100 grams of the potassium-40, since only 11.2% of the potassium-40 produced argon. The rest of the potassium went to calcium-40 (Figure 3.2).

Of the original potassium-40, 100 grams has decayed and 100 grams is still present. Therefore, the rock must be as old as the half-life of potassium. That is, the rock must be 1250 million years old. By knowing the amount of potassium-40 and argon present in a rock, it is always possible to calculate the age of the rock, even though the calculation may not be as simple as it was in our example.

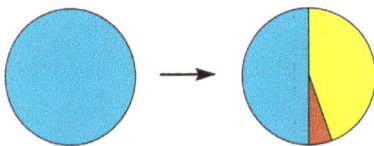

Figure 3.2 Potassium-40 Decay The circle (left) represents a crystal of potassium-40 just after it solidified. The circle (right) is the crystal after one half-life. Half the potassium-40 has decayed to calcium-40 (yellow) and argon-40 (red). [JCB]

Is this method accurate? To answer this question we need to investigate some of the details of the procedure. I will not attempt to address all the potential problems but will discuss a few of the most obvious concerns. First, the argon-40 trapped in the rock must be measured along with the potassium-40 present. With modern instruments this does not pose a serious problem. The argon is measured with an instrument called a mass spectrometer. The measurement involves heating the rock to liberate the argon. The argon gas is then sent through the spectrometer to determine the amount present.

Of course, if the rock had been heated by some natural process (above a few hundred degrees Celsius), the argon would have already escaped, and the measurement will be incorrect. Argon loss is, therefore, of great concern for loss of argon will make the age of the rock appear younger than it is. We usually say that the argon clock is reset to zero when the argon is lost. Often only a small percentage of the argon is lost, and the age determination is not in serious error.

One way of checking for argon loss is to determine the age of several different minerals from the same rock. Remember that igneous rocks often contain several different kinds of minerals. The different minerals will generally have different percentages of potassium and can be used as independent clocks. If the dates from all the minerals do not agree, the dates are said to be *discordant*. If the ages are discordant, we know there must have been some argon loss, and the rock is not well suited for potassium-argon dating. Certain minerals are better suited for potassium-argon dating than others. Biotite for example contains a large percentage of potassium and retains argon well,

making it a useful mineral for potassium-argon dating.

Since about one percent of the Earth's atmosphere is argon, one might wonder if some of the argon in the rocks might not be trapped atmospheric argon. Fortunately, atmospheric argon contains three isotopes, and if there is any atmospheric contamination in the sample, a correction can be made by measuring the abundances of the other isotopes.

In a few instances, certain rocks give ages that are too old. This excessive age is sometimes due to contamination of the sample by pieces of older rocks, but some minerals do contain excess argon that was incorporated into the rock at the time of its formation. This argon is radiogenic argon that comes from the decay of potassium-40 in the deeper interior of the Earth.

Excess argon causes the measured age to appear older than it really is, and since the excess argon is argon-40, it may be difficult to correct for its presence. By too old we mean older than the date when the rock last solidified. Of course the material that composes the rock is as old as the Earth itself, but when the material becomes molten, the gaseous argon generally escapes. If the molten material is trapped below the Earth's surface, however, the argon may not be able to escape as the rock solidifies.

Fortunately, excess argon is not a common problem and certain rock types have been identified as being more likely to contain excess argon. For example, pillow lava is formed when molten lava oozes out of cracks in the sea floor. It solidifies quickly as it contacts the water and tends to trap argon that might be present in the lava. The argon in the lava came from the decay of potassium-40 beneath the Earth's surface and was brought to the surface (the bottom of the ocean) with the lava. Obviously, such oceanic lava cannot be used for potassium-argon dating.

New techniques such as incremental heating allow scientists to better evaluate if excess argon is present or if argon has been lost by the sample. In this method, the sample is heated in increments, and the argon collected for each increase in temperature. As the temperature goes up, the argon released comes from deeper inside the mineral being measured. Excess argon tends to be concentrated on the outside of the mineral, but an increase in argon toward the center of the grain indicates that some argon near the surface has leaked out.

The accuracy of argon dating depends on our ability to accurately measure the argon and radiometric potassium present in the sample. If the potassium and argon are measured by different means, the inaccuracies can be a few percent. A variation on the potassium-argon dating method known as the argon-argon method (or the 40Ar/39Ar method) allows one to increase the accuracy by measuring the potassium and argon by the same method. If the rock sample is irradiated with neutrons from a nuclear reactor, a certain fraction of the radioactive potassium will be transformed into argon-39. The irradiated sample is then heated incrementally (with a laser), and the age is derived from the ratio of argon-40 to argon-39.

How accurate is this method of dating? We know from the writings of Pliny the Younger that Vesuvius erupted in 79 A.D. and destroyed Pompeii. The Berkeley Geochronology Center dated pumice from the event and got a date that was just 7 years off.

Since there are many radioactive atoms that can be used for dating, our confidence in a given age is greatly

enhanced if several independent methods give the same age.

Rubidium-Strontium Dating

The rubidium-strontium (Rb-Sr) method of dating poses different problems than the potassium-argon method. Rubidium-87 is radioactive and decays into strontium-87 by emitting a beta particle (electron). The strontium that results from this reaction is referred to as *radiogenic* strontium. Since there may have been Sr-87 present when the rock crystallized, one cannot simply measure the amount of Sr-87 and Ru-87 for the age calculation. Fortunately, the method used does not require that we know the initial amount of Sr-87 that was present in the sample. The initial Sr-87 abundance is obtained in the analysis.

Isotopes of a particular element differ only in the number of neutrons present in the nucleus. Since the neutrons do not have an electric charge, the only difference in the isotopes is a slight difference in mass. For the heavier elements, this slight difference in mass has little effect on the chemical properties of the different isotopes.

Because of the similar chemical properties of uranium-235 and uranium-238, one of the most difficult jobs in the production of the first uranium atom bomb was the separation of these isotopes. The difficulty of this procedure is what keeps most nations today from making a uranium bomb. Uranium-235 can easily be made into a bomb if enough is available, but uranium-238 will not produce a bomb.

In nature, when a rock solidifies from the molten state, the isotopes of strontium will be mixed in the same ratio throughout the rock. This fact has been verified by analyzing rocks that have formed recently. As the rock ages,

Ru-87 decays into Sr-87. Since we do not know how much Sr-87 or Ru-87 was present in the rock when it formed, we compare them to some stable isotope of strontium. Sr-86 is a stable isotope, and since it is not produced as the result of any radioactive decay process we say it is not radiogenic.

Therefore, if we were to compare the amount of Sr-87 to Sr-86, the ratio would slowly *increase* with time since Sr-87 is increasing but Sr-86 remains constant. However, the ratio of Ru-87 to Sr-86 would slowly *decrease* with time.

If we had just one mineral from a single rock, we could not find the date of formation because we would not know the initial ratio of Sr-87 to Sr-86. However, by measuring several minerals from the same rock, we can determine the initial ratio of Sr-87 to Sr-86 and the age of the rock at the same time.

To understand how this method works, consider the following simplified example. Suppose that when a rock solidified, Sr-87 and Sr-86 were equally abundant in all the crystals that formed in the rock (their ratio is 1:1 or simply 1). As the rock ages, this ratio would increase in those crystals that initially contained some Ru-87. The ratio would increase more in those crystals that initially had more Ru-87.

Referring to Figure 3.3, suppose that three crystals were formed in the rock such that crystal A (upper left) contained no Rb-87. A second crystal B (middle left) contained 2 grams of Rb-87, 1 gram of Sr-87, and 1 gram of Sr-86. A third crystal C (bottom left) was formed so that it contained 1 gram of Sr-86, 1 gram of Sr-87, and 4 grams of Rb-87.

Therefore, the ratio of Rb-87 to Sr-86 (red to green) in the first crystal would be 0 (no Ru-87 present), the ratio in the second crystal would be 2:1

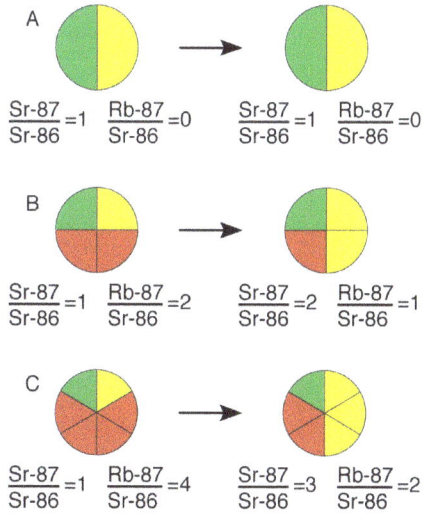

A

$$\frac{Sr\text{-}87}{Sr\text{-}86}=1 \quad \frac{Rb\text{-}87}{Sr\text{-}86}=0 \qquad \frac{Sr\text{-}87}{Sr\text{-}86}=1 \quad \frac{Rb\text{-}87}{Sr\text{-}86}=0$$

B

$$\frac{Sr\text{-}87}{Sr\text{-}86}=1 \quad \frac{Rb\text{-}87}{Sr\text{-}86}=2 \qquad \frac{Sr\text{-}87}{Sr\text{-}86}=2 \quad \frac{Rb\text{-}87}{Sr\text{-}86}=1$$

C

$$\frac{Sr\text{-}87}{Sr\text{-}86}=1 \quad \frac{Rb\text{-}87}{Sr\text{-}86}=4 \qquad \frac{Sr\text{-}87}{Sr\text{-}86}=3 \quad \frac{Rb\text{-}87}{Sr\text{-}86}=2$$

Figure 3.3 Decay of Three Crystals After one half-life, the three hypothetical crystals on the left (A, B, and C) decay into the crystals on the right. Notice that the amount of Sr-86 (green) in a given crystal does not change with time. After one half-life, half of the Rb-87 (red) in each crystal has changed into Sr-87 (yellow). The ratios of Sr-87 to Sr-86, and the ratios of Rb-87 to Sr-86 are shown below each crystal. [JCB]

or 2, and the ratio in the third crystal would be 4.

If we plot the initial ratios of the three crystals in Figure 3.4 (using the ratio on the left as the y value and the ratio on the right as the x value we get the three red dots in Figure 3.4 that lie on a horizontal straight line.

Now, if we wait one half-life, we will find that half of the Rb-87 (red areas in the crystals on the left in Figure 3.3) would be changed into Sr-87 (yellow areas in the crystals on the right in Figure 3.3). The new ratios of Sr-87 and Rb-87 to St-86 are given below each crystal on the right in Figure 3.3. If we now plot these three crystals in Figure 3.4 the dots will no longer lie on a horizontal line. The dashed arrows in Figure 3.4 shows how

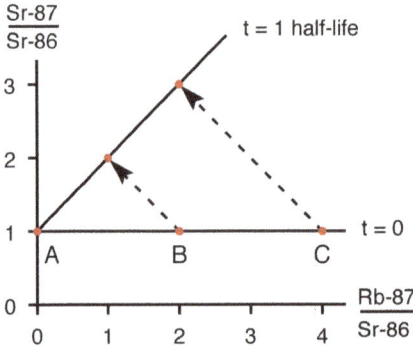

Figure 3.4 Mineral Isochron This plot is of the three crystals in Figure 3.3. At the time of formation, the values fall on a horizontal line. As time passes, the crystals follow the dashed lines. One half-life later, a plot of minerals falls on the sloping line. The steepness of the line gives the age (one half-life in this example), and the vertical intercept gives the initial (Sr-87)/(Sr-86) ratio (one in this example). [JCB]

the crystals changed on the plot as time progressed.

The steepness of the sloping straight line is a measure of the time elapsed since the rock solidified (the age of the rock). The time is one half-life of rubidium-87 in this example. The initial ratio of Sr-87 to Sr-86 was 1, which is where the line intersects the vertical axis.

Notice, that the ratio of Sr-87 to Sr-86 did not change in crystal A with time. That is because it was formed with no radioactive rubidium to start with. If you can find a crystal with no radioactivity, you can determine the initial composition immediately.

To measure the age of a real rock, several crystals from the rock are analyzed and plotted as in Figure 3.4. The sloping line obtained is called a *mineral isochron*, which means *same time*. Crystals with different initial Rb-87 to Sr-86 ratios will lie somewhere on the line. The age is gotten from the slope of the line, and the initial Sr-87 to Sr-86 ratio can be

determined by where the line intersects the vertical axis.

If the rock undergoes thermal metamorphism sometime after its formation, the strontium and rubidium may be homogenized in the minerals. If this rock is then dated using the mineral-isochron method, the date calculated would be the date when the rock was last heated.

However, if instead of minerals, several larger whole rocks from different parts of the same rock *formation* are used to form an isochron, the age of the original formation can still be obtained. This method works because the isotopes are still present in the same ratios, but the isotopes in adjacent minerals were just locally mixed again when the rock was heated. This type of plot is known as a *whole-rock isochron* (Figure 3.5).

The isochron method (either a mineral isochron or a whole-rock isochron) can be used for several radioactive decay schemes other than rubidium-strontium. The isochron method can also be applied to potassium-argon dating. The argon-40

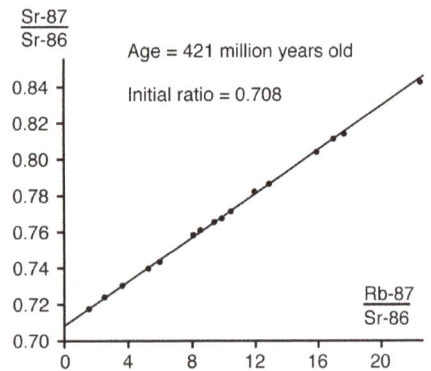

Figure 3.5 Whole Rock Isochron The graph shows the Rb-Sr whole-rock isochron of a 421 million year old rock formation. The data points are from samples of rock taken at different locations in the formation. [data: N.H.Gale et al., credits p 414]

abundance from several minerals in a sample is plotted against the potassium-40 abundance to form an isochron. This isochron method can be used as an additional check on the standard potassium-argon dating technique described above.

Uranium and Thorium Dating

The two most abundant isotopes of uranium, all of which are radioactive, are U-238 and U-235. U-238 decays by a chain of reactions to lead-206 (Pb-206), while the decay chain of U-235 ends with the formation of lead-207 (Pb-207). These two decay chains have different half-lives and, therefore, constitute two independent clocks.

Since there are two independent uranium clocks, they can be used to unscramble much more information than would be possible with just one clock. As we have seen with the rubidium-strontium clock, the age and initial composition can be obtained from a given rock sample. With the two uranium clocks, in some cases there is enough information to determine the age of the rock even when some of the lead has been lost. Being insensitive to lead loss is an important feature since rocks exposed to weathering often loose lead.

There are several ways to calculate the age of a rock using these two uranium clocks. Perhaps the simplest to understand is the lead-lead (Pb-Pb) isochron method. Pb-204 is one of the four stable isotopes of lead, and it is not produced as the result of any radioactive decay process (it is not radiogenic). Its abundance, therefore, is constant in time, and it can be used as a reference when comparing the abundances of Pb-206 and Pb-207.

If the lead isotope content of several different crystals is measured and the ratio of Pb-207 to Pb-204 is plotted against the ratio of Pb-206 to Pb-204 for each of these crystals, a straight line will result. The steepness of the line will give the age of the rock. One advantage of this method is that we only need to know the relative abundances of the various isotopes of lead. Even if the rock has lost some lead recently, the age will not be affected. This is true because all the isotopes are lost with equal probability and, therefore, their ratios will be unaffected by the lead loss.

Another method of using uranium and lead to calculate the age of a rock is to plot the Pb-206/U-238 ratio against the Pb-207/U-235 ratio for several minerals. For a good sample, the points will lie on a straight line from which the age can be obtained. This method is called the U-Pb discordia method and can be used even if there has been a loss of lead sometime in the history of the rock. The correct age is gotten even with these losses because it is the isotope ratios that are used in the calculation, not their absolute amounts.

There are several minerals that tend to form with reasonable amounts of uranium incorporated in their crystals. One that is commonly used for dating is zircon [$ZrSiO_4$]. One advantage of zircon is that uranium often substitutes for the zirconium in the crystal, but because of their larger size, lead atoms are seldom incorporated in the zircon at the time of its formation. Therefore, the initial lead content of the crystal is essentially zero, and the uranium content is relatively high. By simply comparing the uranium (parent) content to the lead (daughter) content, the age can be calculated.

82

Thorium-232 decays to lead-208, and since thorium is often found in the same rocks as uranium, this clock is usually used as a third check when dating a rock using the uranium-lead method. The Th-Pb data can be handled in the same way as the rubidium-strontium data. An isochron is plotted using different minerals in a particular rock or whole rocks are used from several different locations in a given rock formation.

Carbon-14 Dating

Perhaps the most well known dating method is the carbon-14 method. Since rocks do not contain carbon-14, this method cannot be used to date rocks, but it can be used to date pieces of wood, charcoal, bone, or other organic material that contains carbon. Carbon dating is based on the decay of carbon-14 to nitrogen-14 by beta (electron) emission. Since nitrogen-14 escapes from the sample, this method requires that we know the initial ratio of C-14 to C-12 in the sample.

When a plant or animal is alive, carbon is incorporated into its body tissue, and a certain fraction of this carbon will be C-14 (the common isotope of carbon is C-12). When the organism dies, the C-14 trapped in its body will slowly decay to nitrogen-14.

The half-life of C-14 is 5,730 years, which is reasonably short, compared to some of the other radioactive elements we have considered. Because of this short half-life, C-14 can only be used to measure ages less than about 50,000 years. Samples older than about 50,000 years generally do not have enough carbon-14 left to measure using our present methods. The coal deposits in the eastern United States and Europe, for example, have no carbon-14 left since they are a few hundred million years old. Even the coal deposits in the western United States are tens of millions of years old and, therefore, have lost their carbon-14.

Carbon-14 is produced by reactions involving cosmic rays in the upper atmosphere. Cosmic rays are atomic nuclei (mostly hydrogen nuclei) that come from outer space at great speeds. Many of these very high-energy cosmic rays are thrown out during supernova explosions. When a cosmic ray hits the nucleus of an air atom (a nitrogen, oxygen, or argon nucleus), the protons and neutrons are blown out. Carbon-14 can be produced if one of the neutrons blown out of an air nucleus hits a nitrogen nucleus. During the collision, a proton may be knocked out of the nitrogen nucleus and be replaced by the neutron, producing a C-14 nucleus.

Since the percentage of C-14 in the environment is determined by the rate cosmic rays hit the Earth's atmosphere, the reliability of this dating method depends on knowing the cosmic ray flux during the past 50,000 years or so. Assuming that the cosmic ray flux has been constant can lead to errors of almost 20 percent. The correction

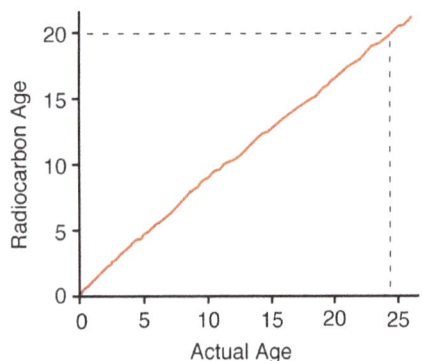

Figure 3.6 C-14 Calibration Curve On the graph, all ages are expressed in thousands of years. Notice that a radiocarbon age of 20 thousand years corresponds to an actual age of about 24 thousand years. [data: P.J.Reimer et al, credits p. 414.]

curve (Figure 3.6) that must be used to adjust C-14 dates of less than about 50,000 years has been determined by analyzing the wood of sequoia trees and bristlecone pines, by analyzing deep ocean sediments, lake sediment, coral samples, and cave stalagmites.

In modern times, the ratio of carbon-14 to carbon-12 has been fluctuating much more rapidly than it did in the past. Nuclear reactions give off many neutrons and these neutrons can produce Carbon-14. Atmospheric explosions of nuclear devices since 1945 have greatly increased the amount of carbon-14 in the environment, especially the atmospheric testing in the late 1950s and early 1960s. The 1963 Limited Test Ban Treaty that banned atmospheric testing stopped the rise of carbon-14 and it has been dropping since then.

The burning of fossil fuels liberates carbon dioxide that contain no C-14, thus reducing the C-14 to C-12 ratio. As stated above, these fuels were produced by plants and animals that died hundreds of millions of years ago, and since their death, all the carbon-14 has been converted into nitrogen.

Before the atomic age, fluctuations in the ratio of carbon-14 to carbon-12 were caused by changes in the cosmic ray flux. The cosmic ray flux is influenced by activity on the surface of the Sun. During times of high solar activity, the number of sunspots increases, as does the amount of gas ejected from the Sun's surface. This gas is a plasma, meaning it is composed of charged particles. As it sweeps out through the Solar System, it interferes with the incoming cosmic rays that are positively charged. This causes a reduction in the number of cosmic rays, and a corresponding reduction in the production of carbon-14. As was described above, these fluctuations in carbon-14

production over the past 50,000 years have been measured by other means.

Using the calibration curve, carbon-14 dating can be a very reliable form of dating when applied to objects less than about 25,000 years old. The dates for objects less than about 8,000 years are not off by more than about 15 years, and the ages of objects between 8,000 and 25,000 years old are accurate to about 150 years. The development of new accelerator techniques and calibrations using polar ice analysis could increase the present range and accuracy of this dating method.

As snow is added to the polar ice caps, air bubbles are trapped in the ice. This air can be analyzed to determine the atmospheric C-14 to C-12 ratio when the ice layer was formed. The ice on Antarctica contains trapped air samples for the past 800,000 years.

Summary of Radiometric Dating

We have investigated only a few of the many methods used for dating, but perhaps the most important point to be made here is that when proper care is taken in picking the sample and carrying out the analysis, the dates obtained by the various methods often agree very well. Not in every case, but in enough instances to convince us that the methods work well.

Parent	Daughter	Half-life (million years)
Potassium-40	Argon-40	1250
Rubidium-87	Strontium-87	48,800
Lutetium-176	Hafnium-176	37,800
Uranium-235	Lead-207	704
Uranium-238	Lead-206	4468
Thorium-232	Lead-208	14,050
Carbon-14	Nitrogen-14	0.00573

Table 3.1 Elements Commonly Used for Radiometric Dating

84

Method	Age (billions of years)
Rb-Sr	3.70
Lu-Hf	3.55
Pb-Pb	3.80
U-Pb	3.65
Th-Pb	3.65

Table 3.2 Radiometric Ages of Amitsoq Gneisses of Western Greenland [data: Baadsgaard, Pettingill and Patchett, Moorbath et al, see credits page 414.]

Why don't the ages always agree? If during its history, a rock has been subjected to changes in pressure, temperature, or various chemical processes, it may have lost some of its radiogenic daughter products, and a particular method of dating may fail.

Fortunately we have many dating methods at our disposal that allow us to measure the ages of different types of rocks, using different dating methods (Table 3.1). Certain rocks retain argon well, while others do not and can, therefore, not be dated by the potassium-argon method. Some rocks can be dated with rubidium-strontium, while others do not contain enough rubidium to employ this method.

When possible, rocks are dated by more than one technique. The use of several methods provides additional support for the age determination. As an example of the agreement that can be obtained when dating rocks, Table 3.2 gives five age determinations of a rock sample from western Greenland. The age of this rock is of great interest because it comes from one of Earth's oldest rock formations.

Relative Ages

Since the ages of most sedimentary rocks cannot be determined directly by radiometric dating, their ages must be deduced from the ages of associated rock formations that can be measured. To understand how this is done, we must investigate some of the processes that are associated with the formation and subsequent distortion of rock layers.

Rocks are laid down in layers, like the layers of a cake. These layers are called strata, and the study of strata and their associated fossils is called *stratigraphy*. Three principles first proposed in 1669 by Nicholas Steno help in the interpretation of stratified rocks. The first of these principles states that the youngest layer of rock or sediment will be deposited on top of older layers. In some rare instances, it is possible for a limited amount of sediment to be deposited beneath an older rock layer, but such cases are easily recognizable. One obvious example of this occurrence is sediment deposited in a cave, where younger sediment is deposited beneath the older rock that makes up the roof of the cave.

The second principle assumes that when sedimentary material is first laid down, the layers are deposited approximately in horizontal planes. There are obviously instances where sediment is deposited on a slope, but these slopes are usually not very steep and can be considered to be approximately horizontal.

The last principle states that when a sedimentary layer is formed, it is deposited in a continuous sheet that either stops at a barrier or ends by thinning out to nothing. To most of us these principles probably sound like nothing more than statements of common sense, but in 1669 they were revolutionary ideas. In retrospect, many great ideas often seem obvious.

Having been initially deposited in approximately horizontal layers, sedimentary layers may be tilted, wrinkled by compression, or partly eroded away. By applying Steno's principles, however, it should be possible to determine the relative ages of the various layers. In most cases, it does not even take a professional geologist to determine the relative ages of different sediments.

Folding is often associated with mountain building, and in extreme cases the rock layers can be completely folded over so that older rocks end up on top of younger rocks. By examining the entire area around the fold, however, the correct interpretation of events can be recognized.

Along with folding, a fault can also distort the formation. Faulting occurs when rock layers are cracked and movement takes place along the fracture. With some types of slanted faulting it is also possible for portions of older layers to end up above younger layers. Again, care must be taken to properly analyze these structures. For complex systems, it may be wise for us to consult a professional geologist to make sure the feature has been properly interpreted.

In a given area, sediments may be deposited during certain times, but erosion may wear down the rocks during later periods. A surface that has been worn down by erosion is called an *unconformity*. Unconformities between rock layers represent time periods when sediment was not being deposited. Unconformities are often formed after an uplift or a fault has elevated a portion of the sedimentary layers to a higher elevation. The deposition of additional sediment is halted and erosion may occur. Later, the area may sink and allow new layers of sediment to be deposited on top of the unconformity.

Although occasionally sediment is deposited at nearly a constant rate (a few grains of sand each day so to speak), a more accurate picture is probably that of an occasional flood, volcanic eruption, or windstorm depositing material in a more sporadic manner. Averaged over thousands of years, however, we tend to think of these little spurts of deposition as being continuous.

The relative ages of fossil layers are easily determined by the relative location of the layers. If a layer is directly on top of another, the one on top is obviously younger. If it is possible to estimate the rate of deposition, one can estimate how much older one layer is than another.

If the ages of certain rocks in a formation can be determined by radiometric dating, the ages of the nearby layers can be inferred. Remember, the ages of sedimentary rocks cannot be measured directly by radiometric methods.

Lava flows or layers of volcanic ash that are occasionally deposited over layers of sediment are extremely important since their age can be measured by radiometric means. The age of a sedimentary layer that has a layer of volcanic ash above and below it can be determined very accurately.

Intrusions of igneous rock are sometimes found in cracks running through sedimentary layers. These intrusions are places where molten rock has squeezed into the crack and cooled to produce a block or slab of igneous rock. Intrusions can be dated and may be used to infer the age of the older sedimentary layers in which it is found.

Many sedimentary layers contain fossils whose ages are of interest to us, and we often need to use relative dating techniques such as those just described to determine the absolute age of a fossils. One of the oldest fossil formations, found in Western Australia, has an age somewhere between 3.4 and 3.6 billion years old. This range was determined by dating igneous rocks above and below the fossils. These and other early fossils will be discussed further in Chapter 7.

Fossil Sequences

In 1793 an English engineer and surveyor named William Smith discovered a distinct order to the way fossils were deposited. For example, fossil fish occurred in layers above those that contained the earliest trilobites (an ancient extinct animal that belongs to the same phylum as the insects and horseshoe crabs). Dinosaur bones were found in layers above those that contained the first primitive fish, and mammoths and other large mammals were found in still higher (more recent) layers. It should be noted that this distinct ordering of the fossil layers, referred to as the geologic column, was discovered long before Charles Darwin published his work on evolution in 1859.

No single location on Earth contains a continuous record of sedimentary deposits. This is because sediments are generally deposited in lower areas, and no single location has been continually low enough to receive sediments throughout even the last 600 million years of Earth's history. There are always many unconformities in any sedimentary sequence, and each

unconformity represents a break in continuity.

To illustrate how Smith was able to establish the sequence of fossil bearing sediments, suppose that in one location we find layer A, perhaps a layer containing a particular type of trilobite, above layer A is layer B containing a different representative fossil, then an unconformity, and finally layer E as shown by the column on the left in Figure 3.7.

At another location we find layer B, which we recognize because of its representative fossils, layer C, and then layer D (middle column in Figure 3.7).

In a third location we find layer B, an unconformity, layer D, and layer E on top (last column in Figure 3.7). The correct order of the continuous sequence is A-B-C-D-E, with A being the oldest and E being the youngest.

In a manner similar to the above example, geologists have been able to assemble a nearly continuous sequence of fossil bearing sedimentary layers for the past several hundred million years (Figure 3.8). Again it should be emphasized that most of this work was completed before Darwin wrote his book on evolution.

As geologists studied the fossil layers of the geologic column, they gave them names (rather than letters as we did). For example, one layer called the Cambrian layer was named after the Cambrian Mountains in Wales. Another, the Jurassic was named after the Jura Mountains of Switzerland and France. Of course, these layers represent different intervals of time and are, therefore, called periods. The Cambrian Period began about 540 mya (million years ago) and ended about 485 mya.

Although the boundaries between most geologic periods correspond to times when certain plants and animals died out or when new forms first

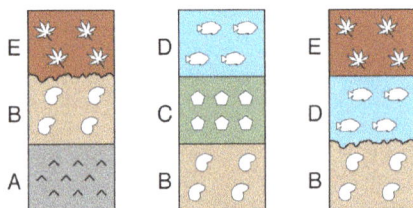

Figure 3.7 Rock Columns *The three figures represent hypothetical rock columns from three different locations. The wavy lines represent unconformities. A is the oldest layer and E is the youngest. [JCB]*

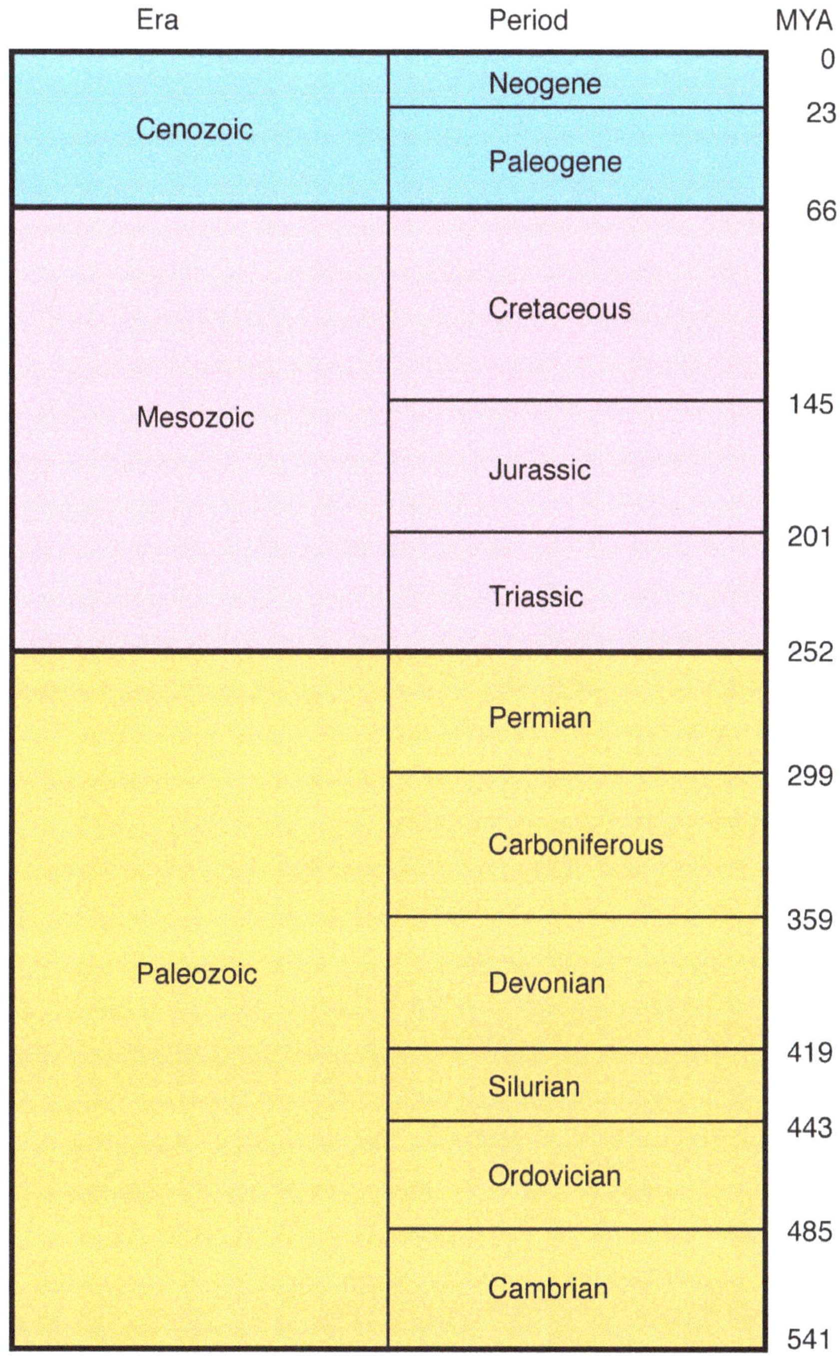

Era	Period	MYA
Cenozoic	Neogene	0
		23
	Paleogene	
		66
Mesozoic	Cretaceous	
		145
	Jurassic	
		201
	Triassic	
		252
Paleozoic	Permian	
		299
	Carboniferous	
		359
	Devonian	
		419
	Silurian	
		443
	Ordovician	
		485
	Cambrian	
		541

Figure 3.8 Eras and Periods of the Phanerozoic Eon There are three periods in the Cenozoic Era. The last 2.6 million years is known as the Quaternary Period, but it is too short to be shown. [JCB; data: International Commission on Stratigraphy, 2014]

appear, a few of the divisions were chosen rather arbitrarily. Some of the original boundaries were defined at unconformities, a poor choice since unconformities represent missing time intervals in the fossil record. In a few cases it is difficult to say just where one period ends and the next begins.

In most cases, geologists have defined boundaries by placing a golden spike (not a real golden spike) at a particular location in the strata at a specific location on Earth. We will investigate some of the details of the fossil record in Chapter 7.

The longest geologic time division is called the *eon*. The last 540 million years has seen a tremendous increase in the diversity of life forms and is known as the Phanerozoic Eon (the visible living eon). This eon is divided into three unequal intervals of time called *eras*. The eras are further divided into *periods* as shown in Figure 3.8.

The time before the Phanerozoic Eon (often called the Precambrian) is divided into three additional eons, the Proterozoic Eon (the former living eon), the Archean Eon (the ancient eon) and the Hadean Eon (named after Hades, the Greek god of the underworld). The Proterozoic extends from the beginning of the Cambrian back to 2,500 mya (million years ago), and the Archean extends from the beginning of the Proterozoic (2,500 mya) back to 4,000 mya.

The name Hadean Eon has been given for the time interval from 4,000 mya back to the formation of the Earth (about 4,600 mya). As we will see in the next chapter, this earliest interval may represent a time when the surface of the Earth had not yet cooled to a point where solid rocks could survive very long on its surface. This interval was also a time when the Moon was being subjected to frequent meteorite impacts, and it seems reasonable to assume that the Earth was suffering a similar fate.

Finally, it should be emphasized that while absolute ages are obtained from radiometric dating techniques, the distinct pattern of fossils in the geologic column can be used to make a more detailed sequence than would be possible using radiometric methods alone.

Since many life forms can come and go during a time span of a few million years, the geologic column can be more finely divided using certain trace fossils than would be possible with radiometric dating alone. Trace fossils are fossils that are found over a wide area and can, therefore, be used to define a specific fossil layer.

Uncertainties in the age of a given fossil layer are due to the uncertainties in the dating procedures and the fact that it is not always possible to find a layer of lava or volcanic ash just above and just below the layer of interest. However, the relative ages of the various layers can be established with much greater precision by the ordering of fossils in the geologic column.

Now that we understand how to determine the age of a rock, our first task will be to find the ages of rocks on the Earth and see if we can determine the age of the Earth itself. Surprisingly, in the next chapter we will discover that the Earth is older than its oldest rocks. How do we determine the age of the Earth?

Chapter 4: The Third Planet

Age and Formation of the Planets

As described in Chapter 2, the Sun, planets, and other members of the Solar System were formed from a cloud of interstellar gas and dust. Since the event occurred almost 5 billion years ago, we must try to piece the story together from the evidence that can be obtained today. Because of the subsequent evolution of Earth, much of the evidence bearing on its formation has been obliterated.

As we explore the Moon and the planets, most of which experienced an evolutionary history similar to that of Earth, our knowledge concerning the details of the formation process will be enhanced. Presently, however, some of our ideas about the early Earth are speculations with a few supportive observations. Because of the limited number of relevant observations, there are often several competing ideas as to the details of the formation of Earth. We will now give a general description of the formation process and attempt to concentrate on those aspects that are well verified.

There is considerable evidence to support the claim that various members of the Solar System formed at approximately the same time. The ages of the oldest meteorites are found to be about 4.57 billion years, and this age is generally taken to be the formation date of the Solar System. Remember, the age we get is the age when the object solidified.

Meteorites are rocks that orbit the Sun and often strike the Earth. Approximately 40,000 metric tons (40 million kilograms) fall on the Earth each year. No meteorite has ever been observed on an orbit, which would indicate that it came from outside the Solar System. A meteorite coming from outside the Solar System would have been pulled in by the Sun's gravity and would be traveling much faster than the ones we observe orbiting the Sun.

It is believed that most meteorites are fragments of small planetesimals, or *asteroids* as they are commonly called, that formed along with the other

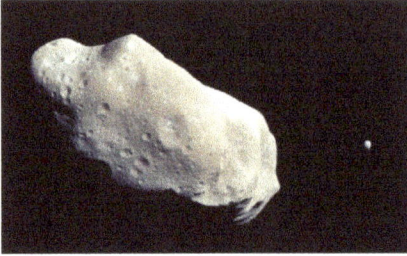

Figure 4.1 Planetesimals *Ida is 60 km long and is the only known asteroid with a satellite, Dactyl (dot at right). Both are left over planetesimals that did not grow into larger bodies. The cratering is evidence of a violent past. [Galileo Project/JPL/NASA]*

planets in the earliest stages of the Solar System. Planetesimals formed by larger rocks sweeping up smaller rocks with which they happened to collide. In the early stages of the Solar System there were large quantities of dust and small rocks orbiting the Sun, so the growth or accretion of these planetesimals could have proceeded relatively quickly (Figure 4.1).

If a larger object grows very quickly by the accretion process, it would become very hot and melt or become partially molten. We can understand this heating process if we note that

generally the objects that collided with the accreting body were moving very fast. The Earth for example moves around the Sun with a speed of about 30 kilometers a second! When two objects collide, much of their energy of motion is converted into thermal energy.

Iron and nickel melt at a lower temperature than silicate rocks and they are more dense, so they would sink to the center of the object. Small objects cool relatively quickly, so as the collisions became less frequent, the smaller planetesimals would have solidified quickly. These objects were never incorporated into larger planets, and today millions still orbit the Sun.

As time passed, collisions broke some of these objects into smaller pieces. Fragments from the center became iron meteorites (Figure 4.2) and the outer silicate layer became stony meteorites. The material at the interface of the iron-nickel core and the stony outer portion would be a mixture, and we believe a rare meteorite called a stony-iron meteorite comes from this interface material (Figure 4.2).

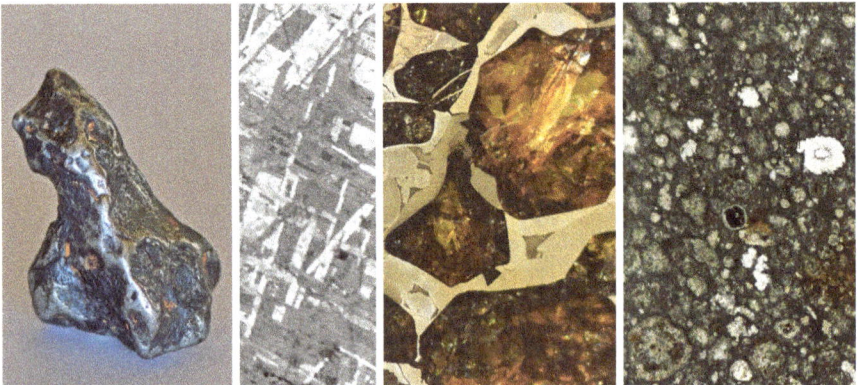

Figure 4.2 Meteorites *(left to right) Iron meteorite; (next), cross section of an iron meteorite showing the unique Widmanstatten structure that was formed as the iron-nickel mixture slowly cooled inside a planetesimal; (next), cross section of a stony-iron meteorite; (last), cross section of a chondrite meteorite showing the chondrules. The cross sections are about 3 cm high. [JCB]*

One type of meteorite known as a *chondrite* is composed of fine grains (called chondrules) of various sizes, up to a millimeter or so in diameter. These chondrules were originally molten or partially molten droplets that cooled and then were accreted to form a larger body. These asteroids did not melt or differentiate further. Chondrites, therefore, represent the oldest members of the Solar System (that is, the first objects in the Solar System to solidify). There composition is very similar to the composition of the Sun except for the gaseous elements like hydrogen and the noble gases (helium, neon, argon).

Since the date we obtain from radiometric methods is the date when the rock solidified, we expect the ages of many meteorites to be very close to the formation date of the Solar System. The ages of hundreds of meteorites have been measured using several different radiometric methods, and we find that most dates cluster around 4.6 billion years. Younger dates indicate that either the parent body took a longer time to cool, or that some of the material was melted later, most probably by the energy generated during a collision with another asteroid.

Direct evidence that the Earth and meteorites are the same age can be obtained by plotting a lead-lead isochron for meteorites. The data points fall on a line whose slope gives the age of the meteorites. If lead from the Earth is analyzed and the lead isotope ratios are plotted on the same graph, we find that the data point falls on the same isochron (Figure 4.3). This *coincidence* suggests that the Earth and the meteorites were formed at the same time, 4.55 billion years ago.

The data point from iron meteorites (that contain no uranium) gives the original isotope ratios. Recall that a

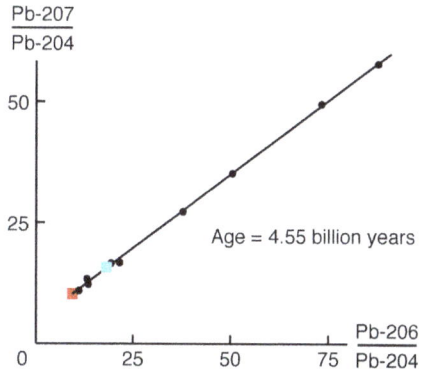

Figure 4.3 Lead Isochron of Meteorites
The red square is from iron meteorites that contain no uranium. Blue square is from modern sediments and young galenas on Earth reflecting current abundances. Other points are from meteorites. [JCB; data: G.Manhes et al.; Murthy and Patterson; York and Farquhart, credits page 414]

sample containing no radioactive material will not change with time and will, therefore, preserve the initial isotope ratios of the sample.

Another way of estimating the age of the Solar System is to measure the age of the Moon's surface rocks. Since the Moon is a fairly small object, it cooled relatively quickly, although not as quickly as the smaller asteroids. If the surface rocks solidified soon after the Moon's formation, they will give an age close to the age of the Solar System. On the other hand, if the rocks cooled more slowly, or were melted later, they will give a younger age than the age of the Solar System.

Since the planets formed by an accretion process, we expect that the rocky planets were originally covered with impact craters. The abundance of craters covering much of the Moon's surface, suggests it has not been disturbed much since its surface formed. Its surface still bears the scars of the accretion process. On the other hand, the Earth's surface is continually

Figure 4.4 The Moon *The dark lunar maria are old circular craters that were blasted out of the lighter highland areas by very large asteroids. They have been filled with magma that came from deeper regions. The maria material is denser than the highland material because it contains more iron. Iron compounds also tend to be darker in color. [UC Observatories]*

being reshaped and there are fewer impact craters on its surface.

As you look at the Moon, even with your naked eye, you will see light areas and dark areas. (Figure 4.4) Seen through a telescope, the dark regions are smooth areas with few craters. Galileo thought they looked like seas, so he named them maria, which is Latin for *sea*. The maria are lower in elevation than the lighter areas, which are called highlands. The highlands are heavily cratered. Now that we have collected rocks from both areas, we know that the maria are made of a different kind of rock than the highlands. The maria material has a higher abundance of iron and other heavy elements and is denser than the highland material.

The lunar maria are giant impact basins that were then filled in by flows of molten material from deeper inside the Moon. Therefore, we would expect the maria rocks to be younger than the

highland rocks. The highlands appear to be the remnants of the ancient lunar crust.

We know that the maria are younger features because they contained fewer impact craters than the surrounding highlands. As meteorites strike the Moon's surface, we expect an increase in the number of impact craters per unit area. However, a lava flow would tend to obscure the craters. Therefore, areas with fewer craters like the lunar maria are regions covered by younger lava flows. The more heavily cratered highlands are the older areas.

Lunar rocks were collected and returned to the Earth from nine different locations by the United States Apollo and the Russian Luna Moon landings. The ages of these rocks have been measured and their ages range from about 3.1 to 4.5 billion years old (Figure 4.5). The older dates are from rocks brought back from the lunar highlands, and they agree amazingly well with the oldest meteorite dates. The younger dates were obtained from rocks found on the lunar maria.

We know when the Moon was formed, but how was it formed?

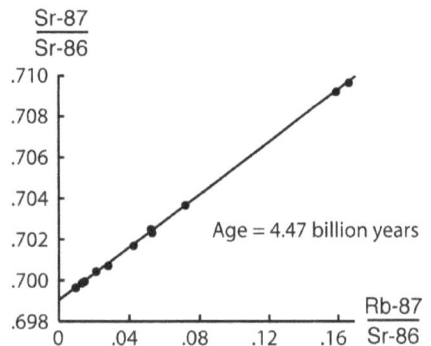

Figure 4.5 Age of a Moon Rock *This rock was collected from the lunar highlands on the Apollo 17 Moon mission. Age was measured by rubidium-strontium isochron method. [data: Papanastassiou and Wasserburg, credits p. 414]*

Certainly an accretion process formed it, but did it accrete along with the Earth out of the same material? Although the surface rocks of the Moon and Earth are very similar, the overall chemical composition of the Moon is different from the Earth.

The Earth contains more of the heavier elements like iron, and the Moon is deficient in water and elements like sodium and potassium that would be easily vaporized in a collision.

To explain these differences, it has been suggested that the Moon formed when the Earth was hit by a large asteroid or planetesimal. The material thrown out by the collision accreted to form the Moon. Since the surface layers of the Earth are more abundant in the lighter elements, the composition of Moon should be more like the composition of the Earth's surface, which it is.

Our current view of the Moon's structure and evolution is based on numerous observations from Earth, experiments performed on the Moon's surface, and analysis of Moon rocks. Although we do not know if it formed by a collision or with the Earth as a double planet, after the Moon formed by accretion, it must have been in a molten or partially molten state. As the meteorite collisions became less frequent, the Moon began to cool, and the different materials began to separate. A crust of less dense material formed, and the more dense iron-rich material sank toward the center. The lighter areas that we call the lunar highlands are remnants of the original crust.

When the solid outer crust was still rather thin, some of the larger planetesimals that struck the Moon were able to crack through this solid outer layer. The resulting craters, some hundreds of miles in diameter, were subsequently filled with denser molten material from the interior. Today we see these lava-filled craters as the darker lunar maria.

After the first billion years or so, the Moon's solid outer layers had cooled and thickened to such an extent that no further significant activity occurred. Except for perhaps a few more craters, the Moon of today looks very much as it did about three billion years ago. In sharp contrast to the Moon, the Earth's surface has changed remarkably in just the past few hundred million years.

Structure of the Earth

If we calculate the average density of the Earth by dividing its entire mass by its volume, we find its average density is about 5.5 times that of water. Of course, some rocks are denser than others, but a representative density for the rocks on the surface of the Earth is about 2.7 times that of water.

Since the density of the surface rocks is so much less than the average density of the entire Earth, the material closer to the center must be extremely dense. Even the great pressure found at the center of the Earth would not be able to squash rock to a density high enough to explain this large average value, and we must conclude that the central portion of the Earth is composed of a different kind of material than the surface.

What is the composition of the material at the center of the Earth? By analyzing the chemical composition of meteorites, the Sun, and other stars, we have found that iron is the most abundant of the high-density elements (Figure 2.24).

When elements are in the form of a hot gas as happens in stars, each element produces its own unique color pattern of light. By studying the

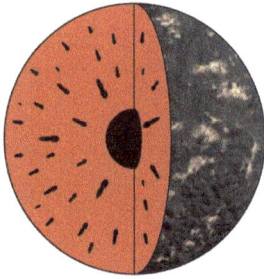

Figure 4.6 Earth Differentiates *Iron melts at a lower temperature than silicate rocks, but it is denser, so it sinks to the Earth's center, forming an iron core. [JCB]*

spectrum of the Sun and stars, we can identify the elements present and calculate their relative abundances. This is how we find that iron is the most abundant heavy element.

The reason for the relatively high abundance of iron can be understood in terms of stellar evolution. Iron is the natural end product of the nuclear reactions that fuel the stars. Based on this indirect evidence and the high density of the material at its center, we believe that large amounts of iron have sunk to the central regions of the Earth (called the *core*), causing the high average density that we observe.

Following the Earth's formation by accretion, it was probably very hot and may have been molten or partially molten. The heat generated by the impact of the incoming meteorites would have melted the surface layers. In a relatively short time, the heat given off by radioactive elements trapped in the interior would have raised the temperature to the melting point of iron. Iron melts more easily than many of the silicate-type rocks, but being denser it would sink toward the center, releasing energy in the process (Figure 4.6).

The liberation of energy as the iron sinks can be understood if we remember that objects generally gain

speed as they fall. This energy of motion is liberated as thermal energy when the object hits the ground, just as a hammer gets warm when it strikes a nail. A dense blob of material sinking in the partially molten Earth would liberate its energy of motion as thermal energy while it descends. This energy, along with the energy liberated by radioactive decay would partially melt the outer layers of the Earth and cause them to circulate. This circulation, which is still continuing today, causes a partial separation of the different materials. The denser materials tend to sink, and the less dense material tends to collect on the surface. Gases and water vapor are carried to the surface by this circulation, which may be how the oceans and primitive atmosphere were formed.

The average density of the Moon is about 3.7 times that of water, indicating that the Moon does not have a significant iron core. This lower average density implies that the Moon was not formed from the same raw materials as the Earth. The composition of the Moon is more similar to the composition of the Earth's outer layers.

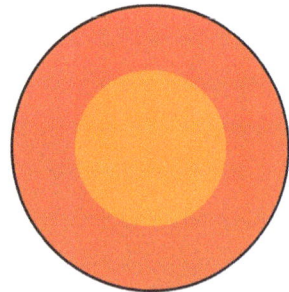

Figure 4.7 Cross-section of Earth *The Earth's radius is about 6,380 km. The outer 5 to 50 km, called the crust, is much thinner than the width of the line in the drawing. The mantle (red) extends down about 2,900 km below the crust. The liquid core (orange) has a radius of about 3,500 km. [JCB]*

If the Moon did form out of the debris from a collision between a large planetesimal and the Earth, the impact must have occurred soon after the Earth formed because the oldest rocks on the Moon are not much younger than the age of the Solar System. However, to explain the low average density of the Moon, the collision must have occurred after the Earth's iron core formed.

The Earth, which has a radius of about 6,380 km, can be divided into three different regions, believed to have different chemical compositions (Figure 4.7). The outermost of these layers is called the *crust*. It is composed of lighter material and varies in thickness from about 5 to 50 kilometers. The continents are areas where the crust is thickest, while the thinnest crust occurs under the oceans.

Next, the *mantle* extends down to almost half the distance to the Earth's center. The region below a depth of about 2,900 kilometers (1800 miles) is called the *core*, which is believed to be composed of iron. The boundaries between these three layers of the Earth are relatively sharp, and they are defined by the behavior of earthquake waves.

We have not drilled into the mantle, so any information about the two inner layers of the Earth must be gotten by other means. Earthquake waves or *seismic waves* are very important tools for studying the conditions inside the Earth. There are two kinds of seismic waves that penetrate deep into the Earth (called body waves), and two kinds that travel only near the surface (Figure 4.8). The surface waves are the ones responsible for most of the damage that is done during an earthquake.

The speed and direction of travel of the various seismic waves are governed by the physical nature of the material through which they are traveling. If a seismic wave enters a region where the density is different, its speed and direction of travel will change.

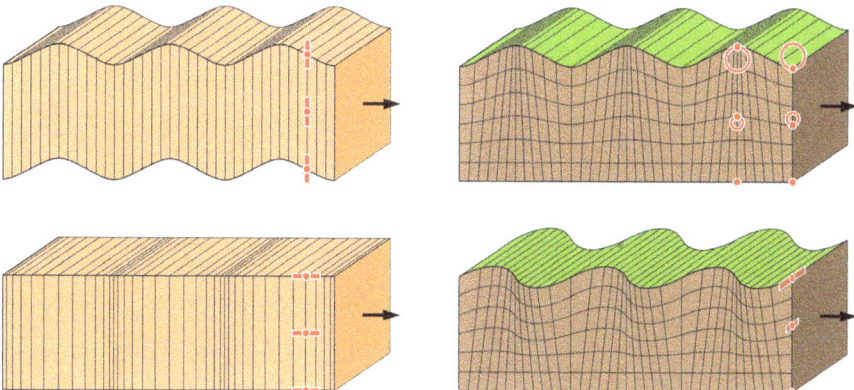

Figure 4.8 Seismic Waves *The red lines and circles show the motion of a particle in the material (red dot) as the wave passes. Waves traveling deep into the Earth (left) are P or compression waves (bottom) and S or shear waves (top). If the wave was coming at you, a P wave would wiggle you forward and backward. An S wave would wiggle you up and down. Surface waves (right) travel only near the Earth's surface (green). Notice that the motion becomes less as you go deeper. In the Raleigh wave (top), the particles travel clockwise in a circle, similar to the motion in a water wave. In a Love wave (bottom), the motion is sideways, perpendicular to the direction of the wave. [JCB]*

Conversely, if an earthquake occurs and we measure the time and location of the body waves as they emerge from the Earth, we can learn something about the nature of the material through which they have traveled. Other scientists have borrowed this technique to create interior views of the human body using CT scanners, PET scanners, and MRI machines.

One type of body wave is called a secondary or S-wave (also called a shear wave). S-waves do not travel through liquids, and since S-waves travel through the crust and mantle, but do not travel through the Earth's core, we conclude that the core must be a liquid of some kind.

The other type of deep-Earth wave is called a primary or P-wave (also called a compression wave). When P-waves enter the core, their speed is greatly reduced, and their direction of travel changes (Figure 4.9). The different physical properties exhibited by the core, coupled with our knowledge that iron is a common dense element have led geologists to conclude that the core is composed of molten iron. P-waves travel faster than S-waves and are, therefore, the first waves from an earthquake to reach a given location.

The inner portion of the core, extending out about 1200 kilometers from the center of the Earth, is a solid. Some geologists believe that the iron in the inner core may contain more nickel while the iron in the liquid outer portion may contain more sulfur. This slight difference in composition along with the higher pressure is believed to be responsible for the change from a liquid to a solid at this depth.

The density of the core is about ten times that of water, and the temperature is calculated to be approximately 4,000 degrees Celsius (about 7,000 degrees Fahrenheit). The interior of the Earth is so hot that if you could imagine cutting the Earth in half, most of the mantle would glow red-hot (Figure 4.7). Toward the center, the glow would become much brighter and be more orange in color.

The boundary between the crust and the mantle is defined by an abrupt change in the speeds of seismic waves. This boundary is known as the

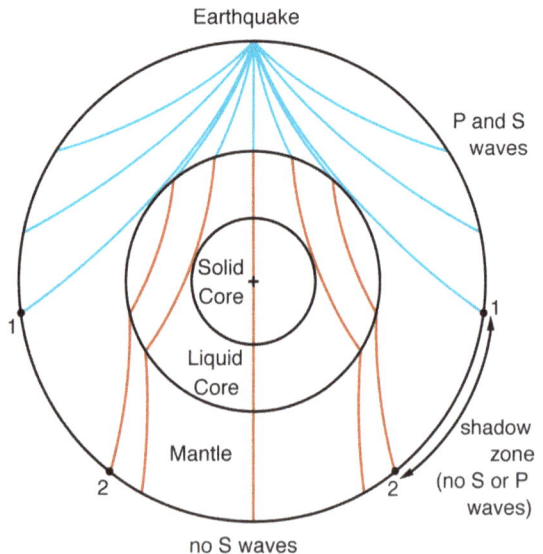

Figure 4.9 Interior View of the Earth Blue lines represent rays (paths of seismic waves) of both S and P waves. Red lines are only P waves (S waves cannot travel through the liquid core). Two blue rays are shown just grazing the core and emerging on the Earth's surface at the points labeled 1. Nearby rays enter the core and the P waves are refracted (bent) strongly. They pass through the core before emerging at points labeled 2. No rays are able to reach the shadow zone region between these two points. Knowing the types of waves and their arrival times at many locations around the world, we can generate an image of the Earth's interior. One ray is shown passing through the solid inner core. [JCB]

Mohorovicic discontinuity or Moho. Both P-waves and S-waves travel faster in the mantle than they do in the crust. The increase in speed is caused by an increase in the density of the material, and this change in density is interpreted as a change in chemical composition. The crust is composed of less dense material than the mantle.

Most rocks are made of silicates (silicon and oxygen) with smaller amounts of other elements such as aluminum, iron, magnesium, calcium, potassium, and sodium. The rocks that make up the continents are quite varied, but are mainly composed of granite (Figure 4.10). The ocean bottoms, however, are more uniform in composition and are composed of *basalt*. Basalt contains about twice as much iron, magnesium, and calcium as granite does and granite contains about 20 percent more silicon than basalt. These abundance differences make basalt denser than granite, about 3.0 compared to 2.7 for granite. Basalt's dark color is created by the iron compounds which happen to be darker colored.

As stated above, the thickness of the crust is not uniform. The lighter continental crust is considerably thicker than the crust under the deep oceans. The ocean crust is 5 to 10 kilometers thick and the continental crust averages about 35 kilometers in thickness. The continental crust is thickest where high mountains are present, being almost 90 kilometers thick under the Himalayas.

Just below the crust is the upper mantle, and it contains an even higher portion of iron and magnesium than the oceanic basalt, giving it a density of about 3.4 times that of water. These density differences explain why the lighter continents float on the underlying mantle material.

Figure 4.10 Granite and Basalt Granite (left) is a course grain rock since it cooled relatively slowly, allowing the minerals to form larger crystals. Basalt (right) cooled more quickly so it is a fine grain rock. Its higher iron content makes it darker in color and more dense. [JCB]

The large blocks of crustal material that make up the continents act like lighter rafts floating on the denser mantle material. Where they push up the highest, they also sink down deeper into the mantle. We observe this effect with blocks of wood floating in water. A block of wood will sink into the water until the weight of the block is balanced by the upward buoyant force of the water. Therefore, a thicker block will float higher in the water, but it will also extend down deeper into the water. Geologists refer to this effect as *isostasy*, which simply means that land formations will sink down until their weight is balanced by the upward buoyant force of the mantle material.

Plate Tectonics

Along with the three regions that we believe differ in chemical composition, the crust and upper mantle can be divided into two layers that differ in rigidity. The crust and outer portion of the mantle are relatively cool, and they have hardened to form a fairly rigid layer that covers the Earth's surface. This layer, which varies from about 40 to 280 kilometers thick, is called the *lithosphere*.

Beneath the lithosphere and extending down to a depth of more than 700 kilometers, is a less rigid layer called the *asthenosphere*. This viscous layer is hotter than the lithosphere and is able to flow (like very thick syrup) beneath the more rigid lithosphere.

In 1912, Alfred Wegener's idea of *continental drift* was not greeted with great enthusiasm by the scientific community because it was based only on evidence of glacial activity in the southern hemisphere and a strong similarity between the contours of the west coast of Africa and the east coast of South America. Since then, however, the evidence in support of continental motion has accumulated at a rapid rate, and today there is no doubt that significant continental motion has taken place.

The lithosphere is broken into about half a dozen large plates, and several smaller ones that form a mosaic covering on the Earth (Figure 4.11). These plates slide over the fluid asthenosphere, carrying the continents and oceans with them. A plate or that portion of a plate that is located under the deep ocean is called oceanic lithosphere, and a plate or portion of a plate that is supporting a continent or continental shelf is called continental lithosphere. Oceanic lithosphere is thinner than continental lithosphere.

The plate boundaries are regions of intense seismic activity. There are four different types of plate boundaries and many interesting features are created at these plate boundaries.

Tectonics refers to the study of the surface structures of a planet, and the mechanisms that are responsible for the formation of these surface features. On the Earth, most of the gross surface features are produced by the motion and interaction of the lithospheric

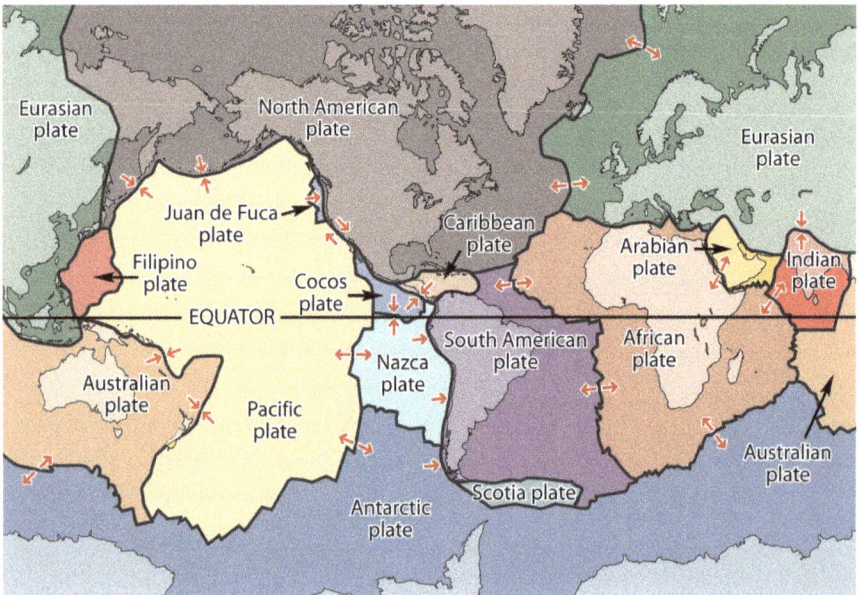

Figure 4.11 Tectonic Plates of the World The lithosphere is broken into several plates that slide over the asthenosphere. The relative motions of the plates are shown with red arrows. [U.S.Geological Survey]

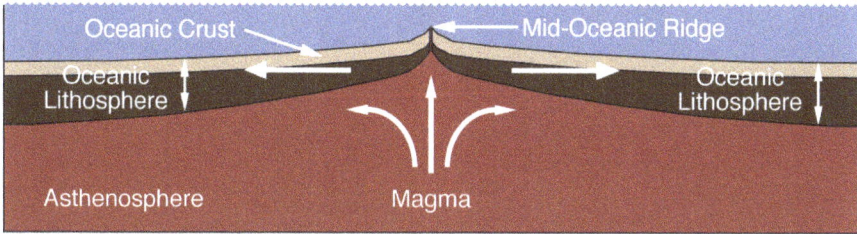

Figure 4.12 Divergent Boundary *As material rises in the asthenosphere, the decreased pressure causes melting. The lithospheric plates separate and new oceanic crust is formed as the lighter materials rise to the surface and solidify. This is happening along the Mid-Atlantic Ridge. [JCB]*

plates, so we refer to this study as plate tectonics. In contrast, the major surface features on the Moon are the result of impact crater tectonics. The Moon, being much smaller than Earth, cooled very quickly, and the rigid outer layer soon became too thick to allow plate motion.

Areas where two plates of oceanic lithosphere are spreading apart are called *divergent boundaries* (Figure 4.12). The plate boundaries in the middle of the Atlantic Ocean, in the eastern and southern Pacific, and in the middle of the Indian Ocean are divergent boundaries. These regions of *sea floor spreading* are interconnected and form a continuous boundary line that is more than 80,000 kilometers (48,000 miles) long. As the two plates move apart, hot molten material from the asthenosphere rises and comes to

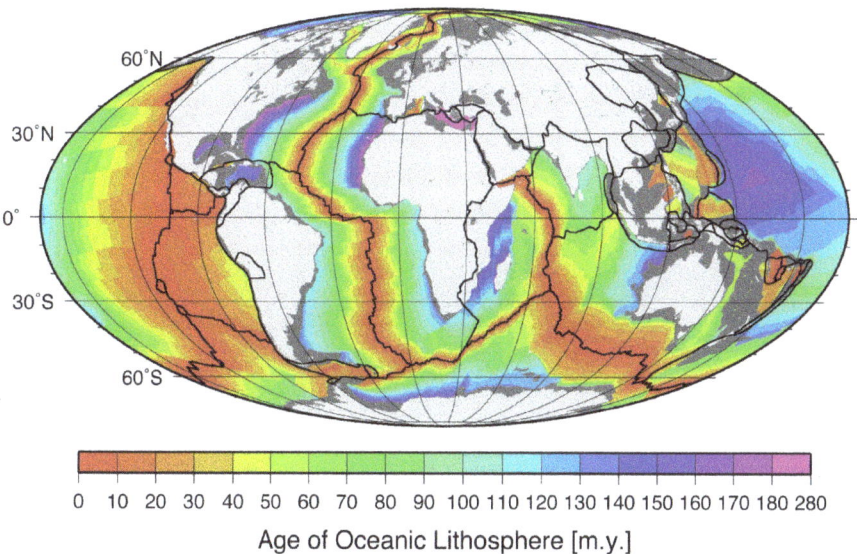

Figure 4.13 Age of the Oceanic Crust *As we expect with sea floor spreading, the oceanic crust gets older as we move away from the divergent boundaries. The oldest oceanic crust is in the western Pacific and the eastern and western North Atlantic. The oceanic crust is very young compared to the age of the Earth! [NOAA/NGDC/R.D.Müller, M.Sdrolias, C.Gaina, W.R.Roest]*

the surface of the Earth's crust (which is located at the bottom of the ocean). As it contacts the cold ocean water, the magma cools and forms new oceanic lithosphere and crust. The lithosphere is thinnest at divergent boundaries.

A chamber of hot magma forms beneath the divergent plate boundary, and since the magma is hotter and, therefore, less dense than the material farther from the boundary, it floats higher on the underlying asthenosphere and pushes up the lithosphere to form a *mid-oceanic ridge* (Figure 4.12). In the Atlantic Ocean, we refer to the divergent plate boundary as the Mid-Atlantic Ridge, while the divergent boundary in the eastern Pacific is called the East Pacific Rise.

The age of oceanic crust reinforces the idea of sea floor spreading. We have measured the ages of rocks from the seafloor and find that they get older the farther they are from a divergent boundary (Figure 4.13).

New oceanic lithosphere is formed on either side of the ocean ridges, but portions of the lithosphere are being destroyed at other places. Destruction can occur were two plates are moving together, a boundary known as a *convergent boundary*. When two plates with oceanic lithosphere collide, one plate will slide under the other (Figure 4.14). These areas are known as

subduction zones. Examples of this type of plate collision can be found south of the Aleutian Islands and east of the Mariana Islands. As the oceanic lithosphere sinks into the hotter asthenosphere, the sinking plate is melted and the lighter portions of the magma rise to the surface forming a *volcanic island arc* along the edge of the non subducted plate. Also, at the plate boundary, the sinking motion of the plate forms a *deep ocean trench*. The Mariana Trench, which contains the deepest spot on the Earth, is such a feature.

Oceanic lithosphere is thinner and denser than continental lithosphere, so if the two collide, the denser ocean plate will slide under the more buoyant continental plate (Figure 4.15). The west coast of South America illustrates what happens when this type of continent-ocean plate collision occurs. As with the collision of two oceanic plates, a deep ocean trench is formed, and a *volcanic mountain chain* like the Andes is created along the edge of the continental plate boundary. As the ocean plate slides under the continental plate, any lighter material that happens to be on top of the ocean plate, such as sediments, island arcs, or fragments of continents, will be scraped off and become incorporated in the continental mass. Much of western North America

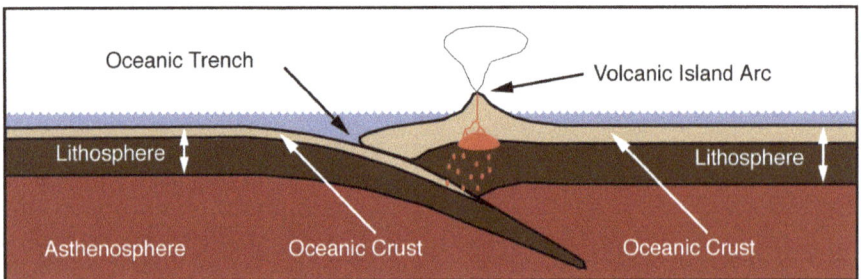

Figure 4.14 Convergent Oceanic Plates When two oceanic plates converge, one dives under the other, forming a deep oceanic trench and a volcanic island arc. An example is the Aleutian Trench and the western Aleutian Islands. [JCB]

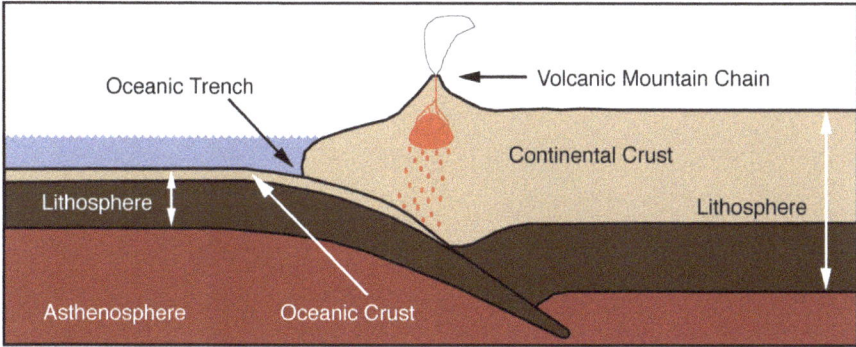

Figure 4.15 Convergent Oceanic and Continental Plates *When an oceanic and continental plate converge, the denser oceanic plate passes under the continental plate, forming a deep oceanic trench and a chain of volcanoes along the edge of the continent. An example is the Peru-Chile Trench and the Andes Mountains along the west coast of South America. [JCB]*

was attached to the continent in this manner.

Eventually, two continental plates will collide, and since both are made of thick, low-density material, neither plate subducts appreciably (Figure 4.16). The plates fuse at the boundary and the compression of the crust forms folded-type mountains such as the Himalayas, Appalachians, and Urals.

The Himalayas began to form about 45 mya when India collided and became sutured to the rest of Asia. The Appalachians climaxed their rather long and complex formation when Africa and North America collided approximately 250 to 300 mya, and the Urals were formed about 250 mya as northern Europe became linked with a continental fragment that would

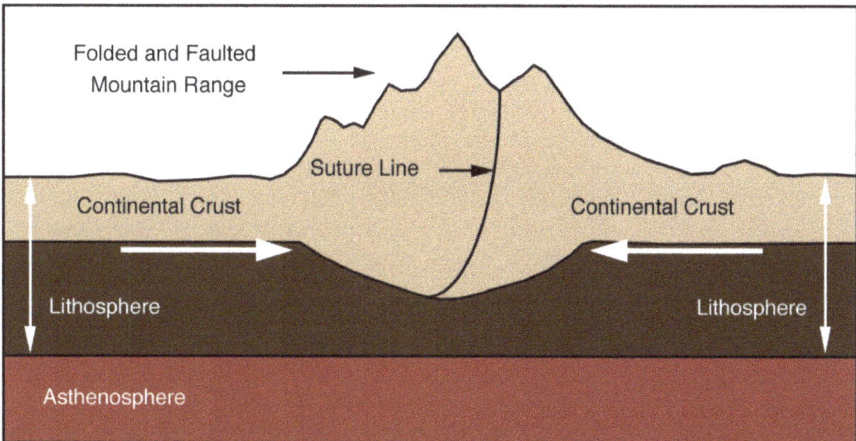

Figure 4.16 Convergent Continental Plates *When two continental plates converge, the lighter crust of both plates remains on top, but is compressed and deformed, creating mountains with many folds and faults. An example is the Himalayan Mountains, forming as the Indian Plate collides with the Eurasian Plate. [JCB]*

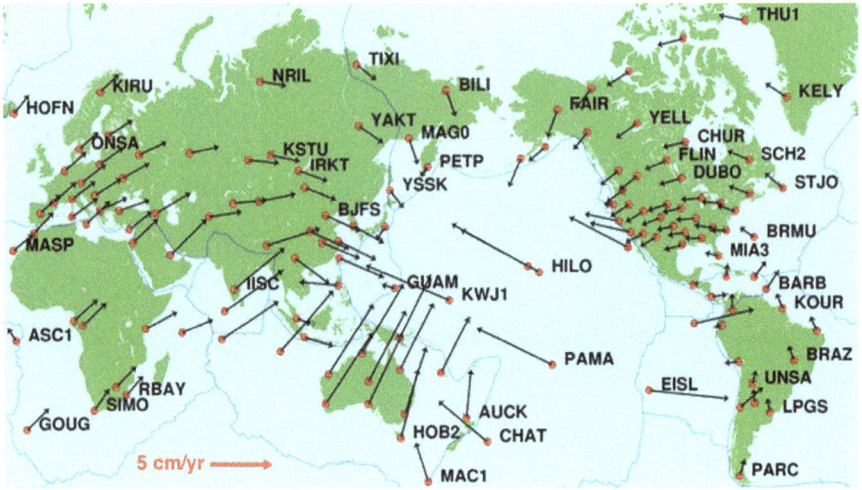

Figure 4.17 Speeds of the Tectonic Plates *The speeds of the tectonic plates are measured by using the global positioning system (GPS). Note the scale for the arrows (in red), near the lower left of the figure. [NASA/JPL]*

eventually become part of northern Asia.

During mountain building, fossil bearing sedimentary rocks can be pushed up, exposing fossil layers that had been buried for millions of years. Because of this process, it is possible to find fossils of marine animals on the tops of mountains.

Plates generally move only a few centimeters a year, but we can measure this small motion directly using the Global Positioning System or GPS (Figure 4.17). The measured plate speeds are consistent with the speeds inferred from geologic evidence.

Paleomagnetism

We can measure where the Earth's tectonic plates are moving now, but we also wish to determine where they were in the past. This information is obtained by studying the magnetism of ancient rocks, a field of study known as paleomagnetism. The Earth has a magnetic field that enables us to use a compass for finding our way. A compass is a small magnet in the shape of a needle that is generally confined to rotate in a horizontal plane. If the needle were free to move in any direction, it would line up with the Earth's magnetic field (Figure 4.18). At the North Magnetic Pole, the north end of the needle would point straight down, making an angle of 90 degrees with the ground. At the South Magnetic Pole the needle would point straight up, and on the Magnetic Equator, the needle would be horizontal to the ground. At intermediate latitudes, the needle would dip down (in the northern hemisphere) or up (in the southern hemisphere) at some angle. This *dip angle* as it is called can be used to determine the approximate latitude of your location.

The Earth's magnetic field is not uniform, but about 90 per cent of the field could be produced by a big bar magnet at the center of the Earth, but tilted 11.5 degrees to the Earth's rotation axis. The poles of this imaginary bar magnet are called the

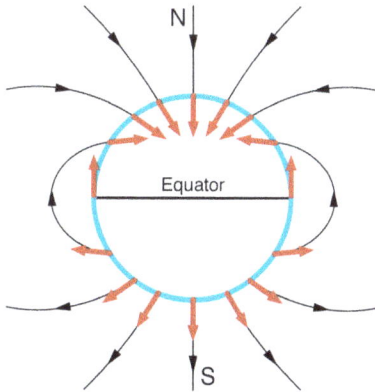

Figure 4.18 Dip Angle *As magma cools, magnetic molecules line up with the Earth's magnetic field (the red arrows). [JCB]*

geomagnetic poles. (Because the Earth's magnetic field is not uniform, the geomagnetic poles do not exactly match the magnetic poles.) The Earth's rotation axis defines the points we call the North and South Poles (the *geographic poles*).

Many atoms and molecules behave like little magnets, but certain minerals have stronger magnetic properties than others. For example, magnetite [Fe_3O_4] is a strongly magnetic mineral that is common in many types of rocks. When a rock is in a molten state, these little magnets tend to line up with the magnetic field of the Earth. As the rock solidifies these little pieces of magnetic rock are frozen in place.

The rock, therefore, retains the orientation angle of the Earth's magnetic field at the time the rock solidified. Even if the rock is later moved, it still retains the original orientation angle it had with the Earth's magnetic field.

In particular, if the plate on which the rock is situated moves to a new latitude or is rotated, the magnetic field of the rock will still reflect the relative position of the magnetic field when the rock solidified. We should, therefore,

be able to determine something about the rock's location when it solidified.

Notice that the position of the rock on the Earth is not uniquely defined by this method. The dip angle frozen into the rock will allow us to calculate the latitude at the time of solidification, but the longitude cannot be determined.

The Earth's magnetic field fluctuates with time and the magnetic poles do not coincide with the geographic poles. Therefore, by using only one rock sample, it is not possible to locate the North Geographic Pole by finding the North Magnetic Pole. However, these two poles are only about 11.5 degrees apart, and measurements made during the past 150 years or so suggest that the magnetic pole wanders about the geographic pole with an irregular period of about ten thousand years. Therefore, when averaged over time periods of several tens of thousands of years, we expect the average positions of the two poles to coincide.

One way of checking the validity of this assumption is to locate the magnetic pole using samples that have cooled in the past few million years. During this time, the continents have not moved significantly, and we find the locations do cluster about the North Geographic Pole of the Earth, not the present North Magnetic Pole (Figure 4.19).

Data from the past 5 million years (not shown) gives a value just 1.2 degrees from the North Geographic Pole (about 140 kilometers). It appears, therefore, that the method can be used to determine the latitude of a land area if we take several rock samples from nearly the same time period and average them.

If we locate the pole using North American rocks of different ages, we find that indeed either the pole has shifted or the continent has moved. If we repeat the analysis using European

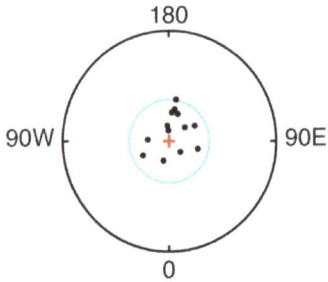

Figure 4.19 Locating the Geographic Pole *The dots show the location of the Paleomagnetic Pole determined from baked clay younger than 2,000 years. The blue circle is 20 degrees from the North Geographic Pole (red plus sign). The mean value of these points is 4 degrees from the North Geographic Pole. [M.W.McElhinny, data from M.J.Aitken, credits p. 414]*

rocks, we find that the pole locations do not match those determined from the North American rocks of the same age. However, if we assume that North America and Europe have been moving apart for the past 80 million years, we find that the poles do coincide!

Considering this type of data, it appears that all the continents were united or approximately united about 240 mya and formed one large supercontinent called Pangaea (Figure 4.20). The fragmentation of Pangaea, and the subsequent motion of its pieces has produced the placement of continents we see today.

When their magnetism was measured, an important discovery was made. Rocks very near the mid-ocean ridge, and therefore, the youngest, did indeed have their magnetic material aligned with the North Magnetic Pole, but the magnets in the rocks a little further from the ridge were found to be pointing in the opposite direction (called a *magnetic reversal*). As the rocks further and further from the ridge were studied, this reversal pattern was repeated (Figure 4.21).

The discovery of this reversal pattern (called *magnetic striping*), demonstrates that the Earth's magnetic field has reversed directions many

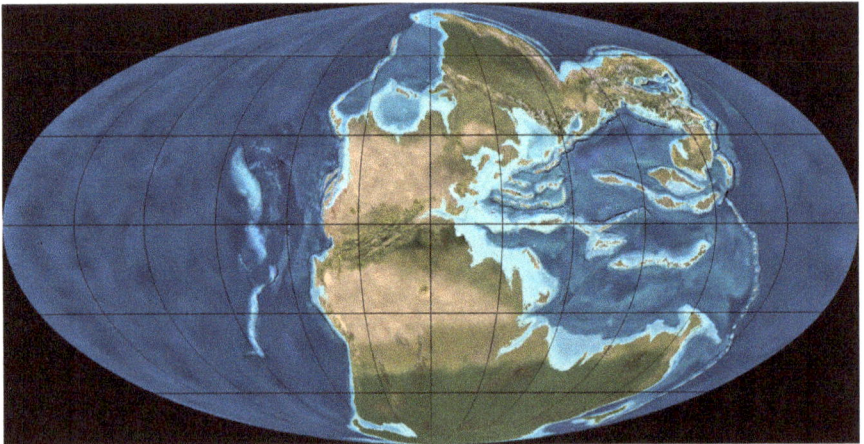

Figure 4.20 Pangaea *The Earth as it looked about 240 million years ago near the Middle Triassic, the dawn of the dinosaur age. North of the Equator was the land mass known as Laurasia, composed of North America just above the Equator and Europe and Asia above and to its right (east). South of the Equator was the land mass of Gondwana. It was composed of South America (on the left), then Africa, then India, Antarctica (at the bottom below India), and Australia (to the far right). [Ron Blakey/Colorado Plateau Geosystems, Inc]*

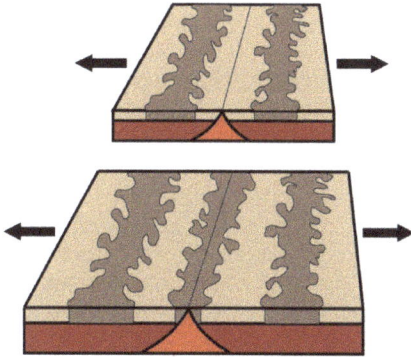

Figure 4.21 Magnetic Reversals At a divergent boundary, magma rises to the surface (bottom of the ocean), where it solidifies as the plates spread apart. Before the last reversal (top), the cooled magma retained the reversed magnetic orientation (light brown). Today (bottom) the newly formed rocks retain the normal magnetic orientation (dark brown). [JCB]

function of time. We see that the reversals occur in a fairly irregular way. There seems to be no way of predicting when the next reversal will occur. The Earth's magnetic field strength also varies in an irregular way. It is currently decreasing in strength, but it could begin to increase at any time. A decrease does not indicate that a reversal is approaching.

The Last 600 Million Years

By measuring the magnetism of rocks with different ages, we can trace the motions of the continents back in time. Because of uncertainties in the measurements, however, the further back we try to trace the movements, the less accurate are of the positions. Remember too, that the longitude cannot be determined from the paleomagnetic data and must be inferred from more recent positions and other indirect data such as the similarity of ancient mountain ranges and similar rock types.

Another difficulty with magnetic determinations is that the older rock samples are more likely to have undergone some form of deformation that might have altered the original magnetic alignment.

The locations of the major land masses have been traced back more than 600 million years, but because of the difficulties mentioned, the geologic community is not in full agreement on the details of the reconstructions. The

times in the past. The reversals are clearly recorded in the rocks on either side of the Mid-Atlantic Ridge and the East Pacific Rise. Although far from regular, the reversals have an average period of about 200,000 years (Figure 4.22).

Some people might find it hard to believe that the Earth's magnetic field could flip upside down, but then it should be pointed out that the Sun has an even stronger magnetic field that reverses direction with a period of approximately eleven years.

Continental rocks also contain a record of the magnetic reversals of the past. The magnetic orientation of rocks of different ages can be used to make a plot of the magnetic reversals as a

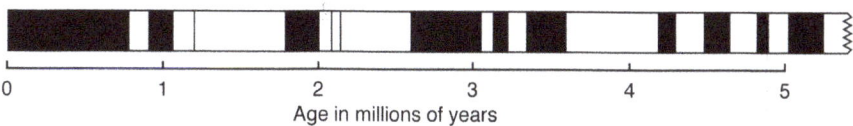

Figure 4.22 Magnetic Reversals The Earth's magnetic field reverses polarity in an irregular way. The graph shows the polarity of the field for the past 5 million years. Normal polarity is black and reversed polarity is white. [USGS]

more recent positions are of course more accurately known than those of 600 million years ago.

About 500 mya, long before the formation of Pangaea, there were several separate land masses that we might call ancestral North America (officially called Laurentia), ancestral northern Europe (called Baltica), ancestral China (which included much of southeast Asia), ancestral Siberia, a piece of central Asia known as Kazakhstania, and a large land mass called Gondwana. Gondwana was composed of South America, Africa, Antarctica, Australia, Madagascar, India, southern Europe, and Arabia.

Laurentia was located very close to the Equator about 500 mya (Figure 4.23). For about the next 300 million years, much of Canada would remain above water, but large shallow seas would cover much of the United States. Southern Ohio and Indiana are excellent places to look for fossils of marine animals that lived during the Cambrian and Ordovician Periods.

Located to the south of Laurentia and across an ocean known, as the Iapetus Ocean, was the landmass of Baltica. During the Cambrian Period, the Iapetus Ocean was widening, but it began to close during the Ordovician Period. About 400 mya Laurentia and Baltica became sutured together to form a single land mass known as Laurussia. The Caledonian Mountains in Britain and Scandinavia were formed along the suture line, and the northern Appalachians were created as the collision progressed.

Much of Gondwana was located near the Equator about 600 mya, but it moved south and about the time the Caledonian Mountains were forming, the African-South American portion of Gondwana was passing over the South Pole. Evidence of the glaciation from this time is seen in southern portions of South America and Africa.

Gondwana continued its motion northward and about 300 mya it collided with Laurussia, which was still located near the Equator. The southern

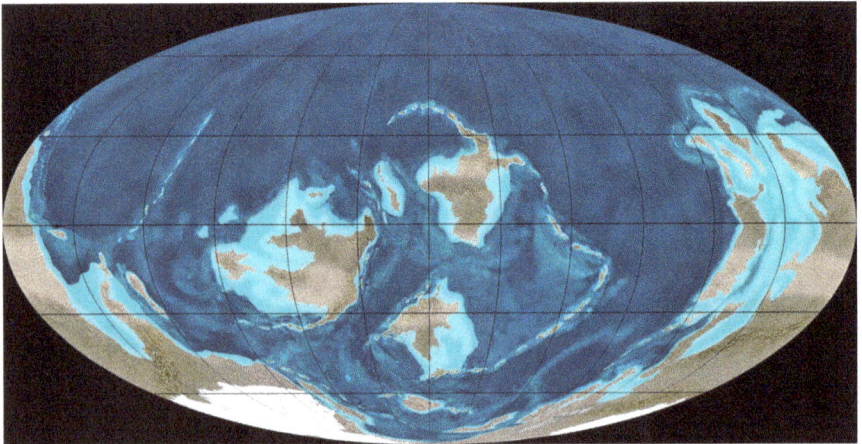

Figure 4.23 Earth During the Middle Ordovician About 470 million years ago Laurentia (ancestral North America) was located near the Equator (left of center) and Baltica (ancestral northern Europe) was located just to the southeast across the Iapetus Ocean. Ancestral eastern Asia (Siberia) was located near the Equator to the east of Laurentia, and Gondwana was centered near the South Pole. [Ron Blakey/Colorado Plateau Geosystems]

Appalachians were formed along the suture line.

Meanwhile, to the north and east, Siberia and Kazakhstania joined to form a single land mass. This combined land mass moved south and collided with Laurussia about 250 mya. The Ural Mountains were formed along the suture line. The combined landmass of Laurussia and Siberia-Kazakhstania is called Laurasia.

China moved into place and Pangaea was assembled about 250 mya (Figure 4.20). With Laurasia extending out to the north and east and Gondwana extending to the south and east, Pangaea resembled a giant letter C centered on the Equator. The body of water that separated the eastern portions of these two landmasses was called the Tethys Sea.

The breakup of Pangaea and the subsequent movements of the various landmasses have played a very important role in shaping the plant and animal life of the different continents. At about the time the first dinosaurs and the first mammals appeared on the Earth, Pangaea began to fragment. The breakup of Pangaea began with rifting between Gondwana and Laurasia (Figure 4.24).

The North Atlantic opened up as North America moved to the northwest, away from Africa and South America. India separated from Gondwana and began its journey northward, and Madagascar drifted away from Africa.

Although Antarctica and Australia remained connected, they began to move away from South America. South America began to pull away from Africa about 150 mya, but northern Brazil and the Ivory Coast remained in close proximity for several million years as these portions of the two continents slid by each other. The Tethys Sea began to close as Laurasia slowly rotated in a clockwise direction.

Africa moved northward, but it remained isolated from the other continental masses between about 100 and 40 mya. It was during this time that many unique animals evolved in Africa like the sea cows, hyraxes, and elephants.

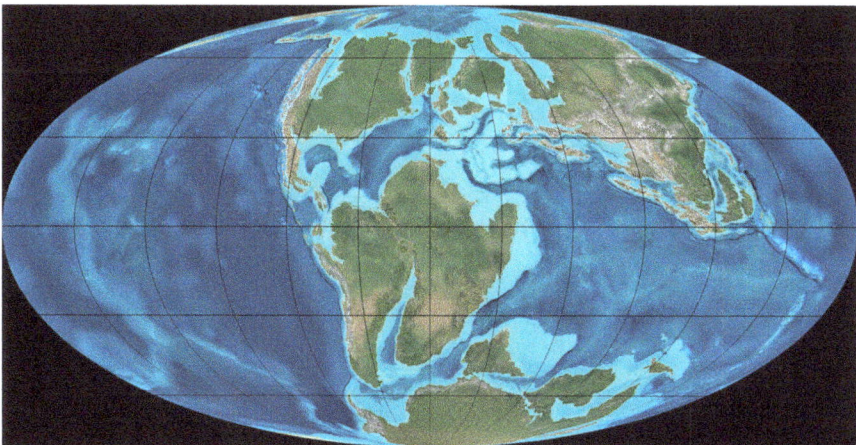

Figure 4.24 Breakup of Pangaea During the early part of the Cretaceous, about 120 million years ago, North America had separated from Africa, and South America was beginning to pull away from Africa. India (just to the right of southern Africa) was breaking loose of Antarctica and moving north. [Ron Blakey/Colorado Plateau Geosystems, Inc]

Near the end of the Cretaceous (when the dinosaurs became extinct) Greenland began moving away from Europe, and a land connection called Beringia joined Asia with Alaska.

About 50 mya, India contacted Asia and the Himalayan uplifting began. At about this time, ice sheets began to form on Antarctica as it separated from Australia and South America and moved over the South Pole.

About 3 mya, the Isthmus of Panama formed, joining North and South America. Until they were connected, South America was an isolated continent for perhaps 100 million years. Australia was isolated for a similar period. This long isolation explains why so many unique plants and animals evolved in Australia and South America. The isolation of Madagascar also produced many unique life forms on this large island.

The motions of the continents have played an important role in the evolution of life and it will be reviewed again in Chapter 7 when we investigate the fossil record and in Chapter 9 when we discuss the biogeographical regions of the world.

The Young Earth

We have gotten a small glimpse of what the Earth looked like during the past 600 million years, but what about the very early Earth? The Earth is 4.6 billion years old, but the oldest rocks that have been discovered are only about 3.8 to 4 billion years old. There may be older rocks that have not been discovered yet, but some geologists believe that various processes obliterated most of the older rocks.

The formations that have endured for the longest time are the continents. Since they are composed of lighter rocks, they are not easily subducted.

The rocks of the ocean bottoms are of course being recycled today as ocean plates are subducted and melted at convergent plate boundaries, and new oceanic lithosphere is created along the mid-oceanic ridges.

During the first billion years after the Earth was formed, the amount of heat liberated by radioactive elements (uranium, thorium, and potassium) was about three times higher than it is today. The higher temperatures that this radiation produced may have caused the plates to move much more violently than they do today. Indeed, all the surface material may have been recycled during these early years, and the continents may not have begun to form until the surface had cooled down somewhat. In any event, we know that by 3,800 mya there were landmasses beginning to protrude above the level of the ocean. We know the land was above sea level because sedimentary rocks are incorporated in these early formations.

Evidence from the Moon suggests that large meteorites continued to impact the lunar surface until about 3,500 mya. Being an even bigger target, the Earth must have suffered even more collisions. These impacts would have erased any evidence of early continental formations.

The composite structure of the present-day continents suggests that they have grown in size through the ages. Island arcs and ocean sediments have been added to the ancient continental cores as the lithospheric plates slide about over the surface of the Earth. The present form of the continents is the result of billions of years of growth, shaping, and recycling. As we will see in the ensuing chapters, the motions of the continents have played an extremely important role in shaping the various life forms that have evolved through time.

The Other Planets

What does plate tectonics have to do with life? Is the Earth somehow unique in its ability to support life? The motion of the plates allows lighter minerals to collect into large masses floating on the dense asthenosphere, thus forming the continents. The average elevation of the continental crust is about 5 kilometers higher on average than the denser basaltic crust. The large quantity of liquid water on the Earth's surface fills these low basaltic basins, forming the oceans. In contrast, our two nearest neighbors, Mars and Venus, show no clear evidence of plate tectonic activity and neither planet has liquid water on its surface.

Although Venus and Earth are nearly the same size and mass, there are important differences. Since Venus is closer to the Sun, you might guess that it gets heated more by the Sun and, therefore, must be hotter than the Earth. That reasoning would be wrong!

Venus absorbs *less* of the Sun's energy than the Earth. Even though it is 70% closer to the Sun and is hit by twice as much sunlight as the Earth, its thick cloud cover reflects 90% of this energy back into space. On the other hand, the Earth only reflects 30% of the incident solar energy back into space. Considering these numbers alone, we would expect Venus to be *cooler* than the Earth.

However, Venus has a thick carbon dioxide atmosphere that creates a strong *greenhouse effect* or *global warming* as some like to call it. The surface of Venus is completely covered by clouds of sulfuric acid. The planet absorbs energy from the Sun and reradiates infrared radiation. Carbon dioxide is the primary greenhouse gas, which absorbs this infrared radiation, and produces extreme global warming

on the planet. The surface temperature is 730 K (850 °F), hot enough to melt lead! There is no liquid water on Venus (Figure 4.25).

Why are the conditions on Venus and Earth so different? Two probable causes are that Venus is closer to the Sun and does not rotate appreciably. On Earth, carbon dioxide and sulfur dioxide dissolve in the ocean, which removes some of these gases from the atmosphere. Plants also use carbon

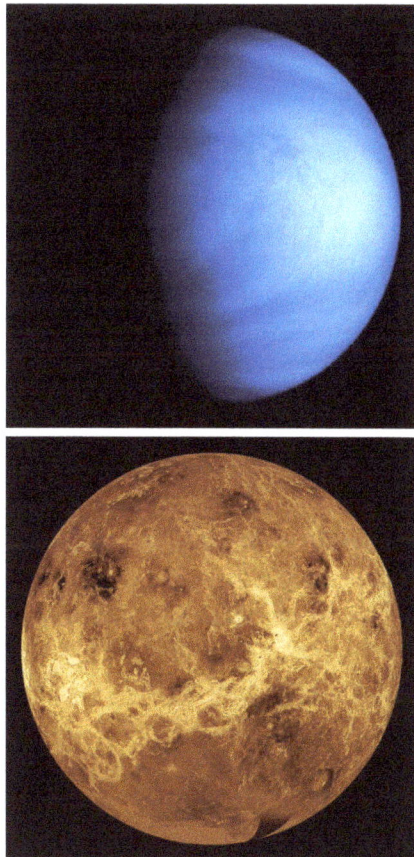

Figure 4.25 Venus In visible and ultraviolet light (top), the surface of Venus is obscured by thick clouds of sulphuric acid. [Galileo Project/JPL/NASA] Radar can penetrate the clouds and reveal the surface features of the planet (bottom). [Magellan Project/JPL/NASA]

dioxide and water to produce sugar which further reduces the carbon dioxide in the atmosphere. In the ocean, some of the carbon dioxide combines with calcium and forms solid calcium carbonate, which precipitates out of the water and forms limestone deposits.

If the ocean eventually becomes saturated with carbon dioxide, it will not be able to absorb much more of the gas. The carbon dioxide abundance in the atmosphere would then increase, enhancing global warming which in turn would cause a further increase in the temperature. Left unchecked, the temperature would continue to rise and the oceans would eventually evaporate. The higher temperature would begin to cook the carbon dioxide out of the limestone, releasing even more carbon dioxide, which would cause even more global warming. Eventually, this process could produce an atmosphere more like the one on Venus.

In the ocean some of the dissolved carbon dioxide combines with water to form carbonic acid, the acid that makes

soda bubbly. As more carbon dioxide is dissolved in the ocean, more carbonic acid is produced and the ocean becomes more acidic.

We are in the process of performing this experiment, by burning fossil fuels that release huge amounts of carbon dioxide. Each gallon of gasoline we burn produces about 20 pounds of carbon dioxide. The ocean is unable to absorb carbon dioxide as quickly as we produce it, and the amount of carbon dioxide in the atmosphere has been steadily rising because we burn fossil fuels and are cutting down the forests.

Today there is 40 percent more carbon dioxide in the atmosphere than there was about 150 years ago (Figure 4.26). In pre-industrial times the concentration as about 280 parts per million (ppm), which means there were 280 carbon dioxide molecules for every million air molecules.

Ice has been accumulating in Greenland and Antarctica for more than 800 thousand years. Air bubbles are trapped with the ice and the air has been analyzed to determine the carbon dioxide content in the air. We find that before the current rise to over 400 ppm, the atmospheric carbon dioxide had not been above 300 ppm for the last 800 thousand years, and the average value was about 230 ppm.

This increase in carbon dioxide, the primary source of global warming, is responsible for the warming trend we are currently experiencing. If this trend continues, one of the first effects will be a warming of the oceans, causing the water to expand and flood the low-lying coastal areas. The ice masses in Greenland and Antarctica have begun to melt, causing even more coastal flooding. If Greenland's ice cap melted completely, the oceans would rise about 7 meters (about 23 feet)!

Does Venus have plate tectonic activity? Although we still have many

Parts per million

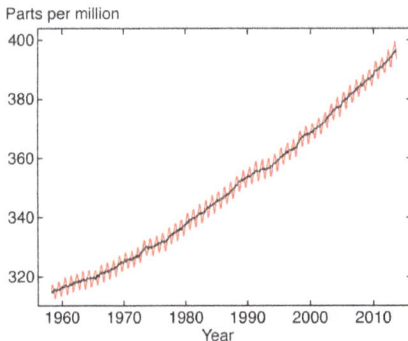

Figure 4.26 Increase in Atmospheric Carbon Dioxide Due to the burning of coal, oil, and natural gas, the abundance of carbon dioxide in the Earth's atmosphere has risen about 40% since the industrial revolution. The small wiggles are yearly variations, higher in spring and lower in fall. [data: Mauna Loa Observatory/NOAA/ Scripps]

questions about plate tectonics, we know that the Earth's hot core produces convection currents in the mantle. The rotation of the Earth may be responsible for deflecting these currents in such a way that they make plate motion possible. Venus has several large volcanoes, which tells us that there is magma rising from its mantle. However, since the planet is not rotating, plate motion may not be possible. There is no strong evidence that plate motion occurs on Venus.

The diameter of Mars is about half the Earth's and its mass is about one-tenth. Its smaller size and mass combine to produce a surface gravity of about 40 percent that of Earth's. A 100 kg person would only weigh 40 kg on Mars. This lower surface gravity means that Mars is unable to hold onto its atmospheric gasses as strongly as the Earth and Venus. Consequently, over the years, Mars has lost a significant fraction of its atmosphere.

The thin atmosphere of Mars is mostly carbon dioxide, and because of a weak greenhouse effect, it is warmer by about 10°C than we calculate based on the sunlight that hits the planet. The warming is not very much because the atmosphere is too thin to absorb much of the outgoing infrared light.

The atmospheric surface pressure is so low that liquid water is unable to exist. Water is present on Mars, but only as water vapor or as ice. Instead of melting, water ice on Mars goes directly to a vapor, behaving much like dry ice (frozen carbon dioxide) does on Earth (Figure 4.27).

Mars is also about 50 percent further from the Sun than we are so it is colder. In addition, its thin atmosphere does not trap the heat so it gets very cold at night. The atmosphere of Mars must have been thicker in the past, and could have supported life then, but it is now a very inhospitable place.

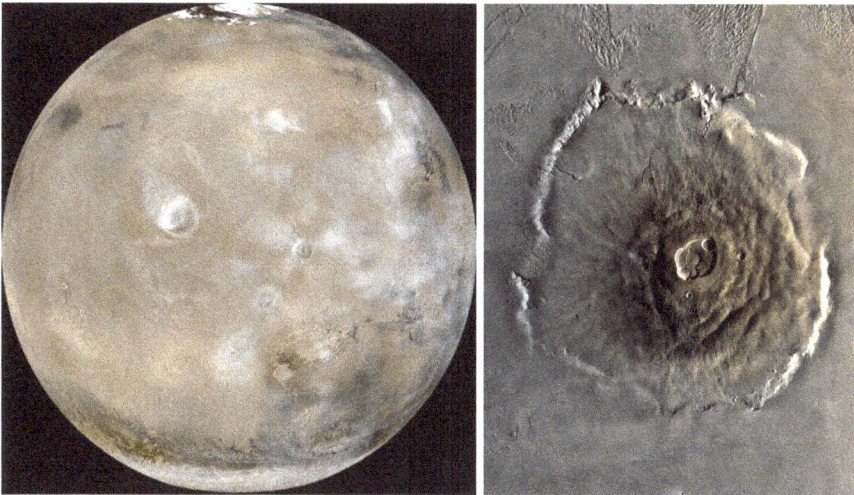

Figure 4.27 Mars and Olympus Mons Mars during summer in the North (left) shows a residual polar cap of water ice is present around the North Pole, and the South Pole is covered with carbon dioxide frost. The white clouds around the middle of the planet are made of water ice crystals. [NASA/JPL/MSSS] The large volcano just to the upper-left of center with the ice clouds drifting to the northwest is Olympus Mons (larger figure at right). Its central caldera is 66 by 83 kilometers. [NASA/Viking Project]

Since Mars is significantly smaller than Earth, it cooled more quickly and its lithosphere is now very thick, too thick to fragment into plates and allow plate motion to occur. Mars has several large volcanoes, indicating that it must have had mantle convection at some time in the past.

One volcano, called Olympus Mons, is the largest volcano known. It is about 600 km in diameter and 25 km high. By comparison, the largest volcano on Earth is Mauna Loa, which forms the island of Hawaii. It is about 120 km wide at its base and rises 9 km above the floor of the Ocean. The relatively thin lithosphere of Earth could not support a volcano as large as Olympus Mons (it would sink down into the asthenosphere). Mars, with its lower surface gravity and thicker lithosphere, is able to support much larger volcanoes.

The next four planets beyond Mars are very large, very far from the Sun, and do not have solid surfaces. Jupiter and Saturn are composed mainly of liquid hydrogen and helium, and although Uranus and Neptune have large rocky cores, their outer layers are composed of liquid hydrogen and slushy ices of water and methane.

While it is possible that Mars or some of the satellites of the outer planets may be able to support some forms of primitive life, the Earth appears to be the only planet in the Solar System with the ability to support advanced forms of life such as insects, higher plants, and vertebrates.

Although science fiction movies are fun to watch, space travel to other planetary systems is highly unlikely. The journey to even the nearest star would last several *generations* and the ship that would have to be capable of supplying all the needs of the crew.

The Earth is the only planet we have that is capable of supporting life. Saving it should be our first priority.

The Early Atmosphere

Today, the atmosphere of Earth is composed of nitrogen (78%), oxygen (21%), argon (1%), and trace amounts of other gasses like carbon dioxide, neon, methane, and helium. Carbon dioxide accounts for only about 0.04% of the atmosphere, but as we have seen, it is a small but extremely important component. Although too much creates detrimental global warming, it is a necessary ingredient for plant growth.

The material out of which the planets, and Sun formed is usually called the solar nebula. As the Earth accreted from the solar nebula, water and gases became trapped inside the growing planet. As the Earth's interior began to circulate, these gases slowly escaped to the surface of the planet. The surface of the early Earth was so hot that the escaping gases were quickly lost. These gas molecules had such high speeds (high temperature) that they would have been able to escape the pull of Earth's gravity.

Our present atmosphere developed in several stages. As the surface of the Earth cooled, the escaping gases began to accumulate, and the escaping water vapor condensed to help form the oceans. Comets are composed of large quantities of water ice and carbon dioxide ice, and some scientists believe that comets colliding with the Earth made significant contributions to the formation of the atmosphere and the oceans.

We expect that the early atmosphere was composed of common gases such as carbon dioxide [CO_2], nitrogen [N_2], and smaller amounts of methane

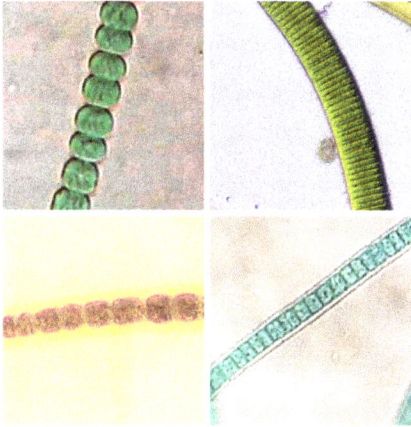

Figure 4.28 Cyanobacteria (clockwise from top left) Anabaena; Lyngbya [NASA]; Tolypothrix [Mathew J.Parker]; and Gloeotrichia [Panek] are cyanobacteria genera. They divide and remain attached to form a thread (~5 micron in diameter), but each cell is an independent organism.

[CH$_4$], ammonia [NH$_3$], and sulfur gases like hydrogen sulfide [H$_2$S]. Although the exact mixture is unknown, the atmospheres of other planets provide some clues. Mars and Venus have atmospheres of carbon dioxide, while Titan, a large satellite of Saturn, has an atmosphere of nitrogen. Although Jupiter and Saturn have atmospheres consisting primarily of hydrogen and helium, they contain significant amounts of ammonia and methane.

Hydrogen and helium are the most abundant elements in the Universe, however, Earth's gravity is not strong enough to hold these light gases. The hydrogen on the Earth is not gaseous hydrogen, but is bound in compounds such as water. The only helium on the planet is trapped in underground pockets.

There is no universal agreement as to the exact mix of gases in the Earth's early atmosphere, but it must have contained little or no oxygen. There is

a great deal of geologic evidence to support this belief. Oxygen is a very reactive gas and would have readily combined with iron, silicon, uranium, and many other substances that were present. However, rocks older than about 1.9 billion years are not fully oxidized. As we will see, the first oxygen-producing organisms were the cyanobacteria or blue-green algae (Figure 4.28). They have been found in fossils that are about 3.5 billion years old.

The percentage of oxygen in the atmosphere began to increase as the oxygen-producers flourished. Between 1.8 and 2.5 billion years ago, the cyanobacteria had produced enough oxygen to raise the atmospheric level of oxygen to about 1 percent. At about that time, formations of oxidized iron known as *banded iron formations* began to appear in sediments (Figure 4.29).

In an oxygen free environment, a great deal of iron could have been dissolved in the ocean water. This dissolved iron began to oxidize and precipitate out as magnetite [Fe$_3$O$_4$] and hematite [Fe$_2$O$_3$] about 2.3 billion years ago. Iron dissolves in water, but iron oxide molecules stick together to form small particles that precipitate out and settle to the bottom.

The iron oxide layers are black, gray, or silver in color. They alternate

Figure 4.29 Banded Iron Formation The silvery layers are the oxidized iron layers (magnetite and hematite), and the reddish layers are the silicate layers. [JCB]

with silicon dioxide (silica) layers to create the banded-iron formations. The silica layers usually contain impurities which give these layers their color. These banded-iron formations contain the major iron reserves of the world today.

Uranium is easily oxidized, but in its fully oxidized state it is only found in sediments younger than about 2.3 billion years, indicating that before that time, there was not enough oxygen in the atmosphere to completely oxidize uranium.

We do not know exactly when the concentration of oxygen became high enough to allow the ozone layer to form, but before life could exist on the land, the ozone layer must have been created. *Ozone* is a form of oxygen that is composed of three oxygen atoms in contrast to ordinary oxygen that contains just two atoms.

The ultraviolet light coming from the Sun is energetic enough to kill organisms, but the ozone layer absorbs much of this ultraviolet light. Because of the presence of this harmful radiation, the earliest life forms would have been confined to the sea where the water could absorb most of the ultraviolet rays. The first fossils of land-living organisms are about 420 million years old, suggesting that perhaps this was near the time at which the ozone layer formed.

We owe our existence to the organisms that produced and help to maintain the Earth's oxygen-rich atmosphere. As we will see, the chemical reactions that they carry out are responsible for much more than just the oxygen we breathe. We will be better able to understand their contributions to our lives after we have investigated some of the chemistry that is the basis of all life. In the next chapter we will study some of the important chemical processes that make life possible on Earth.

Chapter 5: The Molecules of Life

Cells and Super Kingdoms

Although organisms vary a great deal in appearance, there are many similarities that they all share. For example, all organisms are composed of microscopic building blocks called cells. A cell is a small compartment that contains many different chemicals and which carries out the basic functions of life. The smallest life forms are single-celled organisms, while the largest organisms are composed of several hundred trillion cells. (Humans have about 100 trillion (100,000,000,000) cells and a small round worm might have 1000 cells.)

In large organisms, there are many different kinds of cells, each of which carries out a very specialized function. Muscle cells enable us to move, blood cells carry oxygen, nerve cells send messages to different parts of our body, and fat cells store energy. In multicellular organisms, all the cells work in harmony to carry out the processes necessary to sustain the organism. Mammals have about 200 different cell types, insects have about 90 different types, and sponges have 10 to 20 different types of cells.

Many people think of life as being divided into the plant kingdom and the animal kingdom, but when we study their cells, two fundamental groups are recognized. Bacteria and their relatives belong to a group known as the *prokaryotes*. All other living things belong to a group called the *eukaryotes*. The cell structure of these two groups is fundamentally different (Figure 5.1).

Most of the life forms that are familiar to us are eukaryotes. They have cells that are more organized than the cell of a typical prokaryote. For example, the food-producing systems of plants are incorporated into membrane-bounded organelles called plastids. These are small green bodies that contain the plant's chlorophyll. In the higher plants and green algae these organelles are called chloroplasts.

The production of energy for the eukaryotic cell is carried out in membrane-bounded organelles called

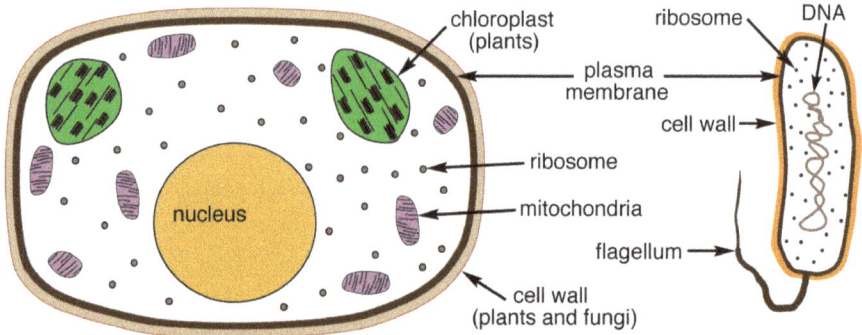

Figure 5.1 Generalized Prokaryotic and Eukaryotic Cells *The eukaryotic cell (left) is a generalized plant cell. Animal cells lack chloroplasts and a cell wall. The prokaryotic cell (right) is not shown at the same scale and should be drawn more nearly the size of the mitochondria or the chloroplasts in the plant cell. [JCB]*

mitochondria, and the genetic information (the brains of the cell) is stored in a membrane-bounded nucleus. Eukaryote, which means *kernel* or *nut*, refers to the fact that the nucleus is enclosed in a membrane. Prokaryotes do not possess these organelles, and the genetic information is not enclosed in a membrane-bounded nucleus.

Along with the structural differences just mentioned, prokaryotes are generally much smaller than the cells of eukaryotes. Prokaryotes range in size from about 1 to 10 microns (a micron is one-millionth of a meter), and the cells of eukaryotes are generally 10 to 100 microns long. Although bacteria do not vary much in size or shape, they do exhibit a great deal of variability in the chemical processes that they use to carry out their basic life functions.

Molecular Bonds

The chemical substances that organisms are made of and the chemical reactions that they carried out have many features in common. The study of these chemicals and their various reactions is a branch of chemistry known as biochemistry. Before we look at the structure of the basic building blocks of life in a little more detail, it will be helpful to briefly investigate the theory of atoms and the forces that bond atoms together.

As an electron moves about a nucleus it is approximately confined to a certain volume of space, which we call an orbital. Although orbitals come in different shapes, the simplest is a sphere as shown in Figure 5.2 (top). The Pauli exclusion principle is part of the theory of quantum mechanics put forth by Wolfgang Pauli in 1921. It states that no more than two electrons can occupy the same orbital. An orbital is said to be filled when two electrons occupy it.

Orbitals that are approximately the same distance from the nucleus are grouped together to form a shell. The maximum number of electrons that can orbit in a given shell varies with the distance from the nucleus. For example, the shell closest to the nucleus is composed of only one orbital and can, therefore, contain only two electrons (this first shell is called the K shell). The second shell (the L shell) is composed of four orbitals and,

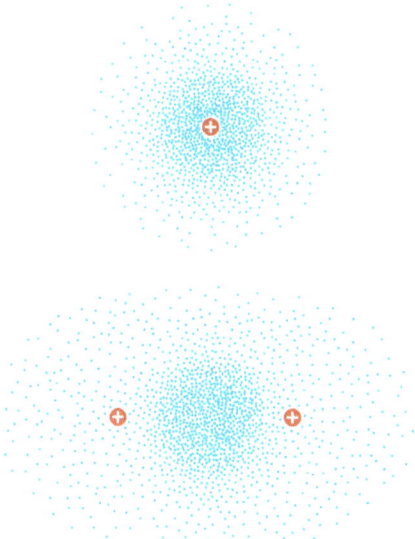

Figure 5.2 Electron Cloud and Covalent Bonding *In the schematic drawing of a hydrogen atom (top), the electron occupies a certain volume of space around the proton (red dot), but spends most of its time where the blue dots are more closely packed. In the hydrogen molecule (bottom), the two electrons spend most of their time between the two protons, creating a net negative charge that bonds the protons together. [JCB]*

molecule are all able to fill their outer shells by *sharing* electrons. As we will see, this sharing of electrons is what bonds the atoms together in a molecule. A few atoms such as helium, neon, and argon already have filled outer shells in their natural state and do not form molecules.

Molecules are two or more atoms that are held together by the electrical attractive forces of the positively charged nuclei and the negatively charged electrons. To understand this attraction, consider a hydrogen molecule that is composed of two electrons and two nuclei (each nucleus is composed of just one proton in the case of hydrogen). The hydrogen atoms fill their outer shell (the K shell) with 2 electrons by sharing each other's electron. The electrons spend most of their time between the two nuclei as illustrated in Figure 5.2 (bottom). This concentration of negative charge between the two positive nuclei is what holds the molecule together. Chemists call this type of attachment a covalent bond because the two electrons are shared by the two nuclei.

Since each hydrogen atom originally had an unfilled orbital (an orbital with only one electron) we can think of these one-electron orbitals as overlapping to form the bond. Covalent

therefore, can contain no more than eight electrons. Figure 5.3 illustrates the K and L shells in a schematic fashion. Notice that the number of electrons that are needed to fill a shell is always an even number since a shell is made of orbitals, and each filled orbital contains two electrons.

When two or more atoms bond together to form a molecule, they do so in a way that allows the shells of the atoms to be filled. For example, oxygen normally has 8 electrons (2 in the first shell and 6 in the second). Since the second shell needs 2 more electrons to fill it to 8, it can combine with two hydrogen atoms as shown schematically in Figure 5.3 to form a water molecule. The oxygen atom and the two hydrogen atoms in a water

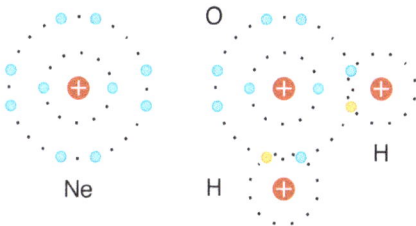

Figure 5.3 Electron Shells *Drawing of a neon atom's K and L shells (left), that are completely filled with 2 and 8 electrons respectively (blue dots). A water molecule (right) is formed when an oxygen atom and two hydrogen atoms fill their outer shells by sharing electrons. [JCB]*

bonds are the strongest type of chemical bonds and are the usual type of bonds found in the molecules that are important in the chemistry associated with life.

Another type of chemical bond is the ionic bond. This bonding is the type that occurs in common table salt. Salt is composed of sodium and chlorine atoms in equal numbers. A chlorine atom needs only one electron to complete its outer shell, and a sodium atom has only one electron in its outer shell (the M shell). By giving up its outer electron, a sodium atom will have a complete outer shell of electrons (the L shell).

Instead of sharing electrons, each chlorine atom completely captures an electron from one of the sodium atoms and becomes negatively charged. The sodium is left with a positive charge since it has lost one of its electrons. The positive sodium atom, or *ion* as it is called, is now attracted to the negative chlorine ion. Notice that in this type of bond an electron was completely captured by the chlorine atom and is not shared with the sodium atom, as in the covalent bond of the hydrogen molecule.

Water and Hydrogen Bonding

Electron orbitals are not always spherical in shape, but can take on one of several possible configurations. Although we will not go into the details here, the theory of quantum mechanics can be used to calculate the shape of these orbitals.

With oxygen, the outer shell (the L shell) is composed of 4 orbitals. The elongated orbitals are equally spaced around the nucleus, forming a structure that resembles a tetrahedron as shown in Figure 5.4a. Two of the orbitals are filled with 2 electrons each, but the

other two contain single electrons. Note that the unfilled one-electron orbitals are always next to each other, and therefore, when two hydrogen atoms attach to the oxygen atom, a lopsided molecule is formed as shown in Figure 5.4b.

When electrons are shared by different kinds of atoms (different elements) as in the case with water, the shared electrons can spend a little more time orbiting one of the atoms (the oxygen atom in the case of water). Since the electrons spend more time around the oxygen than they do the hydrogens, the oxygen atom tends to be slightly more negative while the hydrogen atoms tend to be a little more positive as is shown schematically in Figure 5.4c. Since water molecules have a positive side and negative side, they are said to be polar molecules.

In liquid water the negative oxygen of one molecule tends to attract the positive hydrogens of several other water molecules. This type of attraction is called a hydrogen bond. In general, each molecule of water forms hydrogen bonds with 3 or 4 other water molecules. This attraction explains why water is relatively difficult to vaporize and is a liquid at room temperature.

No other nonorganic compounds are liquids at room temperatures. Mercury and bromine are the only other nonorganic liquids on Earth that are not

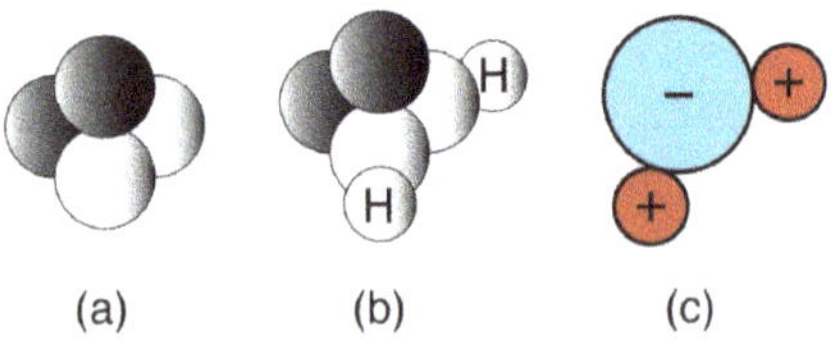

(a)   (b)   (c)

Figure 5.4 Oxygen and Water (a) The L shell orbitals of an oxygen atom showing two filled (shaded) and two one electron orbitals. (b) Two hydrogens share their electron with these one-electron orbitals. (c) Water showing its polar nature. [JCB]

water solutions. It is interesting that carbon dioxide molecules are much heavier than water molecules, and yet carbon dioxide is a gas on Earth. Carbon dioxide is a gas because it is not a polar molecule.

The polar nature of water is responsible for many of the unusual properties that water exhibits. When water crystallizes to form ice, each molecule hydrogen-bonds with 4 adjacent molecules (Figure 5.5). Because of the regular arrangement of the molecules, the crystal structure of water contains more empty space than the liquid form. Therefore, ice is less dense than water, and it floats on top of liquid water. This is extremely important since ice on the top of a lake is able to absorb sunlight and melt quickly in the summer. If ice were denser than water it would sink, and lakes would freeze from the bottom up. Most lakes would freeze solid in winter, and in a very short time much of the water on Earth would be trapped as ice, especially around the North and South Poles.

Compared to many substances, water must absorb a great deal of energy in order for its temperature to change appreciably. Because of this extremely high heat capacity, large bodies of water and the organisms living in the water are maintained at a relatively constant temperature. Water makes up 70 percent or more of most living organisms, and large land-dwelling animals are able to regulate their temperature more easily because of the large amount of energy required to change the temperature of the water in their bodies.

Because so many substances dissolve in water, it is often called the universal solvent. Its solvent properties can be understood in terms of hydrogen bonding. For example, when salt is dissolved in water, the negative ends of

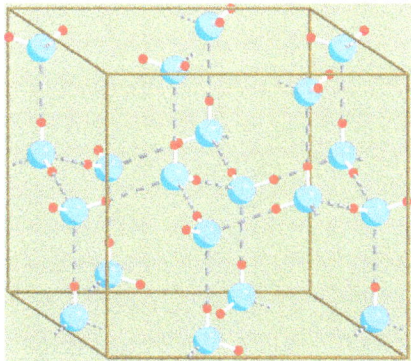

Figure 5.5 Water Ice Because of their polar nature, water molecules in ice bond together leaving empty spaces. Therefore, ice is less dense than liquid water, and ice floats. Blue balls are negative oxygen atoms and small red balls are positive hydrogen atoms. Hydrogen bonds are dashed lines.

several water molecules attract a positive sodium ion, and the positive ends of other water molecules attract a negative chlorine ion.

Each chlorine ion and sodium ion are surrounded by water molecules and are no longer attached to each other when dissolved in water. This type of complete separation does not happen to covalently bonded molecules when they are dissolved in water.

Although hydrogen bonds are weaker than covalent bonds, hydrogen bonds are very important in complex biological molecules. The three-dimensional shape of many large biological molecules is maintained by hydrogen bonding, and two or more large molecules may be hydrogen bonded together to form a larger structure.

When two hydrogen atoms are held together by a covalent bond, the molecule is elliptical in shape as determined by the volume occupied by the orbiting electrons. It is shown schematically in Figure 5.6a. However, molecules are often represented by ball-and-stick models so the positions

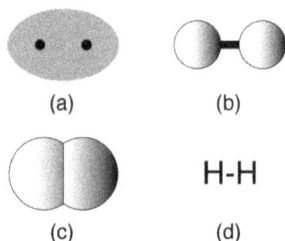

Figure 5.6 Representing a Molecule *A hydrogen molecule can be represented by (a) an electron cloud surrounding the two nuclei, (b) two balls and a stick, (c) a space filled model, or (d) two letter H's connected with a single bond. [JCB]*

of the atoms can be visualized more easily as in Figure 5.6b. These models are especially helpful in studying the positions of different atoms in complex molecules. Space-filled models like Figure 5.6c more accurately represent the space occupied by the molecule and give a better feel for the actual shape of the molecule. In most cases, however, it is easier to represent the atoms by letters and the bonds by lines as in Figure 5.6d.

Carbon

Chemically speaking, the most interesting element in nature is undoubtedly carbon. It has only 4 electrons in its outer shell (the L shell), and to complete this shell it must share 4 more electrons with other atoms. The 4 electrons in carbon's outer shell occupy separate orbitals. As with oxygen, these elongated orbitals are equally spaced around the nucleus and, therefore, form a structure that resembles a tetrahedron (Figure 5.7a). Carbon is one of the few atoms that can join with other carbon atoms to form very long chains. These chains can be linear, branched, or can even form closed circles.

The elements silicon and sulfur are also able to form long chains of identical atoms, but the carbon-carbon bond is about 60 percent stronger than either a silicon-silicon bond or a sulfur-sulfur bond. Because of their weaker bonding, silicon and sulfur chains are less stable, and when heated, they break more easily than carbon chains.

Furthermore, instead of forming chains, silicon and sulfur tend to combine with oxygen because the silicon-oxygen bond and the sulfur-oxygen bond are much stronger than the silicon-silicon and sulfur-sulfur bonds. (Oxygen also happens to be the most abundant element in the Earth's crust.) In contrast, the carbon-oxygen bond is almost identical in strength to the carbon-carbon bond.

Silicon is very similar to carbon in many respects, but its weaker bonding can be understood by considering its size. As atoms get bigger, the electrons are added in shells that are further from the nucleus. Since the outer electrons are further from the nucleus, they are more weakly held to the atom. The electrons bonding two silicon atoms together are further from the nuclei than the electrons bonding two carbon atoms together, and the silicon bond is correspondingly weaker.

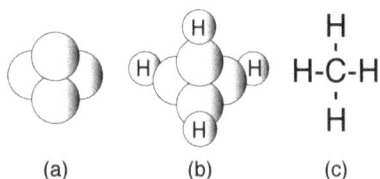

Figure 5.7 Carbon Atom and Methane
(a) Spheres represent the four L-shell orbitals of a carbon atom. Each orbital contains only one electron so they are unfilled. They can be filled if each shares an electron with a hydrogen atom to form methane (b). The carbon forms four single bonds with four hydrogen atoms as represented by (c). [JCB]

When the one-electron orbital of one atom overlaps with the one-electron orbital of another atom, the atoms share the two electrons, and we say a single bond is formed (Figure 5.7b). If two one-electron orbitals of one atom overlap with two one-electron orbitals of another atom so that the atoms share four electrons, we call it a double bond. A triple bond is formed when three one-electron orbitals of an atom overlap with another atom's three one-electron orbitals so that the atoms share six electrons.

Because of the tetrahedron shape of its outer unfilled orbitals (Figure 5.7a), carbon can form single bonds, double bonds, or triple bonds with certain other atoms. Carbon can form single bonds with other carbon atoms or with atoms of hydrogen, oxygen, nitrogen, or sulfur (other very important atoms in the chemistry of life). For example, methane [CH_4] is made when a carbon atom forms four single bonds with four hydrogen atoms (Figure 5.7b).

Double bonds can be formed with other carbon atoms or with atoms of nitrogen, oxygen, or sulfur. For example, carbon can form two double bonds with two oxygen atoms and make carbon dioxide [CO_2]. Triple bonds can be formed with either another carbon atom or a nitrogen atom.

No other element is as versatile as carbon. It is truly amazing that atoms are structured in such a delicate way that only carbon has the correct size and bonding properties to form nature's most complicated molecules. The complex nature of life would not be possible without the unique chemistry of the carbon atom. Because of the complex nature of life, if life does exist on some distant planet, that life must certainly be based on carbon chemistry.

Even though countless biochemical molecules have been discovered in nature, these molecules are composed of only a few dozen basic building blocks. Proteins, the most important substances in living organisms, are composed of 20 different kinds of amino acids. (The word protein comes from a Greek word meaning *primary*.)

Carbohydrates are hydrated carbon compounds, which means they are made of carbon atoms combined with water. Sugars and starches are common carbohydrates. The six-carbon sugar glucose is the basic unit for many complex sugars, starches, cellulose, and glycogen.

Four substances called *nucleotides* form the basic building blocks for the information storage system that all living organisms contain. Lipids are important components found in the membranes of cells. Introductory textbooks in biochemistry tend to be extremely thick, and no attempt will be made to cover the material in great depth. However, to intelligently discuss the aspects of chemistry that bear directly on evolution, we will present a brief introduction.

Energy and Carbohydrates

Living organisms are able to carry out various functions such as growth, movement, and reproduction. To perform these and other functions, the organism must have a supply of energy at its disposal. The ultimate source of this energy lies in the nuclear reactions that occur in the core of the Sun. The energy from these reactions is emitted from the surface of the Sun as light. Some living organisms (like plants and certain bacteria) can convert light energy into chemical energy, a process called photosynthesis.

122

This chemical energy is usually stored as carbohydrates such as sugar and starch. A hydrate is a substance that contains water, and carbohydrates are formed by plants from carbon and water. The carbon comes from the carbon dioxide in the atmosphere and during photosynthesis, the oxygen is split from the carbon dioxide and released back into the atmosphere. The sugar and starch stored in plants are the source of energy for most living organisms.

There are many carbohydrates, but the basic building block of many is the six-carbon sugar *glucose*, which is called a simple sugar. The chemical formula for glucose is $C_6H_{12}O_6$. As can be seen by the formula, this hydrate of carbon is composed of six carbon atoms and six water molecules. Inside a cell, glucose molecules form a ring structure (Figure 5.8).

Fructose is another simple sugar that gets its name from the fact that it is found in many fruits, and in honey and berries. Most fruits have approximately equal quantities of fructose and glucose, although apples and pears have a little over twice as much fructose as glucose. Honey is a mixture of approximately equal amounts of fructose and glucose, with a smaller amount of water, and a few other kinds of sugar. Fructose can be formed from glucose by rearranging the atoms

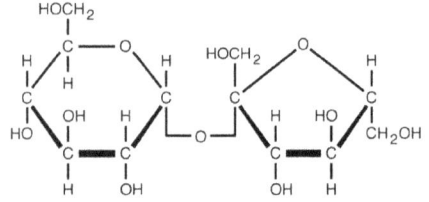

Figure 5.9 Sucrose The common table sugar is a double sugar formed by removing an OH from glucose and an H from fructose and connecting the two as shown. [JCB]

(Figure 5.8). Fructose has the same chemical formula as glucose, but differs only in the way the atoms are connected.

Many complex sugars are formed by connecting two or more simple sugar molecules. For example, two molecules of glucose can be connected to form a double sugar called maltose (a water molecule is released in the process). Sucrose, which is common table sugar, is formed by connecting a glucose and fructose molecule and removing a water molecule (Figure 5.9).

In animals, glucose molecules are connected in large branching chains to form a substance called glycogen. Plant cellulose, a carbohydrate used to give rigid support to plant cells, is formed by connecting glucose molecules in long non branching chains. These chains are hydrogen bonded together to form sheets.

Since the digestive chemicals of mammals are unable to break the cellulose sheets apart, cellulose is not generally used as food by mammals. However, cows and other *ruminants* (mammals that have four stomach compartments and chew their cud) have a type of bacterium in their first stomach that breaks the cellulose down into smaller units so these mammals can use it as food.

Two extremely useful sugars that are used in genetic information storage are

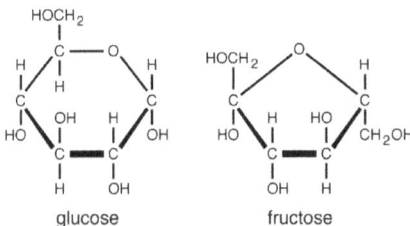

Figure 5.8 Simple Sugars Although glucose (left) and fructose have the same chemical formula, they have different ring structures. [JCB]

the five-carbon sugar ribose and the closely related molecule deoxyribose (Figure 5.10). The prefix *deoxy-* refers to the fact that deoxyribose contains one less oxygen atom than ribose. Indeed, these molecules are more similar than their names seem to imply. It should also be noted that except for the removal of a carbon atom and a water molecule, the structure of ribose is very similar to fructose.

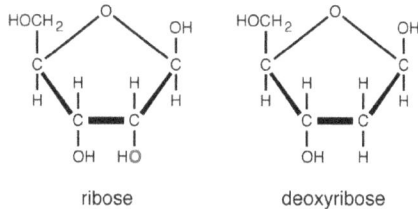

ribose   deoxyribose

Figure 5.10 Five Carbon Sugars Ribose (left) has one more oxygen atom (red) than deoxyribose sugar. Both are important in genetic information storage and information manipulation. [JCB]

Energy can be stored as sugar because energy is released when the sugar is burned (combined with oxygen) to form carbon dioxide and water. Energy is released in a chemical reaction when the final products are more tightly bound than the initial substances.

Consider two hydrogen atoms bound together by a covalent bond as was described earlier. Since these atoms are held together by the electrical attractive forces of the protons and electrons, it would take a certain amount of energy to separate these two atoms. Therefore, when two atoms of hydrogen come together energy must be released. Just as much energy is released when two hydrogen atoms combine to form a molecule as would be required to pull the two atoms apart.

The theory, which says that the energy output must equal the energy input, is called the principle of the conservation of energy. The energy released when atoms combine to form molecules is called the binding energy. By analyzing the binding energy of the initial and the final products of a reaction it can be determined if energy will be given off in the reaction or if energy will be required to make the reaction go.

When hydrogen molecules and oxygen molecules are combined to form water, a large amount of energy is released (remember the Hindenburg). Energy is released because water is a more tightly bound molecule than the molecules of hydrogen and oxygen. On the other hand, when plants make glucose and oxygen out of water and carbon dioxide, energy (in the form of sunlight) must be supplied to make the reaction go. Therefore, the same amount of energy will be released when the sugar is combined with oxygen to form water and carbon dioxide.

Although energy is often stored as sugars or fats, the immediate source of energy for all living cells is an energetic molecule known as *adenosine triphosphate* or ATP for short (Figure 5.11). When sugar is combined with oxygen, the released energy is temporarily stored in ATP. ATP molecules, which have been called the storage batteries of the cell, are charged up by the mitochondria.

The cell uses ATP when energy is needed to perform functions like muscle contraction or the creation of proteins. Specifically, the energy is released when the end phosphate and its three oxygen atoms are removed from the rest of the molecule. The remaining molecule is known as adenosine diphosphate or ADP. The removed phosphate combines with an OH ion and produces orthophosphate [PO_3OH].

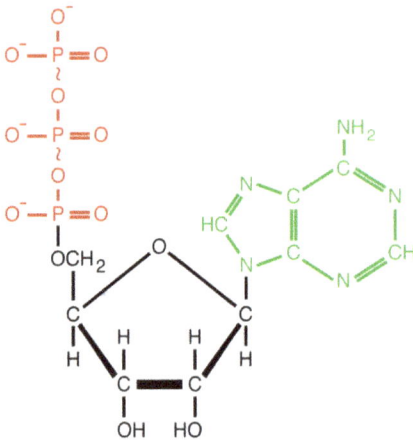

Figure 5.11 Adenosine Triphosphate *ATP is the energy storage molecule used by the cells of all living organisms. Note that ATP is constructed from a ribose sugar (black) with a base called adenine (green) attached to one of the carbon atoms and an added phosphate group (red). The bonds indicated by the two squiggly red lines are the energy releasing bonds that are broken to form ADP and AMP. [JCB]*

ADP can then be split in a similar way to form adenosine monophosphate (AMP) with a comparable amount of energy released. In the mitochondria, energy released in the oxidation of fuel molecules is used to convert AMP and ADP back into ATP where the energy is temporarily stored until it is needed. Although ATP is important because of its energy storage capability, as we will see later, AMP is also an important component of the molecule that stores information for the cell.

Proteins

Proteins are the chemical substances responsible for the major differences we observe in the structure and body chemistry of various life forms. Proteins make us what we are.

Most tissues are composed of different kinds of protein molecules. Tendons, ligaments, skin, and other types of connective tissue are made of fibers containing the protein collagen. Bone is a matrix of collagen and calcium salts, and hair is made of proteins called keratins.

Muscles are made of the proteins myosin and actin. Actin is involved in the different types of motion observed in a wide variety of organisms. Actin accounts for about 10 percent of all the protein in cells.

The protein hemoglobin, which gives blood its red color, carries oxygen from the lungs to the cells. Even the lens of the vertebrate eye is made of crystalline proteins.

Enzymes are proteins that control the rates at which various chemicals react in the body. Digestion is the process by which foods are broken down to a form that can be used by the organism. However, since the body temperature is so low, chemicals that we call enzymes must help the digestion process along.

For example, the protein sucrase is an enzyme secreted into the small intestine that splits sucrose into glucose and fructose. Sucrose molecules (and other complex sugar molecules) are too large to pass through the walls of the intestine, but simple sugars like glucose and fructose can pass through the walls and enter the blood stream, where they are used for energy.

Hormones are substances that regulate or coordinate the bodily functions in multicellular organisms, and many hormones are proteins. The word hormone means to *set in motion*. They are produced in one part of the body, but regulate or set into motion activities in other parts of the body.

For example, the nonprotein sex hormones (testosterone in males and estradiol in females) are produced in the sex organs but are responsible for the regulation of many of the physical changes that occur in other parts of the body during puberty. Insulin, a protein hormone secreted by the pancreas, regulates the amount of sugar in the blood. The protein somatotropin (also known as the growth hormone) causes the liver to produce other hormones that stimulate the growth of bones and muscles.

When bacteria and viruses invade our bodies, some of their molecules (called *antigens*) initiate a response by our immune system. These antigens are usually protein molecules. *Antibodies* are the proteins that our immune system produces to attack these harmful antigens.

Proteins are long molecules that are sometimes folded in convoluted ways. Their shape is maintained by hydrogen bonding. All proteins are constructed from smaller building blocks called amino acids.

Although each amino acid has its own unique chemical properties, they can be grouped into three broad categories. Some are water lovers, some avoid water, and some are indifferent. Many protein molecules are globular in shape, and they reside in an aqueous environment. The amino acids on the exterior surface of the molecule are generally of the water-loving type while those on the inside are usually of the water-avoiding type.

There are 20 different amino acids, but they all have a common structure composed of one nitrogen atom and two carbon atoms (Figure 5.12). The central carbon atom is called the alpha carbon. The 20 different amino acids are made by attaching one of the 20 different side chains (labeled R in Figure 5.12) to the alpha carbon. The 20 different amino acids that are found in living organisms are listed in Table 5.1 along with their abbreviations and

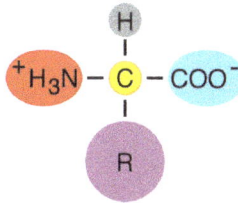

Figure 5.12 Amino Acid The positive amino end (red) and the negative carboxyl end (blue) are bonded to the alpha carbon (yellow). R is the side chain. [JCB]

one-letter symbols. They are grouped according to their general location in globular proteins.

The end of an amino acid with the nitrogen and hydrogen atoms (called the amino end) tends to be positively charged while the end containing the non alpha carbon and oxygen (the carboxyl end) tends to be negatively charged. Protein molecules are made by connecting the negative and positive ends of different amino acids. As each connection is made, a water molecule is removed.

The sequence of amino acids is referred to as the primary structure of the protein. A protein molecule may contain from about 50 to more than a thousand amino acids connected in a long sequence that is often folded into some rather complicated shape. (Very short amino acid chains are usually called polypeptides.) Some protein molecules called globular proteins are approximately spherical in shape, some are irregular shaped, while others called fibrous proteins form long, thin fibers.

When amino acids are attached in a chain, segments of the chain often form a spiral structure called an *alpha helix* (Figure 5.13). This name describes the positions of the alpha carbon atoms which lie along a helix. Hair is composed of long twisted cables of alpha-keratin molecules. Each molecule is a long chain of amino acids

Name	Abbr.	Symbol	Location
Phenylalanine	Phe	F	internal
Methionine	Met	M	internal
Isoleucine	Ile	I	internal
Leucine	Leu	L	internal
Valine	Val	V	internal
Cystine	Cys	C	internal
Tryptophan	Trp	W	ambivalent
Alanine	Ala	A	ambivalent
Threonine	Thr	T	ambivalent
Glycine	Gly	G	ambivalent
Serine	Ser	S	ambivalent
Proline	Pro	P	ambivalent
Tyrosine	Tyr	Y	ambivalent
Histidine	His	H	external
Glutamine	Gln	Q	external
Asparagine	Asn	N	external
Glutamate	Glu	E	external
Lysine	Lys	K	external
Aspartate	Asp	D	external
Arginine	Arg	R	external

Table 5.1 Amino Acids Protein molecules are long chains of amino acids chemically bonded together. There are 20 different amino acids found in the proteins of living organisms. In the table, the amino acids are generally listed in order of their hydrophobicity (how much they dislike water). Amino acids at the top avoid water and those near the bottom like to be in water. The boundaries between internal (water avoiding) and ambivalent, and between ambivalent and external (water loving) are not sharp, but are somewhat fuzzy.

arranged in the alpha-helix structure. This twisted structure gives hair its springiness. When you pull on it, the cable untwists somewhat and allows the hair to stretch. Horns, feathers, nails, and scales are also composed of keratin, which is not surprising since all of these structures have a common evolutionary history. Many proteins, even the globular forms, have segments composed of amino acids that are assembled in the alpha-helix pattern. This pattern appears to be nature's favorite.

Besides the alpha-helix pattern, other structures can be constructed from chains of amino acids, and these different patterns are referred to as the secondary structure of the protein chain. Another common secondary structure is called a *beta pleated sheet* (Figure 5.13). Pleated sheets can be formed when the amino acid chain folds back on itself, and the two strands are hydrogen bonded together by the negative oxygens and the positive hydrogens in the two strands.

Tendons, cartilage, ligaments, blood vessels, skin, and other connective tissue are made of chains of collagen proteins. However, instead of the alpha helix configuration, collagen molecules form a different type of amino acid

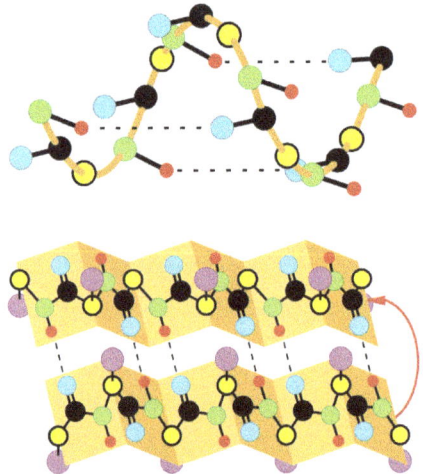

Figure 5.13 Secondary Structures Alpha carbons (yellow) form the alpha helix (top) and hydrogen bonds (dashed lines) of oxygen (blue) and hydrogen (red) stabilize it. Nitrogen (green) and non-alpha carbon (black) are displaced from the helix but are drawn on it to simplify the figure. The side chains are on the outside of the helix (not shown). Beta pleated sheets (bottom), have the alpha carbons on the folds with the side chains (violet) above at the peaks and below at the valleys. The bottom chain folds back on itself (red arrow), and hydrogen bonds stabilize the sheet. [JCB]

chain (secondary structure) that is not as easily stretched. Tendons would not be very functional if they stretched easily when our muscles pulled on them.

Collagen molecules are highly regular chains about 1000 amino acids long. Every third amino acid is glycine, and proline makes up about one-third of the other amino acids in the chain. Skin contains fibrils of collagen that are woven into sheets, and bone is a matrix of collagen that is made into a rigid structure by crystals with a high calcium content. Collagen is a common structural protein and makes up about a third of the protein in vertebrates.

The three-dimensional shape of the protein molecule can either be a long chain as with collagen and alpha keratin, or the protein molecule may be more globular in form. The overall three-dimensional shape of a protein molecule is called the tertiary structure of the protein.

Some protein molecules hydrogen-bond with other protein molecules to form even larger structures. The way two or more protein molecules fit together is referred to as the quaternary structure of the protein. One example of this type of structure is hemoglobin. The hemoglobin in the blood cells of adult humans is composed of two pairs of slightly different protein molecules that are hydrogen bonded together. The two different proteins are called alpha hemoglobin and beta hemoglobin. Even though these molecules are different, their amino acid sequences are so similar that there is little doubt that they evolved from the same ancestral molecule. We will discuss the evolutionary aspect of these molecules in Chapter 10.

DNA, Chromosomes, and Genes

All living organisms possess an information storage mechanism that uses a long molecule known as deoxyribonucleic acid or DNA for short. A DNA molecule is like a long coded message that stores recipes for making different protein molecules. A protein recipe is called a *gene*. The DNA molecule can be thought of as a long necklace made with beads that come in four different colors. Each grouping of three beads stands for one of the 20 different amino acids or the end of the recipe. The code is like a four-letter alphabet that is used to write three-letter words. Each three-letter word stands for one of the 20 amino acids or a period (the end of the recipe). The order of the triplets on the necklace gives the order in which the amino acids must be connected to form a given protein molecule.

The beads used to store the information on the DNA chain are four different kinds of molecules called deoxyribonucleotides (Figure 5.14). Each of the four nucleotides is composed of a deoxyribose sugar with a phosphate attached to one of the carbon atoms (called the 5' carbon), and one of four different molecules called bases attached to one of the other carbon atoms (called the 1'

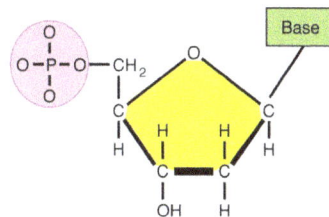

Figure 5.14 Deoxyribonucleotides *The four deoxyribonucleotides are composed of a 5-carbon deoxyribose sugar (yellow), a phosphate group (pink) and one of the four bases. [JCB]*

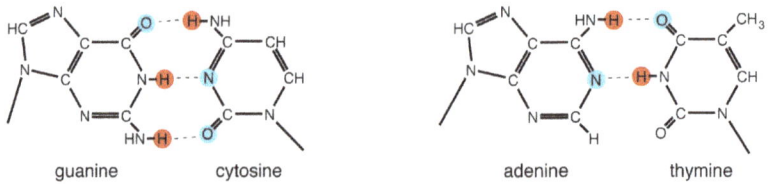

Figure 5.15 Bases *Guanine and cytosine form a complementary pair by hydrogen bonding as shown by the dashed lines. Hydrogen (red) tends to be positive, but oxygen and nitrogen (blue) tend to be negative. Adenine and thymine also form a complementary pair, but their bonding is weaker since they are held together by only two hydrogen bonds. [JCB]*

carbon). One of these nucleotides is AMP (adenosine monophosphate except with a deoxyribose sugar). We have seen it before in association with temporary energy storage.

The bases can be grouped into closely related pairs. Adenine and guanine are called purines, while thymine and cytosine are called pyrimidines. Guanine and cytosine can be hydrogen-bonded together to form a complementary pair (Figure 5.15), and adenine and thymine form another complementary pair. Note that complementary pairs contain one purine and one pyrimidine.

The DNA molecule has the structure of a double stranded helix (like two pieces of string twisted together), and one strand is the complement of the other strand (Figure 5.16). The complementary nature of the two strands is the key to the duplication of the DNA molecule. One strand acts as the template for the formation of the other. When duplication occurs, the double helix is separated and a complement of each strand is generated as different nucleotides hydrogen bond to their complements.

In the DNA duplication process, the complementary DNA chains are generated by a set of three enzymes called DNA polymerases.

All the recipes for the chemical substances that are needed to create the organism are stored in the DNA. Every cell in an organism contains a copy of this information. Bacteria and their relatives (the prokaryotes) are the simplest one-celled organisms, and they have one long circular strand of double helix DNA.

More advanced one-celled and multicellular organisms (the eukaryotes) contain tens to hundreds of times more DNA than bacteria. If the DNA strand of a typical bacterium were laid out in a line, it would only be about 1 to 2 millimeters in length, but the DNA in a human cell has a total length of about 2 meters. Human DNA contains just over 3,000 million base pairs and typical bacteria have about 5 million base pairs.

The DNA of eukaryotes is divided into several separate strands. Each strand is wound around protein molecules (called histones) to form dense rods called *chromosomes*.

The strands of DNA contain the recipes for making the different proteins needed by the organism. Each of these recipes is called a *gene*. Eukaryotes that reproduce sexually have two complete sets of genes, one set coming from each parent. Their chromosomes come in pairs, called *homologous pairs*. One member of each homologous pair came from the mother, and the other member of the pair came from the father. (Humans have 46 different chromosomes or 23 homologous pairs.)

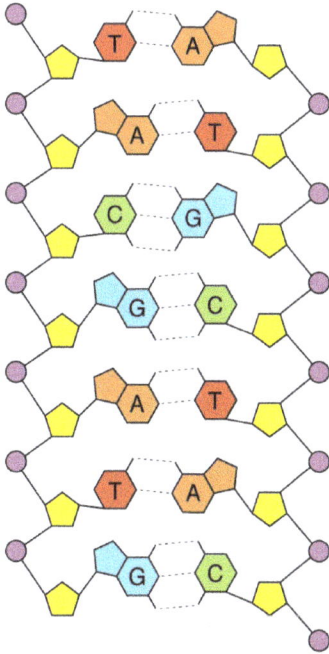

Figure 5.16 DNA The deoxyribose sugars are the yellow pentagons and the phosphates are represented by violet circles. The nucleotides in a given strand are held together by the phosphate bonds. The bases are labeled A, T, G, and C, and the hydrogen bonds are represented by the dashed lines. The helical nature of the bonded strands is not shown. [JCB]

To a large extent, the different genes (called the genetic makeup or the *genotype* of an individual) determine most of the physical and some of the behavioral characteristics of the individual (the physical characteristics of an individual are referred to as the *phenotype*).

The color of a person's hair and skin, their blood type, whether they have blue or brown eyes, their sex, certain behavioral characteristics, and their potential for learning are just a few of the characteristics that are determined by the genes of the individual. Bacteria may have about 4,000 to 5,000 genes while humans have about 23,000 protein coding genes.

Since chromosomes come in pairs (the homologous pairs), eukaryotic organisms possess a pair of genes for each type of protein. These protein recipes may be the same or can be slightly different. (Remember that one member of the gene pair came from the mother and the other gene of the pair from the father.)

If a protein is made that produces a particular characteristic, we say that the gene that produced the protein is dominant. Dominance is observed in eye color. The gene for brown eyes is the gene that is responsible for the production of the brown pigment. One gene is enough to carry out the pigment production, and a person will have brown eyes if they possess one or two genes for brown pigment production. If no pigment is produced, the eyes will be blue. Therefore, we say that brown eyes are dominant over blue eyes. The blue-eyed gene is said to be recessive. Two blue-eyed parents cannot have a brown-eyed child since neither parent carries the gene that produces brown pigment. The situation is a bit more complicated than we have just described. There are several genes involved that control the amount of pigment produced, the tone of the pigment (how dark it is), and the distribution of the pigment in the iris (the pigment is sometimes concentrated near the pupil for example).

Skin color exemplifies a similar process. Skin color is controlled by the production of a substance called melanin. Dark-skinned people produce more melanin than light-skinned people. Certain proteins control the production of melanin. If a light-skinned person with no melanin producing genes and a dark-skinned person with two melanin-producing genes have an offspring, the offspring will possess only one gene for the production of melanin. The offspring will be dark, but since only one of its genes directs the production of melanin, the offspring will generally not be as dark as the darker parent. In actuality, the determination of skin color is more complicated than this description also. In horses, mice, and many other mammals, about three pairs of genes are involved in determining their basic coat color. A similar number of skin color genes in humans produces the observed shades of skin color.

Protein Synthesis

The recipes for making proteins are stored in DNA molecules, but how is the information used to make a protein molecule? Small structures called *ribosomes* are the protein making machines. They are very common in most cells. The ribosomes do not read the information directly from the DNA molecule, but from a messenger molecule (called *messenger RNA*) that is made by using the DNA as a template.

The messenger molecule contains the recipe for making a protein molecule. This messenger molecule is a single strand of *ribonucleic acid* or RNA, and it is synthesized by a set of enzymes called RNA polymerases that read the DNA and make the messenger RNA (abbreviated mRNA).

RNA is very similar to DNA, but is different in two minor ways. The sugar portion is composed of ribose sugar instead of deoxyribose sugar. These sugars differ by one oxygen atom (Figure 5.10).

The other difference is that in place of the base thymine, RNA substitutes the similar base uracil (Figure 5.17). The other three bases, adenine, cytosine, and guanine are the same

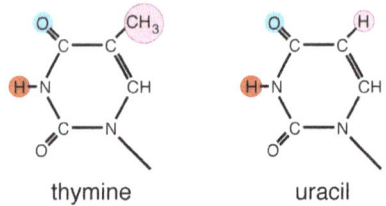

Figure 5.17 Uracil In the RNA molecule, thymine is replaced by uracil. Their difference is highlighted in pink. [JCB]

ones found in DNA. The production of a messenger RNA molecule from a DNA molecule is a process called *transcription*. The mRNA strand contains the triplet groupings of bases (called *codons*) that specify the amino acids in the protein chain.

After the mRNA is transcribed, a ribosome attaches to it and protein production begins, a process called *translation*.

Ribosomes, as their name implies, are composed of a few RNA molecules (called ribosomal RNA or rRNA) and several proteins. Approximately 2/3 of the ribosome is composed of RNA and the other 1/3 is made of protein molecules. The composition and structure of ribosomes depends somewhat on the type of organism. For example, the bacteria and other prokaryotes have smaller ribosomes than eukaryotes. Bacterial ribosomes only contain about two-thirds as many proteins.

It is interesting that bacterial ribosomes can assemble themselves. If the RNA molecules and protein molecules are separated, the pieces will reassemble themselves again to form a ribosome. It also appears that all the protein molecules are not essential. A ribosome will still function even if some of the protein molecules have been removed. Apparently the RNA molecules are more important than the proteins.

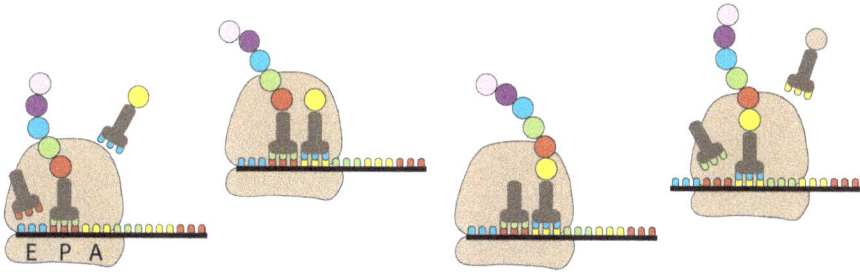

Figure 5.18 Translation (Protein Production) *(from left to right) the empty transfer RNA (brown fork-shaped structure) detaches from the messenger RNA (mRNA) strand (beaded line) and leaves the E-site (exit site) on the ribosome. The growing chain is attached to the transfer RNA (tRNA) in the P-site. The next amino acid (yellow circle) is carried in by the appropriate tRNA molecule which binds to the mRNA codon at the A-site on the ribosome (second picture). Next (third picture), the growing protein chain is bonded to the new amino acid, and the ribosome moves along the mRNA (last picture) to the next codon. The empty tRNA exits from the E-site and the cycle is repeated until a stop codon is reached. [JCB]*

The RNA in the ribosome (the rRNA) helps to hold the messenger RNA (mRNA) in place. When the mRNA is in place, the three letter words (codons) are read (translated), and the proper amino acid is brought into place to build the protein molecule. Amino acids are brought to the ribosome by other RNA molecules (called *transfer RNA* or tRNA). They attach to the mRNA codon and the correct amino acid is added to the growing protein chain (called a polypeptide chain).

A transfer RNA molecule (tRNA) has a three base sequence that is complementary to one of the three base codons on the mRNA. This three base complementary sequence is called an *anticodon*. The amino acid that is specified by the codon is attached to the other end of the transfer RNA. Since there are 20 different amino acids, there must be at least 20 different kinds of tRNA molecules. However, most organisms have more than 20 kinds. Transfer RNA molecules are folded molecules of RNA and are sometimes represented schematically as cloverleaf structures. In reality, however, tRNA molecules are more L shaped.

There are three sites on a ribosome (sites A, P, and E) that can hold tRNA molecules. The P site (short for the peptidyl binding site) is where the growing polypeptide chain is located. After the growing amino acid chain is attached to the next amino acid, the ribosome slides down the mRNA molecule and the empty tRNA leaves from the E-site (exit site). A tRNA brings in the next amino acid to the A-site (short for the aminoacyl binding site). The growing chain of amino acid molecules is terminated when a stop codon is reached (Figure 5.18).

The Genetic Code

One of the most important biochemical discoveries of the 1960s was the deciphering of the genetic code. Figure 5.19 shows the 64 possible mRNA codons and the corresponding amino acid for which they code. Since there are 64 codons, but only 20 amino acids and one stop code, many amino acids are specified by more than one codon.

Stronger 3-3 Hydrogen bonds	Mixed 2-3 Hydrogen bonds	Mixed 3-2 Hydrogen bonds	Weaker 2-2 Hydrogen bonds
Pro CC (C/U)	Ser UC (C/U)	Leu CU (C/U)	Phe UU (C/U)
Pro CC (A/G)	Ser UC (A/G)	Leu CU (A/G)	Leu UU (A/G)
Ala GC (C/U)	Thr AC (C/U)	Val GU (C/U)	Ile AU (C/U)
Ala GC (A/G)	Thr AC (A/G)	Val GU (A/G)	Ile/Met AU (A/G)
Arg CG (C/U)	Cys UG (C/U)	His CA (C/U)	Tyr UA (C/U)
Arg CG (A/G)	stop/Trp UG (A/G)	Gln CA (A/G)	stop UA (A/G)
Gly GG (C/U)	Ser AG (C/U)	Asp GA (C/U)	Asn AA (C/U)
Gly GG (A/G)	Arg AG (A/G)	Glu GA (A/G)	Lys AA (A/G)

Pyrimidines: C,U   Purines: G,A

Figure 5.19 The Genetic Code for mRNA Codons *The columns give the strength of codon-anticodon bond formed by just the first two bases (not the third), when a tRNA anticodon bonds with its mRNA codon during protein synthesis (see Figure 5.18). When codons in the first column bond with their tRNA anticodon, two C-G pairings are formed with 3 hydrogen bonds each (see Figure 5.15). So, the first two bases are bound by 6 hydrogen bonds, a strong bond. On the other hand, in the last column the codon-anticodon bond involves two U-A pairings which are only bound by two hydrogen bonds each, making 4 hydrogen bonds for the first two bases, a weaker bonding. For the second column, the first base is held by two hydrogen bonds and the second is more strongly held by 3, making a total of 5. In the third column, the bonding is reversed. For the yellow boxes, changing the third base does not change the amino acid. For the white boxes, changing the third base from one purine to another or one pyrimidine to another does not change the amino acid. Protein synthesis starts with AUG (Met).*

When two or more codons specify the same amino acid, the bases in the first and second position are always the most important. Changing the last (third) base does not change the amino acid being specified in half the cases. In all cases, interchanging the pyrimidines uracil and cytosine in the third position does not change the amino acid, and in all but 2 cases, interchanging the purines adenine and guanine in the third position does not change the amino acid.

There are other regularities that have been noticed in the genetic code. For example, the second base appears to be the most important. If the second base is uracil (upper right quadrant), one of the internal (water avoiding) amino acids is specified. A cytosine in the second position (upper left quadrant) always specifies an ambivalent amino acid, and unless the first base is uracil, an adenine in the second position (lower right quadrant) specifies an external (water loving) amino acid.

Although the genetic code appears to be almost universal, several variations are known. For example, in most mitochondria, UGA codes for Trp instead of stop and AUA codes for Met. AG(AG) are stop codons in vertebrate mitochondria, but are serine codons in most invertebrate mitochondria. Notice that these changes make the code simpler and even more symmetrical as far as the third base position is concerned. As we will see later, mitochondria may have evolved from ancient bacteria, and the observed variation in their code could be a relic from the distant past when the code was somewhat simpler.

Mutations

We are now able to understand certain kinds of genetic mutations, which are events that produce new variations in nature. Darwin noticed that the individuals within a given population exhibit a great deal of variation. They vary in color, height, weight, there are subtle differences in body chemistry, and many other variations that may affect the individuals' chances of survival. Darwin did not know what caused these differences, but we now know that most are due to differences in the genetic make-up (the DNA) of the individuals.

A sudden change of one base in the DNA of an organism is called a *point mutation*. Point mutations may be caused by copy errors when the DNA is duplicated or by chemical reactions causing one base to change enough that it is copied as a different base. Since genes are protein recipes, a change in a particular gene may show up as a change in its corresponding protein.

We can classify point mutations into three general types. *Substitution* is where one base is changed into another. For example, thymine might be changed to cytosine. Since an amino acid is specified by a three-base sequence on the DNA molecule, changing one base could result in the production of a slightly different protein. However, since several codons can specify the same amino acid, every substitution in the DNA does not necessarily result in a protein change.

Even if a substitution produces an amino acid change, the change may be effectively neutral. Many of the amino acids in a protein molecule are not crucial to the effectiveness of the protein. For example, one water-loving amino acid might function just as well as another water-loving amino acid at a particular spot in a given protein sequence. The amino acid sequence of many proteins is very nearly random. Experiments have been done in this area, and one study of a bacterial protein indicated that only about 8 of the 268 amino acids in the chain were absolutely critical to the functioning of the protein.

Horse and human hemoglobin have virtually the same three-dimensional structure even though 43 of the 287 amino acids in the alpha and beta chains are different. Often, changing one or more amino acids in a protein molecule does not radically change the effectiveness of the protein.

A potentially harmful point mutation is one that causes a codon to change to a stop codon (called a *nonsense mutation*). For example, one type of hemophilia is caused by a mutation in the blood clotting protein that changes arginine to the stop codon. The fragmented protein is non functional.

The second type of point mutation is the *insertion* of one or more bases into the DNA chain. Unless some multiple of 3 bases is inserted, this type of mutation results in a portion of the DNA molecule being read in a completely different manner. The stop message may not be read, or a stop could occur earlier. Shortened chains usually result in nonfunctioning proteins. However, many organisms have two or more genes that direct the production of a given protein, and if one gene becomes nonfunctional, the other may produce enough of the protein to assure the survival of the individual. In some cases, however, the results may be severe enough to cause the death of the individual.

The third type of point mutation is the *deletion* of one or more bases. As with insertion, if some multiple of 3 bases is deleted, the individual may not be adversely affected. However, the deletion of one or two bases usually results in a nonfunctioning protein. Deletions and insertions are relatively rare point mutations. About 96 percent of the point mutations are substitutions. Deletions account for about 2 or 3 percent of the point mutations and about 2 percent of the point mutations are insertions.

The word mutation usually conjures up an image of some deformed individual, and indeed they can occur. Most of the time, however, a mutation merely results in the production of a slightly different protein. This variant protein may be better suited to its task, it may not be as functional, or it may perform much the same as the original protein. If the individual with the mutation survives, however, more variety will have been introduced into the population.

Studies of the frequency at which mutations occur indicate that there are about 30 mutations in each human sperm. In other words, every person in the world carries many mutated genes. In the vast majority of cases these mutations are not easily detected. Obviously all mutations are not as harmful as many people have been led to believe.

How does a reptile become a bird? The answer is really quite simple to understand. Over time, the recipe for a

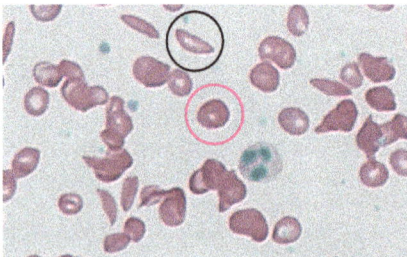

Figure 5.20 Sickle Cells *Blood sample of a person with sickle-cell anemia showing a normal cell (circled in red) and a sickled cell (circled in black). [Graham Beards]*

reptile is simply changed by many mutations into the recipe for a bird. The details of when each of these mutations came about would be quite difficult to determine.

It is now easy to see why different forms of radiation and certain chemicals produce mutations. Many forms of radiation carry enough energy to alter a portion of a DNA molecule. If the DNA molecule in an egg or sperm cell is affected, the offspring will carry that mutated gene. (A mutation that results in a mutated sex cell is known as a *germinal* mutation while a mutation in some another tissue cell is called a *somatic* mutation.)

Certain chemicals can also initiate reactions in DNA molecules, resulting in mutations. Cancer cells are often the result of somatic mutations that can be induced by certain chemicals called carcinogens or by radiation of the type associated with medical tests and the waste products from nuclear reactors.

The most extensively studied protein is human hemoglobin, the molecule in red blood cells that carries oxygen from the lungs to the cells. Over 300 variant hemoglobins have been discovered and probably more than 600 exist.

One particularly interesting point mutation, which is relatively common in certain parts of Africa, is responsible for a condition called sickle-cell anemia. If both chromosomes of the homologous pair contain the sickle-cell mutation (*homozygous* for the sickle-cell trait), the individual's blood cells form in a sickle shape, and the hemoglobin molecules cannot carry oxygen effectively (Figure 5.20). Consequently, the person usually dies at a young age.

However, if only one of the homologous chromosomes contains the sickle-cell gene (*heterozygous* for the sickle-cell trait), only about half the hemoglobin molecules will be of the sickle-cell type. The oxygen carrying ability of the blood is reduced and the person tends to be anemic, but usually does not die.

Generally, such a dramatic alteration in the properties of a protein molecule would not be expected by changing only one amino acid. However, the particular amino acid that is affected by the sickle-cell mutation is in a critical position on the outer surface of the molecule. The point mutation causes glutamate (an external or water loving amino acid) in position six of the beta hemoglobin chain to be replaced by valine (an internal or water avoiding amino acid). This change produces a sticky area on the outer surface of each hemoglobin molecule that causes the molecules to combine in long bundles. These long bundles distort the blood cell and produce the sickle shape.

Carrying the sickle-cell mutation appears to confer a rather distinctive disadvantage on the individual. Why has nature not severely selected against this mutation? The answer is that there is an advantage conferred by the sickle-cell hemoglobin that can be greater than the disadvantage. Individuals with the sickle-cell mutation are resistant to malaria because the malaria parasites cannot live in the distorted red blood cells.

A frequency study of the sickle-cell gene shows that it is most abundant in areas where there is a high incidence of malaria. The sickle-cell mutation is an illustration of a mutation that confers an advantage when only one sickle-cell gene is present, but a very severe disadvantage when two sickle-cell genes are present. In malaria-infested areas of Africa, natural selection has operated to preserve this mutation.

As we saw in the last section, there are slight variations in the genetic code. Why are there different genetic codes?

One possibility is that the genetic code itself evolved early in the history of life. As we saw, changing the third base usually does not change the protein recipe, or changes it in a minor way. (One exception is a change that produces the stop codon.) In addition, changing the first base usually causes one amino acid to change to a similar one. (For example, one water-loving amino acid to another water-loving amino acid.)

The regularities in the genetic code suggest that the code was not the result of random chance, but must have evolved to its present state. Of all the millions of possible genetic codes, computer simulations have found that the current code is one that minimizes the effects of mutations and reading errors. This means that most mutations do not produce significant changes in the protein recipe. Although the mutated protein might be slightly different, it usually still functions.

By minimizing the effect of a mutation, the code also increases the probability that a mutation will be beneficial. The reason for this increase is because the mutations that do occur tend to have a smaller effect on the protein. Mutations that have a smaller effect are more likely to result in a protein that is an improvement. Mutations that produce large effects are more likely to be harmful.

The genetic material in a fertilized egg contains the DNA recipes needed to make the organism. This recipe book contains all the protein recipes, but it also has contains the information needed to produce the substances that guide the development of the organism. That is, information on how the various proteins are to be assembled into the organism. Proteins guide certain stages in the development of an organism, but it now appears that some DNA codes for RNA molecules that also help orchestrate the development of the organism.

Equivalent proteins are proteins that perform the same function in different organisms. For example mouse hemoglobin and human hemoglobin are equivalent proteins, and bird keratin and lizard keratin are equivalent proteins. Many proteins in one organism have equivalents in other organisms. Most of the proteins found in humans have equivalents in mice. As expected, the recipes for equivalent proteins are different from one organism to the next, but they still serve similar functions

Although equivalent proteins may have different recipes, these differences alone do not seem to be large enough to account for the observed differences we generally see in two organisms. It is very likely that the differences between two organisms is more than just the differences in the structural proteins of the two organisms. A dog is not just a cat with different hair proteins, different blood proteins, muscle proteins, bone proteins, etc. The differences in structure must be due to differences in the architectural information. For example, cats have retractable claws, but dogs do not. To produce this difference, there must be information in the DNA that produces proteins and/or RNA molecules that guide the development of the claws of the dog and the cat.

Only a fraction of the DNA of a eukaryote is used to make proteins. Humans have about 23,000 protein coding genes, but less than two percent of our DNA encodes for these protein molecules. The purpose of the remaining DNA is not fully understood. Even though it is not used to make proteins, most is transcribed into RNA. In the eukaryotic cell, RNA is transcribed in the nucleus, but translation (the process of making proteins) occurs in the cytoplasm. Since prokaryotes have no nucleus, RNA is transcribed and immediately translated.

This nonprotein coding DNA has been called junk DNA, but most is not junk at all. Prokaryotes do not have very much of this junk DNA. It is possible that the excess DNA is necessary to orchestrate the development of more complex organisms. Perhaps the evolution of this regulatory system is what allowed eukaryotes to evolve into complex organisms. This is an area of current research that will undoubtedly produce some very interesting results.

We stated earlier that genes are protein recipes, but since we now know that some genes are recipes for RNA molecules that never get translated into proteins, we need to revise our definition. Genes are not just protein recipes, but can also be recipes for RNA molecules. We have known for some time that the DNA must contain recipes for the set of transfer RNA molecules and the RNA molecules in ribosomes, but it now appears that other RNA molecules might also play a large role in the development of the organism.

Dollo's Law

In any scientific discipline, many little principles are discovered, and often they are given the inflated status of a law. As we have already seen, many of these laws are not even true. For example, Ohm's law of electrical resistance is very useful, but it is only an approximation for some substances. In evolution, there is a principle of irreversibility known as Dollo's law, which states that the evolutionary process is not reversible. For example, mammals evolved from fish ancestors, but the process could not be reversed. A mammal could not evolve back into a fish. Mammals have evolved swimming forms such as whales and dolphins, but a dolphin is not a fish.

In retrospect, Dollo's law is really just a statement of probability. The evolutionary process involves so many mutations, that after a population evolves far enough in one direction, it would be virtually impossible for the process to exactly reverse itself and have a portion of the population evolve back to an earlier form. For example, the ancestors of horses had feet with five toes, but over millions of years their descendants have evolved feet with a single toe. (The toenail of this single toe is the horse's hoof.) Because of the many protein changes (mutations) that must have taken place to cause this change, it would be impossible for the process to exactly reverse itself. It would be impossible for the descendants of a group of horses to evolve back into a group of five-toed mammals like the ancestors of the horse.

Fish evolved into amphibians and then into reptiles (like dinosaurs) and in the process, fins evolved into feet with toes. On the molecular level, hundreds of mutations must have been involved in the process. Later, a group of reptiles evolved back into aquatic creatures and in the process, the feet with toes evolved back into a type of fin or flipper. Although these flippers superficially resembled the fins of a fish, they were quite different structurally. For example, their bone structure was quite different. Statistically it would have been virtually impossible for the mutations that led to the feet to be exactly reversed when the feet evolved into flippers.

On the molecular level the aquatic reptiles and fish must have been quite different. Today, dolphins superficially resemble sharks, however, they are very different anatomically and are quite different on the molecular level. The DNA of dolphins is much more similar to human DNA than it is to shark DNA.

Methods

The reactions that are carried out in cells generally involve chemicals containing the element carbon. Carbon compounds are called organic compounds because it was once thought that only organic or living organisms could produce these compounds. We now know that organic compounds follow the same chemical principles as inorganic compounds and organic chemicals are now routinely produced in the laboratory.

A great deal of progress in biochemistry has been made in the past several decades. In 1951, Frederick Sanger determined the amino acid sequence of insulin, and in 1977 he sequenced the first viral DNA. James Watson and Francis Crick discovered the double helix nature of the DNA molecule in 1953. The decipherment of the genetic code was completed in 1966, due in large part to the efforts of Marshall Nirenberg. Keiichi Itakura made the first synthetic gene in 1977.

Today we have well-established methods for determining the amino acid sequence of proteins as well as the base sequences of DNA and RNA molecules. The techniques and procedures that were used to reach our present level of knowledge are too numerous to discuss in detail here, but a few will be briefly described to illustrate some of the clever experiments that have been performed.

The first step in breaking the genetic code involved making an RNA molecule that only contained the base uracil. When this poly-U molecule was added to cell extracts containing ribosomes, chains of phenylalanine were produced. Therefore, UUU must be the codon for phenylalanine. In a similar experiment, poly-A produced chains of lysine, demonstrating that AAA is the codon for lysine. By synthesizing other mRNA molecules with different triplet sequences, the other codons were eventually identified.

One method of determining the sequence of amino acids in a protein (called protein sequencing) is the Edman degradation process (devised by Pehr Edman). In this process, the amino acid at the amino end of the chain is labeled by attaching a molecule of phenyl isothiocyanate to it (the name of the molecule is irrelevant). This labeled amino acid is then split off the chain and identified (Figure 5.21). Repeating the process allows the next amino acid in the chain

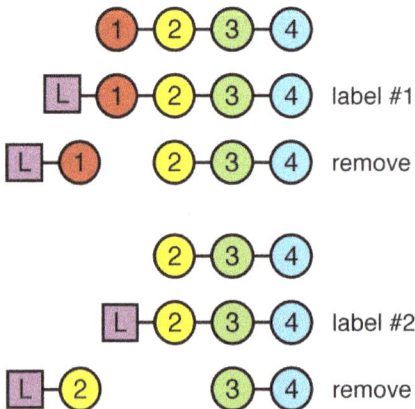

Figure 5.21 Edman Degradation Process
This process is a procedure for sequencing a protein. The first amino acid is labeled, removed, and identified (top three pictures). Next, the second amino acid is labeled, removed, and identified. The process continues to the end of the chain. [JCB]

to be identified. The complete sequence is determined in this manner.

Sequencing of a single protein requires many time consuming steps, but today, *DNA sequencing* can be accomplished with relative ease. Therefore, instead of sequencing a protein, it is easier to sequence its corresponding DNA chain.

The Sanger method of DNA sequencing (developed by Frederick Sanger) is relatively easy to understand. It involves growing partial segments of the DNA chain that is being sequenced. (DNA polymerases are used to grow the partial segments of DNA.) The growth of each partial chain begins at one end of the template DNA (the end known as the 3' end). A radioactive phosphorus marks the beginning of the chain.

The growth of the chain is stopped at some point by adding a modified nucleotide instead of the normal one. (The modified nucleotide lacks an OH group at the 3' position.) The importance of this modified nucleotide is that it stops the chain growth at that point. In this way, strands of various lengths are produced, each ending at a specific nucleotide position. If some modified adenine nucleotides are added to the growing solution, all the chains will be terminated at an adenine position. If some modified guanine nucleotides are added, the chains will be terminated at a guanine position, and so on.

The method involves growing chains in four different containers, each of which contains a small amount of one of the four modified nucleotides. Therefore, after many chains are grown, a given container will contain sequences of varying lengths, but the termination of each sequence will occur at a position corresponding to the modified nucleotide added to the container.

For example, consider the sequence CATGAGTACT. There are ten possible sequences that could be grown starting at the C end (the 3' end): C, CA, CAT, CATG, CATGA, CATGAG, CATGAGT, CATGAGTA, CATGAGTAC, and CATGAGTACT. After the sequences are grown in the four different containers, the container with the modified adenine would contain the sequences CA, CATGA, and CATGAGTA, all the possible sequences ending in A.

The container with the modified guanine would contain the sequences CATG and CATGAG, all the sequences ending in G. The container with modified thymine would have the sequences CAT, CATGAGT, and CATGAGTACT and the container with modified cytosine would have the sequences C and CATGAGTAC.

Once the four classes of sequences have been prepared, the next job is to separate the sequences according to their length. This is done using specially prepared gels by a process known as *gel electrophoresis*. The four different solutions are placed on four different locations at the top of the gel, and a weak electric field is applied to the gel (Figure 5.22). The electric field causes the DNA sequences to move down through the gel, but their speeds depend on the size of the molecule. The smaller the sequence, the further it moves down through the gel.

In our example the sequence containing just C is the smallest and therefore will move the furthest down the gel. The dark line at the bottom of the gel in Figure 5.22. This process produces four columns in the gel corresponding to the four different terminations, and a separation within each column that depends on the size or length of the sequence (Figure 5.22). The correct sequence of the DNA

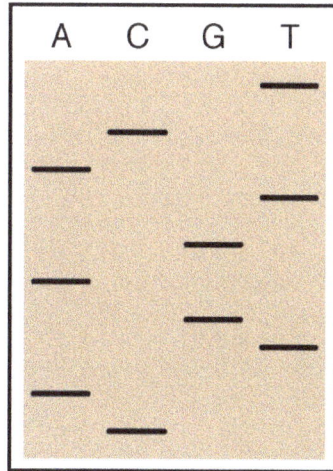

Figure 5.22 Gel Electrophoresis The *letters across the top give the last base of the DNA sequences in the respective columns. The dark line at the bottom of the gel was produced by the shortest sequence which was C (it ends in C so it is in the C column). The next dark line from the bottom was produced by the sequence CA (it is in the A column since it ends in A), etc., on up the gel. The entire sequence read from the bottom to the top is: CATGAGTACT. [JCB]*

molecule is simply read up from the bottom of the gel: CATGAGTACT.

The three-dimensional structure of proteins, DNA, and many other substances is determined by X-ray analysis. Many proteins can be purified and made into crystals. The crystals are then studied using X-rays, a high-energy form of light with a very short wavelength. When X-rays reflect off the atoms in a crystal, they form a distinct pattern that can be used to determine the three-dimensional arrangement of the atoms in the crystal.

The Origin of Life

We now know a great deal about the chemistry of life, but how did life first form on the Earth? The details of the origin of life are not known at this time, but some of the pieces may have been discovered.

As we have seen, amino acids are important components of life's building blocks. In 1953, an experiment by Stanley Miller showed how amino acids and other important molecules could have been produced on the early Earth. Miller showed that if a mixture of nitrogen, water vapor, methane, and traces of ammonia is sparked in an electrical discharge chamber, many important molecules are made. These gases were likely present in the Earth's early atmosphere, and lightning would have supplied the electrical spark.

Since the Miller experiment in 1953, many similar experiments have been performed using different gas mixtures and alternate energy sources. These energy sources include ultraviolet light and high temperatures as would be expected in association with volcanic lava. The results of these various experiments are similar to those of Miller as long as oxygen is not included as one of the gases. Because it is so reactive, we would not expect significant quantities of oxygen to have been present in the Earth's early atmosphere. Essentially all the oxygen in the present atmosphere was formed as a byproduct of photosynthesis.

Table 5.2 lists the amino acids that were produced by sparking a mixture of methane, nitrogen, water vapor, and

traces of ammonia. The numbers give the relative yields of each.

The mixture of amino acids and other organic compounds produced in Miller-type experiments is often referred to as an organic soup. Many scientists believe that a significant amount of this organic soup was produced on the early Earth, and that life eventually formed from it. The details of exactly how this life-forming process might have happened, however, are not known at this time.

For comparative purposes, it might be of interest to look at the frequency distribution of amino acids in protein molecules. Figure 5.23 shows a distribution that was obtained from 74,000 proteins which contained 26.8 million amino acids. The amino acids that were listed in Table 5.2 are shown as red bars.

As can be seen from the graph, ten of the twelve most abundant amino acids are those produced in the spark experiment. This high correlation would suggest that we are not dealing with mere coincidence. It should also be pointed out that many amino acids are made in the spark chamber that are not found in protein molecules. The full significance of these findings is not understood. It is to be hoped that some future discoveries will shed more light on this subject.

Astronomical evidence indicates that many organic molecules were made in the solar nebula out of which the Earth and other members of the Solar System were formed. Amino acids, sugars, and other organic compounds have been found in meteorites. In addition, many fairly complex molecules have been discovered in interstellar gas clouds. It appears that organic compounds, that were once thought to be unique to living organisms, are easily made in outer space and were present in the

Alanine (A)	790	Glutamate (E)	8
Glycine (G)	440	Serine (S)	5
Aspartate (D)	34	Isoleucine (I)	5
Valine (V)	20	Proline (P)	2
Leucine (L)	11	Threonine (T)	1

Table 5.2 Amino Acid Synthesis The amino acids were made by sparking a mixture of nitrogen, water vapor, methane, and traces of ammonia. Relative abundances are shown. [data: S.L.Miller, see credits p 415]

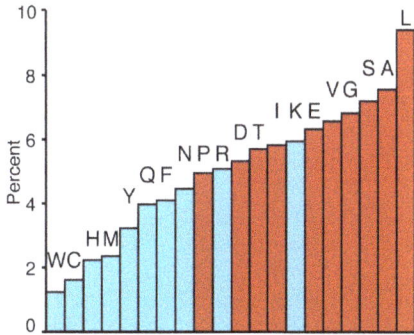

Figure 5.23 Amino Acid Abundances *The percentages are based on 74,000 protein sequences. The one-letter symbols (Table 5.1) are shown above each bar. Amino acids made in sparking experiments are drawn as red bars (see Table 5.2). [JCB; data: SwissProt release 36.0, compiled by: Andrea Beckel-Mitchener]*

solar nebula when planet Earth was formed.

Life must have started with the formation of a molecule that could reproduce itself. Since proteins are such important components of life, many scientists believe the first self-replicating molecules must have been proteins. However, as we have seen, you cannot make proteins without DNA and you cannot make DNA without proteins. So which did came first?

DNA is relatively stable, but its close cousin RNA is more versatile. Many scientists believe that RNA played an important role in the early evolution of life. Perhaps RNA led the way for proteins. Many investigations have involved RNA, and several interesting results have been discovered. RNA is used in many structures such as ribosomes and messenger RNA. Maybe it could do everything. We have discovered several RNA enzymes that can catalyze reactions, and ribosomes contain an RNA enzyme that catalyzes the making of the proteins.

RNA molecules have also been discovered that can copy other RNA molecules. All these discoveries seem very promising, but some problems still need to be solved. For example, some type of energy source is required and this may require some additional helper molecules.

One other question is how the RNA on the early Earth was formed. As it turns out, two of the RNA nucleotides (the pyrimidines) have been produced out of chemicals that are likely to have been present on the early Earth.

The formation of like is still not solved, but much progress has been made, and there is little doubt that someone will eventually succeed in showing how it might have happened

An explanation for the zoo of elementary particles looked hopeless in the 1950s and early 1960s, but in 1964, the quark model cleared up the mystery. Solving mysteries like this is what science is all about.

At present, the means by which life originated on the early Earth is one of the great unsolved mysteries of our time. There has been much speculation about the earliest life-forms and their chemical evolution. Papers have even been published that claim to calculate the probability of forming life on the early Earth. It should be pointed out, however, that calculating probabilities for an event that we do not fully understand would seem to be an exercise in absurdity. If one does not know the process, how does one begin to calculate the probability of it happening?

The origin of life involved the formation of life from nonliving matter. Some may have the idea that Pasteur and others proved that living organisms could not arise from nonliving matter. Before Pasteur's experiments, many people believed that organisms arose spontaneously from inanimate material.

For example, they believed that maggots came from rotten meat, and bacteria arose from beef broth. Pasteur and others showed that complex organisms do not arise from nonliving matter.

However, even bacteria are very complex compared to what the earliest life-forms must have been like. Neither Pasteur, nor anyone since has shown that the first simple life-forms could not have arisen from nonliving matter.

The distinction between the living and the nonliving is not always absolutely clear. It is sometimes very difficult to accurately define life. Would we define a self-replicating molecule as a living thing?

A virus is composed of a short strand of DNA or RNA (generally less than 50 genes) enclosed in a protein coat. To reproduce, a virus must enter a cell and use the cell's machinery (the ribosomes and transfer RNAs). By itself, a virus cannot manufacture proteins or duplicate its own genetic information, and many scientists do not classify viruses as living organisms.

Some scientists believe that viruses evolved from fragments of DNA or RNA and represent a later development in the history of life, while others believe they may be the remnants of very ancient life-forms. Regardless of how viruses originated, they do illustrate the sometimes difficult problem of differentiating between the living and the nonliving. In the next chapter, we will look briefly at the great diversity in the living world.

Chapter 6: The Living Earth

Introduction

Although Darwin was not aware of genes or the chemical changes that are associated with mutations, he did know that the individuals in a population exhibit a great deal of variation. We now know that different protein recipes or genes produce these individual variations. As we saw in the last chapter, variations in genes are introduced into the population as a result of mutations. As time passes and more of these mutations become established in the population, the population changes. That is, the population evolves. Darwin called these changes, descent with modification.

Darwin proposed an explanation (called natural selection) for the reason certain traits or characteristics become established in a population. Those traits that make an individual better suited for survival to a reproductive age will be passed on to the next generation. Traits that make an individual less suited to survive and bear young are less likely to be passed on, and the genes responsible for these traits will tend to disappear from the gene pool. As more changes are added to the population, organisms may become significantly different in appearance from their ancestors.

Often different portions of a population are isolated, and two or more forms evolve from the same parent stock. Isolation is believed to be one important way branches form on the evolutionary tree of life.

If all living organisms evolved from earlier life forms in the manner described above, there should be many clues to the interrelationships that exist in the living world. For one thing, organisms that are closely related should have many anatomical features in common with their close relatives and fewer features in common with their more distant relatives. It was these anatomical similarities among groups of organisms that suggested the idea of evolution. For example, all birds have feathers and a beak, but all mammals have hair and mammary glands.

Darwin and other early evolutionists were not aware of the chemical basis of evolution, but if evolution is responsible for the variety of life forms on Earth, the chemistry of closely related organisms should also be very similar. This prediction has been verified in numerous cases. For example, if chimpanzees and humans evolved from a common ancestor just 5 to 8 million years ago (5 to 8 mya), we would expect their proteins to be very similar, and indeed they are. The 287 amino acid sequence that forms human hemoglobin is identical to the sequence in chimpanzees. The horse, a more distant relative of humans, has hemoglobin molecules that differ from human hemoglobin by 43 of the 287 amino acids.

Although analyzing proteins would undoubtedly be the best way of determining how closely organisms are related, anatomical features are generally used because the vast numbers of organisms make chemical analysis impractical. Since proteins determine the anatomical features of an organism, a family tree generated from protein analysis should closely parallel a family tree based on anatomical features. As we investigate the classification scheme, we will look into some of the anatomical features that are generally used to determine the relationships between organisms.

Taxonomy

When we are presented with a large number of different objects, our natural inclination is to classify these items in some useful way. Putting things into boxes, even if they sometimes do not fit as well as we would like, helps us remember their similarities and differences. Classifying different objects sometimes helps us see patterns and relationships that might otherwise go unnoticed.

There are many ways to classify things, but some methods turn out to be more informative than others. Stars are often classified by the color of the light they emit. This classification system resulted in the discovery that blue stars are generally, more massive and brighter than red stars.

The arrangement of living organisms according to how closely they are related is known as *taxonomy*. We could classify all life forms by weight or color, but such classification schemes have obvious weaknesses and have not led to any enlightening discoveries. A scheme based on weight would mean that organisms that grow would be put into different categories as they gain weight. We would often find parents and their young offspring classified differently. Some plants and animals change color with the seasons, so a classification scheme based on color would mean these organisms would be put into different categories depending on the time of year.

Body shape, although often useful, is sometimes a poor characteristic to use in classification. Whales and porpoises are very fishlike in appearance, however, they share more anatomical structures with mammals than they do with fish. Young organisms often have body shapes that are quite different from those of their mature parents. Consider the caterpillar that is eventually transformed into a butterfly.

While there is no single anatomical feature that is universally used for the purposes of classification, it is generally agreed that certain combinations of traits indicate close relationships. For example, the general bone structure of vertebrates, the teeth of mammals, the bills and feet of birds, or the reproductive structures of plants are often used for classification. Organisms that share many common features are grouped together, and a rather elaborate classification scheme has evolved over the years. Chemical analysis has shown that most plants or animals in a given group have similar DNA sequences.

It may sound like taxonomy is not very scientific, but as we study various organisms and compare their similarities and differences, it becomes obvious that certain organisms share many more features with each other than they do with other organisms. The degree of similarity between two different organisms is interpreted as a measure of how closely they are related to each other. For example, a Tiger shares many more features with a Lion, than either does with an elephant. We, therefore, conclude that a Tiger is more closely related to a Lion than either is to an elephant. This conclusion was reached centuries before the idea of evolution was proposed.

Now that it is possible to analyze proteins and to sequence DNA molecules, we are indeed finding that the proteins of Tigers are more similar to the proteins of Lions than they are to the proteins of elephants. Family trees based on protein or DNA sequencing have been constructed that strongly parallel the family trees based on anatomical similarities.

The smallest natural unit of classification is generally considered to be the taxonomic *rank* of species. A species is defined to be a group of organisms that are capable of interbreeding and producing fertile young. Being reproductively isolated means that even if two different species happen to mate, no young will be produced, or if young result, they will be sterile. Horses and donkeys are considered to be different species

because their young (called mules) are sterile.

Horses and cows are different species because no offspring are produced when the two mate. The reason for this incompatibility can be traced to the proteins of the two species. If the protein recipes from the mother and father differ by too much, the various proteins produced in the fertilized egg cell will not be able to function together, and the cell will die.

Lest we be led to believe that species are always well defined, we should remember that over time, a species can diverge and evolve into two or more new species. We see this happening today. Most species do not form a uniform population over their entire range, but are often divided into slightly different groups called subspecies or races. If these subspecies remain isolated over a long enough time, mutations in the groups will eventually cause the groups to evolve into different species. Occasionally experts decide that two subspecies are different enough to justify splitting the groups into two distinct species.

Sometimes, however, the experts (often different experts) decide that two species are so similar that they lump them together as one species. Changes like this are expected since evolution proceeds slowly and there is not an exact moment when one species splits into two species. Experts may have different opinions on whether a split has occurred or not.

Several species that share many common traits are placed together into a higher taxonomic rank called a genus. For example, the Lion, Tiger, Jaguar, and Leopard are considered to be so similar that they are all put into the same genus (*Panthera*). Giving both its genus and its species name uniquely identifies a particular species. The two-part name is the scientific name of the species. This naming system is known as *binomial nomenclature*. The genus name is capitalized while the species name is not. Both are expressed in their Latin forms and are written in italics. For example, the scientific name of the Lion is *Panthera leo*, the Tiger is *Panthera tigris*, and so on.

Often species have common English names, and we will use the convention that common names of species like Greater Panda and African Elephant will be capitalized. Generic names will not be capitalized like white pigeon (where we are describing the color of a bird) or elephant (there are three different species of elephants). This convention makes it clear if we are talking about a species or a group of several species. For example, elephants hyraxes, manatees, the Dugong, and

the Aardvark belong to a large group called the Afrotheria.

Scientists use scientific names to avoid confusion since the same species may have several different popular names, some of which may of course be in different languages. Notice that whether two species are placed in the same or in different genera is a rather subjective decision. Genera and higher groupings are man-made categories whereas the species classification is a somewhat more natural grouping.

Genera whose members share many common traits are placed into the same family. For example, the cat family (Felidae) contains mammals with long pointed canine teeth, four toes on their hind feet, and retractable claws. Members of this family include the genus *Panthera* as well as the genera that contain the House Cat (*Felis*), the

Kingdom:	Animalia	Plantae	Fungi
Phylum: or Division:	Chordata	Magnoliophyta	Ascomycota
Class:	Mammalia	Liliopsida	Pezizomycetes
Order:	Lagomorpha	Liliales	Pezizales
Family:	Leporidae	Melanthiaceae	Morchellaceae
Genus:	*Oryctolagus*	*Trillium*	*Morchella*
Species:	*cuniculus*	*undulatum*	*conica*
Common name:	European Rabbit	Painted Trillium	Black Morel

Figure 6.1 Hierarchy of Taxonomic Ranks The box above the pictures gives the taxonomic ranks of the European Rabbit, Painted Trillium, and the Black Morel. [rabbit, JJ Harrison; trillium, D.Gordon E.Robertson; morel, Beentree]

Mountain Lion and Jaguarundi (*Puma*), the Cheetah (*Acinonyx*), the lynxes and Bobcat (*Lynx*), the Ocelot (*Leopardus*), and several others.

Similar families are grouped into orders. Because of the similar structure of their teeth, the cats, the dogs, the bears, and certain other flesh-eating mammals are placed in the order Carnivora. Members of this order are called carnivorans. (The term carnivore can refer to any meat-eating organism, even a plant like the Venus Flytrap.)

Similar orders are grouped into classes. All organisms that have hair and possess milk-producing glands to nourish their young belong to the class that is named Mammalia (popularly called the mammals).

Similar classes are grouped into the same phylum. All organisms that have a flexible supportive rod (called a notochord) running down their back at some stage in their development belong to the phylum Chordata. In mammals and other familiar members of the chordate group, the notochord is replaced by the backbone as the organism develops. Botanists generally use the term division instead of phylum for classifying plants at this level.

Finally, groupings of similar phyla are called kingdoms. Chordates, along with insects, spiders, crabs, clams, snails, squids, starfish, jellyfish, sponges, earthworms, and many other worm-like creatures belong to the kingdom Animalia (the animals).

The words describing the various groupings such as species, genus, family, class, etc., are called the taxonomic *ranks*. The hierarchy of ranks is summarized in Figure 6.1 along with several examples. The common names are also included.

Note that there are seven categories in the classification scheme, however, this number is purely arbitrary. Indeed, often terms such as superfamily (a

grouping containing more than one family), subfamily (a group containing a portion of a family), subphylum, superorder, subspecies, etc., are used when convenient. Also keep in mind that since all life forms are the descendants of a common ancestor, we should not be surprised to find that it is not always easy to put organisms into the classification system just defined. There are many examples of organisms that seem to fit between the boxes, but that is what we would expect if the life forms we see today evolved from earlier life forms.

The various classification boxes are called taxonomic ranks. Taxonomists try defining taxonomic ranks such that the grouping contains an ancestor and all its descendants. Such a grouping is called a *clade*. For example, if all the mammals (and no other animals) evolved from a single ancestor, the mammals and their ancestor would form a clade. Since a plot of the evolutionary sequence of organisms looks like the branches on a tree, it should be obvious that a system where all the groupings are clades is not possible (Figure 6.2).

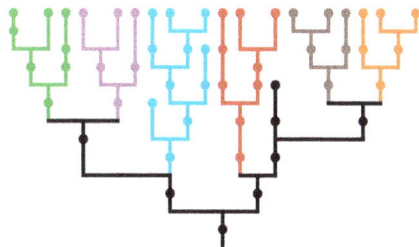

Figure 6.2 Clade Difficulties *This family tree (cladogram) has been divided into 6 clades (different colors). The dots represent species, but several species (black dots) do not fit into any of the clades. If we lump the green and violet clades into a clade of higher rank, the higher rank clade will include a species (black dot below the two groups) that is not in either of the lower rank clades. [JCB]*

As expected, the classification of various groups is in a nearly constant state of flux. Even experts do not always agree, and they may have trouble trying to decide if a grouping should be called a family or a subfamily. As more information is gathered, their opinions may change. More detailed DNA analysis will help greatly in the classification process.

In addition, classifying plants is not quite the same as classifying birds, and the same people do not often work with several different groups of organisms.

As we study fossils it is often extremely difficult to tell the difference between some mammals and certain reptiles, or to tell early fossil birds from certain dinosaurs. With time, mammals and reptiles became quite distinct, but when mammals were first evolving from reptiles, it was very difficult to tell the difference.

The characteristics that define the various groupings have been changed occasionally when new discoveries are made. For example, mammals were once defined as animals that give birth to live young, have hair, and possess milk-producing glands. However, after the discovery of several egg-laying mammals (the Platypus and the spiny anteaters), the part about bearing live young had to be dropped.

One interesting aspect of the classification system of organisms is that smaller groups can be nested completely within the larger groupings. Groups do not overlap with other groups. For example, the group we call cats is part of a larger group (the carnivorans belonging to the order Carnivora), which is a subgroup of the mammal group, and so on. In addition, the cat group does not overlap with the dog group. This grouping can be illustrated schematically by a Venn diagram (Figure 6.3).

The pattern is just what we would expect if the life forms of today evolved from a common ancestor. We are seeing the tips of an evolutionary tree.

When we try to classify unrelated objects such as stars, crystals, or geometrical figures, a quite different pattern emerges. We find that it is generally impossible to classify these objects into groups that nest within groups. In general, the groupings overlap (Figure 6.3). Because they are unrelated, and it is impossible to classify them in a meaningful way such that the groups do not overlap.

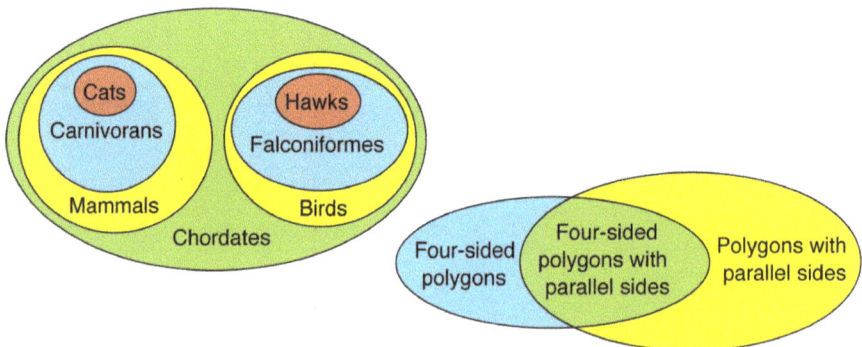

Figure 6.3 Venn Diagrams of Related and Unrelated Objects For related objects, criteria can always be chosen such that smaller groups are totally contained within larger groups (left). That is not possible for a group of unrelated objects (right). Even with carefully chosen criteria, one group will overlap another. [JCB]

For example, we could group geometrical figures according to the number of sides they have. If we then group the figures with parallel sides, we find that they do not all lie within the group with four sides. Many six-sided figures also have parallel sides, as do many five-sided figures, and so on.

In contrast, all mammals belong to the group of animals that have hair and mammary glands, and if we form another group of all organisms that have four sharp gnawing front teeth that grow throughout the organism's life (we might call them the rodents), we find that every one of these rodents is a mammal. There are no fish or reptiles or birds with teeth of this type.

As we learn more about the various organisms that inhabit planet Earth, our classification scheme will undoubtedly have to be modified. DNA sequencing is a powerful new tool that is currently being used to more accurately determine how various organisms are related.

The number of different species living on the Earth today is not known, and the estimates vary widely. About half may be prokaryotes and half eukaryotes. Of the eukaryotes, only about 1.6 million species of have been described, and about a million are insects. Estimates of the total number of insect species range from a few million to more than 20 million. We have just begun to inventory the vast array of organisms that inhabit this planet.

Since there are several classification schemes being proposed at this time, we will investigate one called the six kingdom system. Remember that the number of boxes we choose to make in a given rank is somewhat arbitrary. In the six kingdom system, the prokaryotes are divided into two kingdoms and the eukaryotes are divided into four kingdoms.

The four eukaryotic kingdoms are the Fungi, which contains the molds, mushrooms, and other fungi, the Plantae, which are the familiar plants, the Animalia, which are the familiar animals, and the Protista, which includes most of the one-celled eukaryotic organisms. Since the multicellular organisms evolved from one-celled organisms, there are protists that are similar to the multicellular organisms in the other three eukaryotic kingdoms.

There are two major groups in the prokaryote kingdom. They are the Bacteria, to which most of the familiar bacteria and the cyanobacteria belong, and the other group is the Archaea.

The Archaea are very common in the ocean and make up a significant fraction of the biomass of plankton. They are also found in extreme environments such as hot springs and highly acidic or very salty water (Figure 6.4). Their numbers dominate in the sea floor sediment. Even though they are abundant and surely very important, we only have a limited understanding of them.

Archaea differ from both Bacteria and eukaryotes genetically, which has

Figure 6.4 Grand Prismatic Spring This hot spring in Yellowstone National Park supports colonies of cyanobacteria and archaea. Brilliant colors are produced in times of heightened growth. [M.Ziinkova]

led some taxonomists to suggest that the three groups should be placed in separate superkingdoms or domains. The three domains would be the Bacteria, the Archaea, and the Eukarya. This classification system would have three domains and many kingdoms such as plants, animals, fungi, etc. Considering the uniqueness of the Archaea, it would at least make sense to expand the kingdoms to six by placing the Archaea in a separate kingdom (Figure 6.5). Remember, the groupings are arbitrary to some extent.

We would like to become familiar with some of the members of these kingdoms, but our discussion will be necessarily brief since an in depth treatment of all the kingdoms would be a rather colossal undertaking in itself.

Within the kingdoms, the number of phyla and divisions is somewhere around 90 and depends somewhat on the judgment of those doing the grouping. Therefore, even attempting to introduce the major phyla of the living world would exceed the bounds of this endeavor. We will, however, introduce some of the phyla that will be important for our later discussions.

Before we discuss the groups of organisms, however, we will briefly discuss viruses, a group of objects that do not fit into the classification scheme of living things.

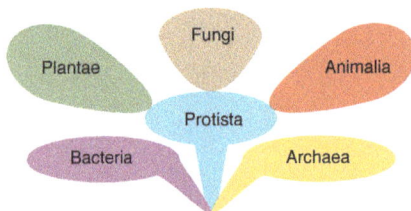

Figure 6.5 Six Kingdom Classification System The Protista, Plantae, Fungi, and Animalia belong to the Eukarya domain. [JCB]

Viruses

Viruses do not seem to fit into any of our classification boxes. Indeed, in many ways viruses appear to lie in the region between the living and the nonliving. All members of the six kingdoms are organisms composed of a cell or group of cells, but viruses do not have cells.

Although an object may possess certain properties that we recognize as characteristics of living organisms, the differentiation between the living and the nonliving is not always clear-cut. Living organisms can use certain chemicals in the environment to grow and sustain their organization. However, crystals are able to grow, and we do not consider them to be alive. All living things also can reproduce. They are able to make other organisms similar to themselves.

Viruses are well known as the agents of diseases such as AIDS, ebola, polio, smallpox, measles, mumps, rabies, influenza, serum and infectious hepatitis, herpes, certain types of leukemia, and most colds and flu. However, viruses do not fit into our classification system because they are not composed of cells and cannot reproduce on their own.

Viruses contain nucleic acid that is enclosed in a protein shell, and some even have a membrane covering. The shell is made of several proteins arranged in a repeating pattern that often produces a symmetrical structure (Figure 6.6).

Depending on the type of virus, the nucleic acid is either a single or a double-strand of DNA or RNA. Some viruses have strands that are circular, but others have linear strands. The amount of nucleic acid also depends on the type of virus, and it reflects the number of genes (protein recipes) present. This number varies from less

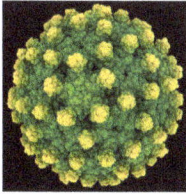

Figure 6.6 Hepatitis B Virus *This model is of a virus that produces liver disease and kills over a half million people each year. The disease is preventable with a vaccine. This virus is about 40 nanometers across. Typical bacteria are about 100 times larger. The coloring of the model is arbitrary. [Antonio Šiber/antoniosiber.org]*

than ten to perhaps a few hundred. Recall that typical bacteria, the simplest living organisms, may have a few thousand genes.

Although viruses contain the genetic information necessary for their reproduction, they do not possess ribosomes or the other machinery necessary for the production of proteins. Therefore, they are not able to replicate themselves without the aid of a host cell.

A virus is just an encased set of genes, without the means of reproducing itself. Viruses must invade a cell and take over the cell's protein replication mechanism to reproduce themselves. Using the genetic information contained in the virus, the host cell produces viral components, and new viruses are assembled from the components. Viruses can assemble themselves with no assistance from the host cell, and the pieces can even assemble themselves in a test tube.

The largest viruses are the pox viruses such as the one that causes smallpox. They are about the size of the smallest bacteria and are the most complex of the viruses. Among other things, they possess a double outer membrane and a double strand of linear DNA. To reproduce, most DNA viruses must invade the nucleus of the host cell

where the DNA is transcribed into messenger RNA by the machinery of the cell. However, the pox viruses can transcribe their own DNA into messenger RNA, and this process is carried out in the host cell's cytoplasm (the fluid surrounding the nucleus).

In a retrovirus (like AIDS), the genetic information is stored as a messenger RNA, and the reproduction of the virus is the reverse of the process for a DNA virus, hence the name *retro*. The retrovirus mRNA is used to make DNA, which is then inserted into the host DNA. The viral DNA becomes a part of the host's own DNA. This DNA is then used to produce the new viral components. This process is the reverse of the way proteins are usually made.

When exposed to a virus, our body begins duplicating the parts of the virus. In most cases our immune system recognizes the virus as a foreign substances and begins creating specific antibodies that attack and destroy the virus. However, viruses evolve very quickly and a single mutation can change one of the proteins in its shell, producing a new strain. If you are infected with this new mutated virus, the antibodies that your immune system created may not recognize the new virus. Therefore, your immune system must start from scratch and create new antibodies to fight the new virus. This is why we get the flu so many times. It is not the same flu we caught before, but a newly mutated flu virus.

Viruses are usually classified as nonliving objects, because without the help of the machinery of the host cell, they are not able to reproduce themselves. Outside the host cell, they are incapable of carrying out any of the functions that we generally associate with living organisms. However, whether viruses are living or not depends ultimately on just what our definition of life happens to be.

Viruses mutate and evolve very quickly and in the process they produce new genes. When these genes are transferred to the host organism they may be permanently incorporated into the genome of the organism. This process could produce new genes that are useful to the host organism. Although we do not know much about the frequency of this evolutionary process, it could conceivably be very important.

Viruses can also transfer genetic from one host organism to another, a process called *transduction*. Bacteria can also transfer genes in this fashion and organisms that are not closely related can occasionally end up with similar genes because of this process.

How did viruses originate? There are several ideas, one of which suggests that viruses represent the remnants of an ancient form of life that preceded the cellular organisms. If this idea is correct, the viruses of today have survived by adapting to a parasitic way of life. Although we are more familiar with the viruses that cause diseases in humans, viruses can infect most living organisms.

Viruses that infect different species of bacteria are called *bacteriophages*. They are the most abundant viruses, and coincidentally they infect the most ancient forms of cellular life.

As we look at other organisms in the living world, it is interesting how many of the very early life-forms have survived by adapting to a parasitic way of life. Disease causing bacteria are just such examples. As a parasitic organism evolves, many of its body parts become modified or are even lost.

Tapeworms do not possess a mouth or a digestive tract, but absorb food directly through the surface of their body. Their ancestors must have had these structures just as their close relatives do today, but after many generations, their mouth and digestive tract were lost as an adaptation to their parasitic way of life. Since they live in an environment containing food that has already been digested, tapeworms do not need a mouth or digestive tract.

Often, structures that are not needed tend to be greatly reduced or even lost in future generations. Other examples of this process are the loss of sight by cave-dwelling creatures, and the reduction in size of vestigial organs such as the human appendix.

Tapeworms are so modified that today they can only survive in the nutrient-rich environment of certain vertebrate intestines. Likewise, viruses may be the modified ancestors of an earlier type of life that was much simpler than the cellular life forms that dominate the world today. Just as tapeworms no longer have a digestive tract, viruses may have lost the ability to reproduce by themselves because it was not essential for maintaining their parasitic way of life.

A competing idea about the origin of viruses proposes that they were originally cellular forms of life that were severely modified after they adapted to the parasitic way of life. Essentially, a virus is thought of as a cell stripped of everything except the bare essentials, the necessary genetic information and a protective coat of protein. Again, the severe loss of structures is explained as the result of adapting to a parasitic way of life.

A third idea about virus origins suggests that they may have evolved from fragments of genetic material from the host cell. We do not have enough information at this time to choose between these ideas. As more data is collected, someone may formulate a new idea that provides a better explanation.

In the mean time, we find viruses of evolutionary interest because they seem to partially bridge the rather hazy gap between the living and the nonliving. Their rapid evolution is also an excellent example of how mutations can change the genetic make up of a population.

The Prokaryotes

If viruses are not the remnants of an ancient form of noncellular life, then the prokaryotes are the most ancient forms of life about which we have any direct knowledge. As we will see in the next chapter, fossil evidence indicates that members of this group were flourishing in the oceans at least 3,500 million years ago (3,500 mya).

Prokaryotes are the bacteria, the cyanobacteria, and the Archaea. The prokaryote cell is generally much smaller and simpler than the eukaryote cell. Perhaps the most obvious structural difference is the lack of a membrane-bounded nucleus in the prokaryote cell. In eukaryotes, this membrane separates the nuclear material from the cytoplasm, the fluid that makes up the rest of the cell.

In the eukaryote nucleus, the DNA is wound around protein molecules called histones and organized into several separate strands called *chromosomes*. The chromosomes become very distinct rod-like structures when the cell divides. In the prokaryote cell, however, the DNA consists of a single loop that forms an indistinct mass near the center of the cell.

Besides the nuclear differences, the cytoplasm of a eukaryote cell contains certain structures or organelles that the prokaryote cell lacks. *Mitochondria* are organelles found in all eukaryote cells and *chloroplasts* are present in most plant cells. Many eukaryote cells have

flexible whip-like structures called flagella, cilia, or undulipodia. These undulipodia are complex structures composed of eleven filaments called microtubules. Some prokaryotes also have whip-like structures called flagella, but their flagella are much simpler structures.

All prokaryote and eukaryote cells are bounded by an outer membrane, which serves as a container for the cytoplasm and regulates the passage of certain chemicals. This membrane is composed of a double layer of lipid (fat) molecules in which various protein molecules are imbedded. Many of these proteins function as one-way gates that allow certain useful chemicals to pass into the cell and expel those chemicals that are not needed.

Along with this membrane, many cells have a more rigid outer covering called a cell wall. Animal cells and a few bacteria lack cell walls, but other organisms have cells that are bounded by various types of cell walls. As we stated earlier, plant cells have walls composed of cellulose. This wrapping of cellulose provides the rigidity that is required to support a plant. Animal cells, lacking this type of support, are very flexible. The cells of vertebrate animals are attached to bones, which supply the needed support for the animal. Insects have an external skeleton of chitin that supplies their support.

Most bacteria have a cell wall that maintains the characteristic shape of the organism, but it is not made of cellulose. Different types of bacteria display a variety of cell walls, and the chemical differences in the walls can be used for classification purposes. Certain bacteria, called Gram-positive bacteria, have cell walls that will absorb a stain (called Gram stain) and take on a deep purple color. Other

154

bacteria have a cell wall that does not absorb Gram stain well, and these organisms take on a light pink color when stained. The prokaryotes that have the most similar body chemistry are the most closely related.

Prokaryotes do not show a great deal of variation in shape. There three basic shapes which are the sphere or oval (called a coccus bacterium), a rod-shape (a bacillus bacterium), and a helical or spiral-shape (a spirillum bacterium). Shape is not generally a very good indicator of how closely species are related although several spiral-shaped bacteria comprise a natural grouping of closely related forms known as the spirochetes. (Figure 6.7)

Bacteria reproduce asexually by a method called *binary fission*. During fission, the DNA is replicated, and the

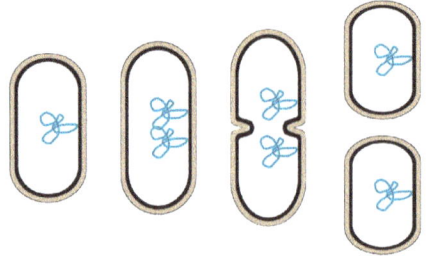

Figure 6.8 Binary Fission The DNA is duplicated and the circular strands are attached to the cell membrane. The cell grows and the membrane pinches the cell in two. In some instances, the cells do not separate and a chain of cells is formed. [JCB]

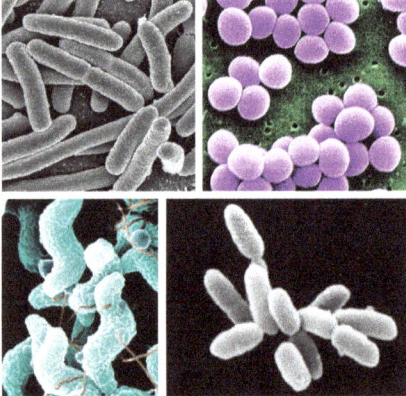

Figure 6.7 Prokaryotes (clockwise from top left) Scanning electron microscope images of Escherichia coli, a rod-shaped Gram negative bacterium found in the intestines of many animals. [Rocky Mountain Labs/ NIAID/NIH]; Staphylococcus aureus, a Gram positive bacterium that causes "staph" infections. [CDC/Janice Haney Carr]; Halobacteria, an Archaea that lives in places with a high salt concentration like the Dead Sea and the Great Salt Lake.; Campylobacter jejuni, a spiral-shaped bacterium that causes food born illness. [USDA/ARS/EMU/DeWood, Chris Pooley]

two circular strands of DNA are attached to different places on the cell membrane. As the membrane grows, the two clusters of DNA separate, and the cell membrane constricts near the middle of the cell, pinching the cell in two (Figure 6.8).

Transfer of genetic material has been observed in some bacteria. This sexual process involves the growth of a tube from one bacterium to another and the one-way transfer of genetic material. Only one bacterium receives DNA, and usually only a portion of the DNA is transferred. The new genetic material replaces parts of the old DNA, and the leftover DNA is then dissolved, leaving the organism with the same amount of genetic material as it started with before the sexual process began.

Undoubtedly, most people identify bacteria with various types of illnesses. Indeed, the list of diseases is quite impressive. Typhoid fever, diphtheria, cholera, leprosy, gangrene, syphilis, gonorrhea, tuberculosis, tetanus, botulism, bubonic plague, and Rocky Mountain spotted fever are just a few of the more well known diseases for which bacteria are responsible.

On the other hand, without the important products manufactured by bacteria, higher forms of life could not exist. Certain bacteria are the only organisms capable of converting gaseous nitrogen into compounds that can be used by other organisms. These nitrogen fixing bacteria are found in the nodules on the roots of legumes such as peas.

Organic matter would soon engulf the Earth if it were not for the bacteria. They play an integral part in the decay process by breaking organic matter into simpler substances that can then be used by other organisms. Many bacteria produce useful antibiotics such as streptomycin, erythromycin, and terramycin.

Cyanobacteria were so named because of their cyan or blue-green color. They are able to convert sunlight into usable energy by the process of photosynthesis, and oxygen is given off as a waste product. Over the past several billion years, these organisms have formed the oxygen atmosphere of the Earth (Figure 4.28).

Not only is oxygen used by higher organisms for respiration, it makes an even more important contribution to the living conditions on Earth by forming a layer of ozone (a molecule composed of three oxygen atoms). The ozone layer is located 20 to 30 km above the Earth's surface, and it shields the Earth from the deadly ultraviolet light rays emitted by the Sun. Before the formation of the ozone layer, life on the land was not possible. The earliest life forms lived in the oceans where the water partially protected them from the Sun's ultraviolet light.

Today the ozone layer has been damaged by the release of certain chlorofluorocarbons (molecules made of chlorine, fluorine, and carbon). By a series of chemical reactions, these chlorofluorocarbons remove the extra oxygen atoms from ozone molecules, changing them into ordinary molecular oxygen.

Considering the conditions under which they can live and the chemical reactions that they are able to perform, the prokaryotes appear to be a much more diverse group than the eukaryotes. Many live under conditions similar to those conditions that must have prevailed during the early stages of the Earth's formation. As evidenced by their fossil record, the prokaryotes are an extremely old group of organisms. Their great chemical variability is also evidence of a long evolutionary history.

The first prokaryotes evolved before the atmosphere contained any significant amount of oxygen. After the evolution of the cyanobacteria, the oxygen content began to rise, and some organisms evolved mechanisms to protect themselves from this very reactive gas. Oxygen must have been poisonous to most early forms of life. Bacteria that are able to live in the presence of molecular oxygen are called *aerobic* bacteria.

Many bacteria were never able to evolve protective devices against the highly reactive nature of oxygen, and the ones that survive today exist only in those places that are void of oxygen. These organisms are called *anaerobic* (meaning *without air*) bacteria. Still other bacteria live best under conditions of reduced oxygen, and others can live under both aerobic and anaerobic conditions. The existence of anaerobic bacteria reinforces the idea that bacteria were common before the oxygen atmosphere was formed. These anaerobic bacteria managed to survive only in limited habitats where oxygen is not found.

The Archaea, contains members so unique that they are usually placed in a separate kingdom. Chemically, they are as distinct from other prokaryotes, which are simply called the Bacteria, as the Bacteria are from the eukaryotes. In particular, the cell membrane is composed of a different type of lipid, the cell wall has a different composition, and the RNA in the ribosomes has a much different ordering of the four bases.

Since ribosomes are such important structures for the production of proteins, it seems unlikely that ribosomes could have changed very rapidly. The significant differences in their ribosomal RNA are interpreted to mean that the Bacteria, the Archaea, and the eukaryotes diverged at a very early time. The observed differences are the result of the slow accumulation of random changes in the RNA base sequences.

The largest group of Archaea is the methane producers (the methanogens). These organisms cannot live in the presence of molecular oxygen (they are anaerobic), and they produce methane [CH_4] as a byproduct of their energy production. They are found in stagnant lakes and in the stomachs and intestines of many animals where they produce methane gas. Since they are anaerobic, they probably were more abundant before the evolution of the oxygen producers. Methane is a green house gas, and some scientists have suggested that the methane producers may have kept the early Earth from freezing.

Geologic evidence suggests that the oxygen producers replaced the methane producers about 2,300 mya. Glacial deposits from this time contain uraninite and pyrite, which are only produced under conditions of very low oxygen levels. However, above these deposits is a layer of red sandstone containing hematite, a mineral only formed under oxygen rich conditions. This evidence suggests that the oxygen producers became more abundant at this time, causing a decrease in the number of methane producers.

Along with the anaerobic methane-producers, there are other Archaea that live in very strange environments. Some can live in hot springs at temperatures up to 80 degrees Celsius, others live in hot acid springs where they convert sulfur into sulfuric acid, and still others are only found in extremely salty lakes such as the Great Salt Lake and the Dead Sea.

Salt loving Archaea known as halobacteria are able to use light energy to form ATP (the molecule that stores energy for the cell). These organisms do not use the familiar pigment chlorophyll for this purpose, but a protein that works in a manner somewhat similar to the visual pigment rhodopsin, that is found in the vertebrate eye. The mechanism for capturing and using the light energy is very simple in comparison to the mechanisms used by other bacteria or the mechanism used by green plants.

Generalized photosynthesis can be described by the following equation:

$$2H_2A + CO_2 \rightarrow CH_2O + 2A + H_2O$$

Where "A" can stand for an atom of sulfur, or an atom of oxygen, or it could be nothing. The type of photosynthesis most of us are familiar with is the type carried out by the green plants. In this particular process, "A" in the above equation stands for oxygen. Green plants combine carbon dioxide with water to form a simple sugar and free molecular oxygen. Several types of bacteria can also carry out photosynthesis, but the methods and pigments used are much more varied than the process that is used by all the green plants.

The purple sulfur bacteria use hydrogen sulfide (where "A" is sulfur in the above equation) to produce carbohydrates and sulfur. As can be seen by studying the above equation, these bacteria do not release oxygen in the process of photosynthesis. Many are anaerobic and can, therefore, not tolerate oxygen. These anaerobic forms with their simplified mechanism of photosynthesis are believed to be organisms that were more common in ancient times, before the cyanobacteria produced the atmospheric oxygen.

Chlorophyll is the familiar pigment associated with photosynthesis. However, there are several types of chlorophyll that differ slightly from one another. Although all have a high degree of similarity, the differences might be significant from an evolutionary point of view. Two types of bacteria known as the green bacteria and purple bacteria contain several forms of chlorophyll. These forms are only found in bacteria. Cyanobacteria use a different type called *chlorophyll a*, one of the two forms of chlorophyll found in green plants.

A photosynthetic bacterium known by the genus name of *Prochloron* is interesting because it contains chlorophyll a and chlorophyll b, the same two chlorophylls found in green plants. The *Prochloron* species that have been discovered are found living in association with marine organisms. No truly free-living forms have been discovered.

As we will see later, there is strong evidence to suggest that chloroplasts in green plants are the ancestors of a type of photosynthetic bacteria. It is believed that these bacteria invaded the cell of another organism, and they began living together in a mutually beneficial relationship. The genus name *Prochloron* reflects this type of relationship although current studies suggest that a cyanobacteria ancestor most likely became the chloroplast.

The Bacteria have been divided into about a dozen groups. Some of these are the Gram-positive bacteria, the spirochetes, the cyanobacteria, the green bacteria, and the purple bacteria. The purple bacteria include such well-known species as *Escherichia coli* (often just called *E. coli*) and *Salmonella typhi* which is responsible for typhoid fever.

There are many unusual types of bacteria. Chemoautotrophic bacteria are capable of living on inorganic matter alone (salts, air, and water). No other organisms are capable of this type of life. Some bacteria form multicellular structures, and still others mimic the fungi in appearance.

From an evolutionary standpoint, there are many things that make the prokaryotes a particularly interesting group of organisms. They are much smaller and simpler than the eukaryotic organisms. For example, the average protein of a prokaryote is about two-thirds the size of an average eukaryote protein (about 200 amino acids for prokaryotes compared to about 300 for eukaryotes). Bacteria possess many of the features that one would expect to find in organisms with a long evolutionary history that predates the formation of an oxygen atmosphere. Many are anaerobic, and many have survived by adapting to a parasitic way of life.

Bacteria exhibit much variation in the bacterial photosynthetic processes, and we find bacteria that are able to capture light energy by much simpler means than the mechanism used by green plants. Many bacteria can survive by using only inorganic matter, an ability that none of the higher organisms can claim. Since only bacteria can perform many vital chemical reactions, higher organisms are dependent upon the bacteria for their survival. Higher forms of life could not have existed before the appearance of the bacteria.

Most bacterial research has been done on those bacteria that produce diseases in humans or on those that are directly related to our well being. As a result of this emphasis, the study of forms that may be of more interest from an evolutionary point of view may have been missed. Undoubtedly as more work is done in this field, many new and interesting discoveries will be made which may further illuminate the important chemical innovations that first evolved in the bacteria.

Reproduction

Since the more abundant nuclear material in the eukaryotic cell is organized into several chromosomes, eukaryotic cell division (called *mitosis*) is more complicated than binary fission.

Although mitosis is a continuous process, it has been divided into several stages. During interphase, the chromosomes duplicate themselves, but remain connected to each other at a point called the *centromere*. These two copies of the same chromosome are called *sister chromatids* (Figure 6.9). After the duplication process, the chromosomes form into dense rods known as mitotic chromosomes (Figure 6.10). The sister chromatids are still connected at the centromere during this stage which is known as prophase. The condensed mitotic form of the chromosome is the one familiar to most people because it can be seen in an ordinary microscope.

While the chromosomes (still double) are condensing into the mitotic form, the nuclear membrane dissolves, and a system of microtubules or spindle fibers is set up. It may be remembered that each eukaryotic flagellum is constructed of several microtubules. During the next stage (called metaphase), the chromosomes line up near the middle of the cell, and they become attached to the spindle fibers at their centromeres. At this time, the connections at the centromeres are broken, and one chromosome from each previously connected pair moves to opposite sides of the cell. This stage is called anaphase.

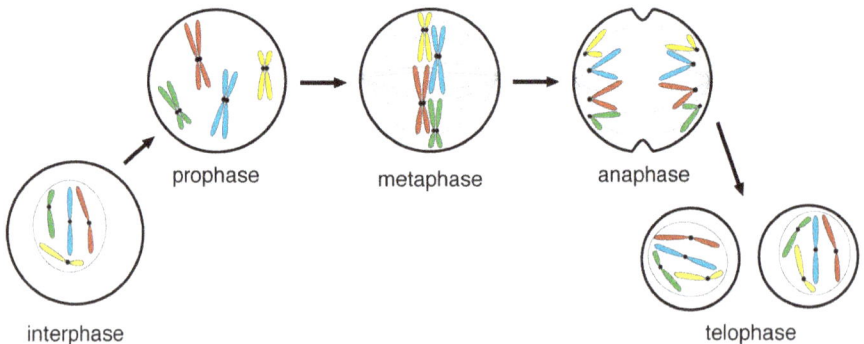

Figure 6.9 Mitosis Interphase is the stage where chromosomes are duplicated. The chromosomes are unwound and the long threads that are hard to see in the interphase and later telophase stages, but for clarity, they are shown as compact forms here. The sister chromatids remain connected at the centromere, and they line up along the central plane of the cell during metaphase. The duplicated pairs separate and are pulled to opposite sides of the cell during anaphase. The DNA unwinds and the cell divides during telophase. [JCB]

Figure 6.10 Mitotic Chromosome *An electron micrograph of a duplicated human chromosome during metaphase. The sister chromatids are connected at the centromere (constricted region). DNA strands are seen around the edges. Chromosomes are about the size of a typical bacterium (5 microns). [Hans Ris/Cell Image Library]*

Finally during the last stage (telophase), the chromosomes arrive at opposite ends of the cell and new nuclear membranes form around each of the two separate bundles of chromosomes. During this stage, the cell is divided into two separate cells. In cells that contain rigid cell walls such as plant cells, a new membrane and wall grow between the separated nuclei, but in animal cells that do not contain cell walls, the cell membrane pinches together near the middle of the cell until the two cells are separated.

Notice that the net result of mitosis is that two cells are formed, which usually contain identical genetic information. Generally we expect the nuclear material to be identical, but the duplication process is not always absolutely precise, and copy errors may occasionally arise. Many single-celled organisms reproduce by this method of mitotic division, and new growth cells are formed in multicellular organisms by this process.

Sexual reproduction involves the formation of a *zygote* from the union of two cells, called germ cells or *gametes*. Each parent supplies one gamete. In the higher animals, the gametes are the *sperm* and the *egg*, and the zygote is the fertilized egg. If two ordinary cells were combined, the offspring would have twice as many chromosomes as either of its parents. This doubling does not happen, because the cell division that forms germ cells is modified such that the germ cells end up with half the number of chromosomes present in ordinary cells. This modified form of mitosis is called *meiosis* (Figure 6.11).

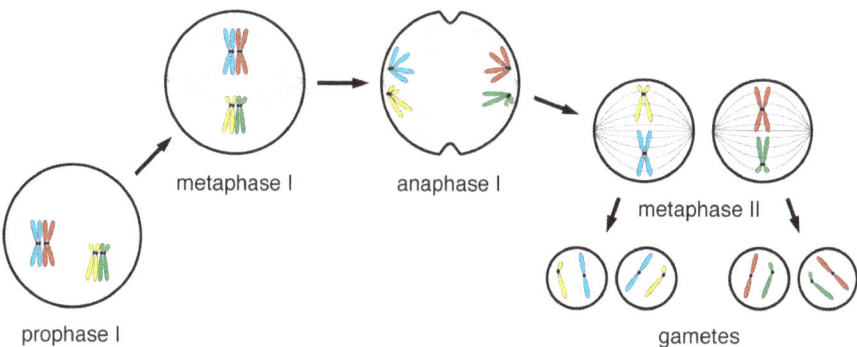

Figure 6.11 Meiosis *Meiosis is a modified form of mitosis followed by regular mitosis. After the chromosomes are duplicated they remain connected at the centromere. During prophase I, the homologous pairs come together and lie along side each other forming a tetrad. Crossing over, which exchanges some genetic material can occur at this time. During metaphase I, the tetrads line up along the central plane of the cell. The pairs split and go to opposite sides of the cell during anaphase I. This part of the process produces two cells, each with one duplicated member from each homologous pair. From telophase I (not shown here), each cell undergoes ordinary mitosis, producing four gametes. [JCB]*

Meiosis involves two slightly different types of division and results in the formation of four germ cells. (Germ cells have half the number of chromosomes as other body cells.) The second division is identical to mitosis. The first division looks much like mitosis, but with a few differences. Eukaryotes have an even number of chromosomes, which form pairs called *homologous* pairs (one member of the pair came from the mother and one came from the father). During prophase I, the chromosomes (which have already been duplicated) line up with their homologous partners. Notice there are 4 chromosomes since each member of the homologous pair was duplicated before the pair was joined. This configuration is called a triad.

Next, the pairs of duplicated chromosomes line up near the middle of the cell during metaphase I, but the duplicated chromosomes do not separate. Instead, one member of each homologous pair moves to opposite sides of the cell, and two cells are formed with half the number of chromosomes. At this stage, the duplicated chromosomes are still connected. The final division occurs just like mitosis, except that each cell now has only half the usual number of

chromosomes. Although this process produces four cells, there are only two different cells. This is because the last division is just like mitosis, which produces identical cells. Therefore, the net result of meiosis is that two copies of two different cells are formed. Each cell has half the normal number of chromosomes.

Cells with half the normal number of chromosomes are referred to as *haploid* cells, and cells with the full complement of chromosomes are called *diploid* cells. Haploid cells contain only one member of each homologous pair, while in diploid cells both members of the homologous pairs are present. Remember that the homologous pairs of chromosomes are not generally identical, because the corresponding genes (protein recipes) on different members of the pair are generally different.

A very important evolutionary process is often associated with meiosis. Pieces of the homologous pairs may be exchanged during prophase I, when the homologous pairs of chromosomes are joined. One common method of exchange is called crossing-over. The exchange of pieces of chromosomes is important because it allows the genes on different homologous pairs to be mixed (Figure 6.12). For example, one chromosome might contain the gene for blond hair and the gene for blue eyes while the other member of the homologous pair might have the genes for black hair and brown eyes. Crossing-over makes it possible to produce a new chromosome containing the gene for black hair and the blue eyes. The other chromosome would contain the gene for blond hair and the gene for brown eyes.

Notice that the process of meiosis is more complicated than one might expect. The same result could have been accomplished by just separating

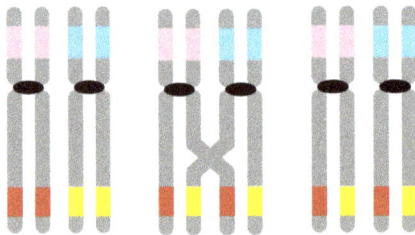

Figure 6.12 Crossing Over During prophase I, the duplicated homologous pairs line up and form tetrads (left). Crossing-over (middle) produces new chromosomes (right) with some genes exchanged between homologous pairs. [JCB]

the homologous pairs without also duplicating them first. If duplication did not occur first, the process would result in two germ cells instead of four. However, meiosis is interesting from an evolutionary point of view. The first division is just a slightly modified form of mitosis, and the second division is identical to mitosis. Rather than being a completely different process, meiosis is just a modification of mitosis.

Meiosis is an important process for several reasons. Along with creating haploid germ cells, the homologous pairs are separated at random. This random mixing of the homologous pairs means that many possible combinations can be assembled in the germ cells, resulting in a great deal of variation in the offspring. The more diverse the offspring, the greater is the chance that one of the offspring will be able to survive in a changing environment. The offspring that are able to survive and reproduce will pass their advantageous characteristics on to the next generation. These variations are believed to be the main advantage of sexual reproduction.

When beneficial mutations occur, these changes can be mixed in many more ways if the organisms reproduce sexually than if sex is not involved. For a species without sex, two mutations cannot both spread through a population unless the second mutation happened to occur in an organism that contained the first mutation. Therefore, there are usually just three possible combinations from two mutations. Individuals can have just the first mutation, other individuals can have the second mutation, and still other individuals can have neither mutation.

For a species that reproduces by sex, recombination can bring several beneficial mutations together. For example, two mutations will yield four possible homozygous combinations. (*Homozygous* means the individual has two copies of a particular gene.) Some individuals could have no mutations, some could have one mutation, some could have the other mutation, and some could have both mutations (Figure 6.13 top row). There are 6 more heterozygous combinations (second row). (*Heterozygous* means the individual has only one copy of a particular gene.) For three mutations there are 8 homozygous possibilities (rows 1 and 3 of Figure 6.13), and 22 additional heterozygous possibilities. Therefore, for three mutations, there is a grand total of 36 different possible combinations (all rows of Figure 6.13).

Figure 6.13 Mutations If an organism reproduces sexually and if crossing over occurs, two mutations (colored lower case letters), can produce 4 homozygous combinations (top row) and 6 heterozygous combinations (2nd row). With 3 mutations, there are 4 additional homozygous possibilities (3rd row) for a total of 8, and 22 additional heterozygous possibilities (last 3 rows) for a total of 28, giving a grand total of 36 combinations. [JCB]

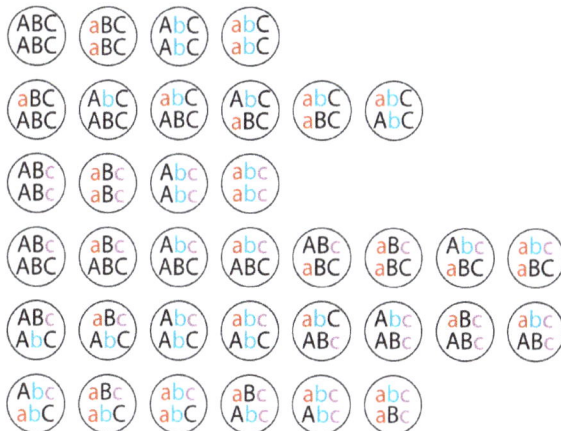

Another important advantage of sexual reproduction is that each individual has two copies of each gene. Therefore, if a harmful mutation occurs in one of the genes, the other copy can continue to produce useful proteins.

Endosymbiotic Origin of the Eukaryotes

As we have seen, a eukaryotic cell is quite different from a prokaryotic cell. The division between these two great superkingdoms is much greater than are the divisions separating the four eukaryotic kingdoms, (the protists, fungi, plants, and animals).

Since eukaryotes have organelles that appear similar to certain bacteria, it has been suggested that mitochondria and plastids (called chloroplasts in higher plants and green algae) were originally free-living bacteria that somehow became incorporated in host cells (Figure 6.14). Perhaps a large cell engulfed a photosynthetic bacterium similar to a cyanobacteria and instead of digesting the bacterium, the two

organisms began a life of cooperation together. The plastid would have supplied food for the host cell, while the host cell provided the necessary chemicals and a suitable environment for the plastid.

A similar origin for the mitochondria is envisioned. Mitochondria are the energy production centers of the cell (they produce ATP), and they are similar to certain purple bacteria. They are found in all eukaryotes, but plastids are of course found only in plants and algae.

Organisms that live together for the mutual benefit of both are said to have a *symbiotic* relationship. Therefore, the idea that plastids and mitochondria were once independent organisms that invaded a host cell is referred to as the endosymbiotic origin of the eukaryotes. Evidence for this origin of the eukaryotes is very strong.

What evidence is there to support the idea? First, both plastids and mitochondria are not manufactured by their host cells, but reproduce by binary fission as do bacteria. When a cell divides, each new cell receives some of the plastids and some

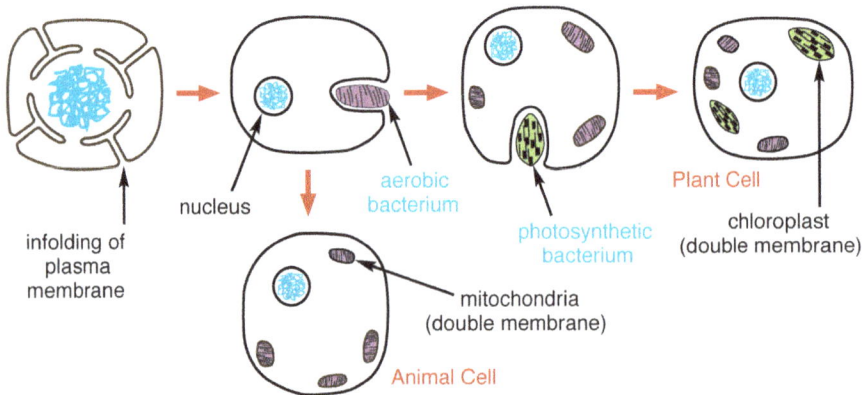

Figure 6.14 Endosymbiotic Origin of the Eukaryotes It is believed that the nuclear membrane formed by infolding of the plasma membrane around the DNA in some ancestral prokaryote (upper left). Next, an ancestral eukaryote engulfs an aerobic bacterium which evolves into the mitochondria. Next, a photosynthetic bacterium is engulfed and evolves into the chloroplasts. [JCB]

mitochondria. If the plastids or mitochondria are removed from a cell, the cell is not able to construct new organelles.

In addition, both plastids and mitochondria contain DNA, usually in circular form as is the DNA of bacteria. These organelles are able to produce some of the proteins that they need to perform their respective functions, but not all of them. They also produce their own machinery for the manufacture of proteins. That is, they produce their own ribosomes and transfer RNA molecules, all of which are quite different from those used by the rest of the cell.

Molecular studies show that the DNA and protein molecules of plastids are more similar to the corresponding molecules of certain photosynthetic bacteria than they are to the molecules made from the nuclear DNA of their own cells. For example, the ribosomal RNA molecules produced by the chloroplasts of the higher plants studied are more similar to the corresponding rRNA molecules of bacteria than they are to the corresponding rRNA molecules made by the nucleus of their own cell.

While this is strong evidence for an endosymbiotic origin, remember that higher plants did ultimately evolve from bacteria. It is possible, although not as likely, that the plastids somehow became organized in a membrane bounded structure just as the nucleus did, and they evolved separately. The plastid DNA could have evolved more slowly than the nuclear DNA, explaining the close similarity between the molecules of plastids and bacteria.

Based on the DNA evidence, most scientists believe that the plastids in green algae and higher plants evolved from a cyanobacteria-like ancestor. However, the plastids in red algae are significantly different from the plastids in higher plants. The plastids in red algae must have evolved from a different photosynthetic ancestor. At least two endosymbiotic events may be needed to explain the two different plastids.

The endosymbiotic origin of mitochondria is also based on DNA studies. The DNA evidence supports the idea that mitochondria evolved from some type of free-living purple bacteria (or proteobacteria) that was engulfed by a host cell. If the mitochondria are the result of a single symbiotic origin, their subsequent evolution has been very diverse. There is a great variety in the structure, composition, and even the genetic code used by mitochondria.

Protista

There is much jostling with the classification system of the one-celled organisms and the name protist may eventually be replaced with several other categories. We will use it for convenience.

The Protista kingdom includes the one-celled organisms and the very simple multicellular forms of life. Since multicellular organisms evolved from this group, we find some protists that are plant-like, some that are more animal-like, and still others that have properties more like the fungi.

A few members of this kingdom illustrate the difficulty that can be met when trying to differentiate between the plants and animals. We normally think of plants as organisms that possess chloroplasts and are able to use light energy for the production of food. However, a plant-animal protist whose genus name is *Euglena* has this ability, and yet it moves about and feeds in a manner we normally ascribe only to an animal. If the chloroplasts are removed from a *Euglena*, the organism cannot produce food photosynthetically, but is able to live as a food-ingesting organism. There is a separate species of *Euglena* that does not possess chloroplasts and feeds on organic material like a typical animal.

As happened with the bacteria, it appears that many one-celled organisms have survived by adopting a parasitic way of life. Some of the better-known human diseases that are caused by members of this kingdom are malaria, amebic dysentery, African sleeping sickness, elephantiasis, coccidiosis, and giardiasis. Many others are parasites in nonhuman organisms.

Some protists live in partnership with other organisms, an example of symbiosis. The protists that live in the guts of termites exemplify this way of life. These organisms help in the digestion of cellulose, which is one of the primary ingredients in the termite's diet. The cellulose-digesting protists found in termites are members of an important group of protists that possess flagella or undulipodia. The group is called the flagellates.

Some flagellates have chloroplasts and appear plant-like, but some do not contain chloroplasts and are more animal-like. Because they are similar to many of the cells in higher organisms, most scientists believe that the ancestors of the multicellular plants and animals were members of this group of protists.

Several groups of protists are of interest because of their presence in the fossil record (Figure 6.15). One of these groups, the foraminiferans, contains organisms that form a shell (usually called a *test*) made of calcium carbonate [$CaCO_3$]. These tests are often preserved in fossil sediments. Foraminiferans are animal-like organisms that feed on algae, other protists, and small animals. The remains of foraminiferans have been present in the fossil record since Early Cambrian times (approximately 540 mya),

The radiolarians are animal-like protists, named for their radial symmetry. They form intricate skeletons made of an opal type of silica. Fossils of radiolarians have been found in Middle Cambrian rocks.

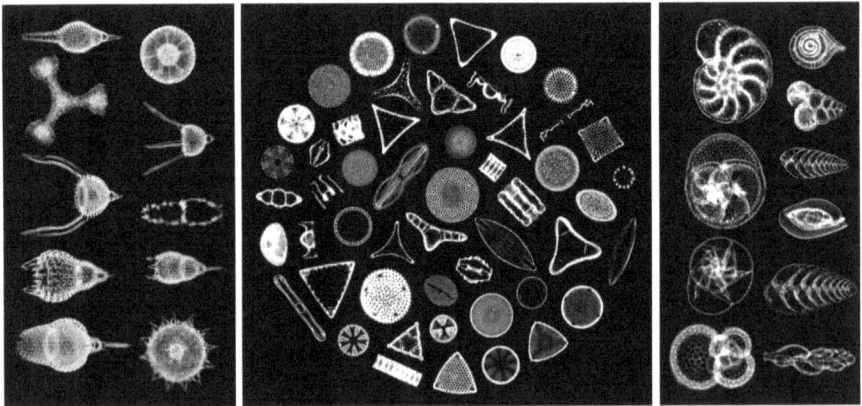

Figure 6.15 Radiolarians, Diatoms, and Foraminiferans Radiolarians (left) have silica tests; diatoms (middle) have silica coverings; foraminiferans (right) have calcium carbonate shells. [USGS/Randolph Femmer]

Diatoms, a group of plant-like protists, are photosynthetic organisms that have a silicon dioxide (silica) covering called a *frustule*. A frustule is composed of two overlapping halves that fit together like a pill box. The frustules of diatoms have been found in fossil sediments since the beginning of Cretaceous times (approximately 140 mya). These sediments often form thick deposits called diatomaceous earth.

Diatoms share many features with the brown algae, and both have been placed in the same group (called chromists). Photosynthetic members of the group are the only organisms that use chlorophyll c, a form that is different from the molecule that green plants use. Molecular studies suggest that the chromists diverged from the other eukaryotes fairly early and some taxonomists feel they should be placed in a separate kingdom.

Some protists superficially resemble certain fungi. These forms include a very strange group of organisms called the slime molds. Slime molds are divided into two groups, the cellular slime molds and the true slime molds (sometimes called the plasmodial slime molds).

The cellular slime molds are particularly interesting because of their unusual life cycle. These unicellular organisms live independently and reproduce by simple mitotic division for part of their life, but as the food supply runs low, they aggregate and form a slug-like mass of individual cells that can move about as a unit. Eventually the slug settles down, and the cells form into a fruiting structure. In some species, the cells that were once identical free-living organisms take on different forms during the slug and the fruiting stage. The fruiting structure produces small protected cells called spores that produce a new generation of free-living unicellular organisms.

Fungi

This kingdom includes yeast, molds, and mushrooms of various kinds. It is estimated that there are several million species, and they are currently grouped into seven phyla. Their classification is based mainly on the nature of their reproductive structures.

The most well known fungi belong to two phyla, Basidiomycota, called the club fungi, and the Ascomycota which are commonly called the sac fungi (Figure 6.16). The club fungi are

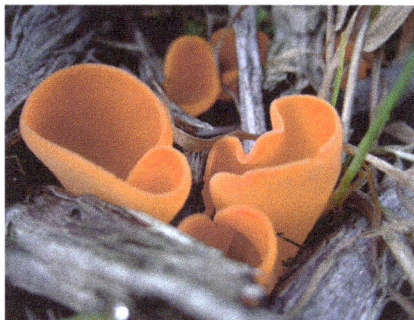

Figure 6.16 Colorful Fungi (top) Fly Agaric or Fly Amanita (Amanita muscaria) is a toxic Basidiomycota. [H.Zell]; Orange-peel Fungus (Aleuria aurantia) is an Ascomycota. [Aiwok]

named after the club-shaped cells or basidia that produce the spores. Common examples are the mushrooms, puffballs, and the plant diseases called rusts and smuts.

The sac fungi is currently the largest Fungi phylum with over 60,000 species. They are named after the asci that are produced around the rim of a cup-like reproductive structure. Each ascus produces spores. Widely known members of the phylum are the morels (Figure 6.1), truffles, yeasts, and the genus *Penicillium* (the fungus that produces penicillin).

One phylum, the Glomeromycota, is not well known, but its members are extremely important, and higher plants may not have evolved without them. They penetrate the roots of most land plants, and they form a symbiotic relationship with the plant. The plant supplies the fungus with food such as simple sugars, and the fungus helps the plant uptake water and vital nutrients like phosphorus, sulphur, and nitrogen compounds. Evidence of this symbiotic relationship has been found in fossils of the earliest plants.

An important group of fungi forms symbiotic relationships with different species of algae. These symbiotic organisms, called lichens, are often found on bare rock outcroppings (Figure 6.17). Lichens are very important because they secrete chemicals that help to breakdown rocks, the first step in the production of soil. About one-fifth of all species of fungi live in this symbiotic way, and almost half of these belong to the Ascomycota phylum.

Many people probably think fungi are plants, but there are several important differences between these two kingdoms. The most obvious difference is that fungi lack chlorophyll and, therefore, cannot manufacture food as plants do. Fungi obtain their food from decaying organic matter by excreting enzymes that breakdown complex organic molecules into smaller molecules that are then absorbed by the organism.

In fungi, adjacent cells often are not separated by a cell wall. Cells of this type are called *multinucleate*. Finally, most of the fungi have cell walls composed of chitin, which is a nitrogenous polysaccharide (a complex sugar with nitrogen atoms). The green plants have cell walls composed of the complex sugar cellulose. Cellulose is a long chain of glucose molecules.

There are many beneficial fungi such as the edible mushrooms, yeast, certain molds used in making cheese, and *Penicillium*. Fungi also play a vital role in decomposing organic matter so the byproducts can be used by other plants.

A few of the less respected members of this kingdom include the fungi that are responsible for athlete's foot and ringworm. Many fungi are also parasitic on plants, and some cause a great deal of damage. A few famous examples are the fungus that causes Dutch elm disease and the fungus that was responsible for the Irish potato blight.

Figure 6.17 Lichen *This unidentified lichen is growing on bare rock. [Ildar Sagdejev]*

Alternation of Generations

All life forms reproduce to create new organisms similar to themselves. In the case of bacteria, they merely undergo binary fission (Figure 6.8). The reproduction of many organisms involves some type of fertilization. As plants progress from one generation to the next, they go through a cycle known as alternation of generations.

To illustrate this cycle suppose we begin with a plant that we will call a *sporophyte*, a spore producing plant (Figure 6.18). (The word sporophyte is a combination of the word *sporos* meaning *seed* and *phyton* meaning *plant*.) Each cell of a sporophytes has a full complement chromosomes. At a certain stage in the life of this sporophyte, it produces cells by meiosis that contain only half the number of chromosomes as the normal cells of the sporophyte. We call these haploid cells *spores*. Since the plant is a spore producing plant, we call it a sporophyte.

These haploid spores are dispersed in various ways, and each one can grow into a new haploid plant, which we call a *gametophyte* (meaning gamete plant or gamete producing plant). This haploid plant may look like the sporophyte plant or it could look very different. At a certain stage in the life of the gametophyte, the plant produces special cells not by meiosis, but by ordinary mitosis. Therefore, the cells are haploid cells with the same number of chromosomes as the haploid gametophyte plant. We call these haploid cells *gametes*. So the gametophyte is a gamete producing plant. Notice that gametes contain half the number of chromosomes as the sporophyte plant did.

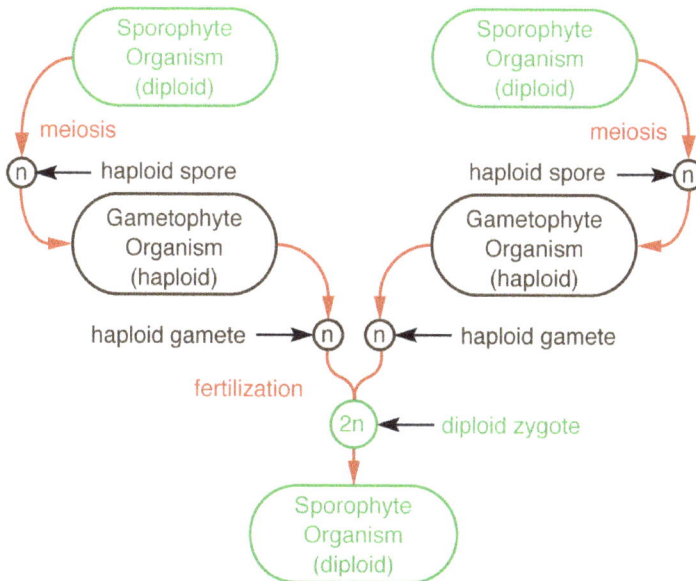

Figure 6.18 Alternation of Generations (Diploid structures are shown as green and haploid structures as brown) Diploid sporophyte plants (top) produce haploid spores by meiosis. Spores grow into haploid gametophyte plants which produce haploid gametes by mitosis. Two gametes (generally from different gametophytes), combine to form a diploid zygote. The zygote grows into a diploid sporophyte plant and the cycle begins again. [JCB]

Spores and gametes are both haploid cells. The only difference is that sporophytes produce spores by meiosis while gametophytes produce gametes by ordinary mitosis.

The haploid gametes are dispersed, and a gamete can unite with another gamete, even a gamete from a different gametophyte. When the two haploid gametes unite, the combined cell, which we call a *zygote*, has twice the number of chromosomes (called diploid) as a gametophyte plant. When this diploid zygote grows into a plant, the plant will be a diploid sporophyte plant and the cycle begins again. In plants this cycle is called alternation of generations.

The alternation of generations is the basic plan of reproduction by fertilization in the plant world, but it has been modified in various ways by different groups of plants. In many plants and algae we find macroscopic sporophytes as well as gametophytes. In many other plants, we find that the gametophyte generation comes in two forms which produce either male or female gametes. In some cases one type of gamete is not dispersed until after fertilization, in which case we identify it as the female. The more mobile gamete is called the male.

One of the most basic versions of the alternation of generations is illustrated by the green algae *Ulva* (commonly called sea lettuce). In *Ulva*, the sporophyte plant looks just like the gametophyte plant. Some of these gametophytes produce female gametes and others produce male gametes. The female gametes are somewhat larger. In higher land plants, the sporophyte looks quite different from the gametophyte plant.

The word spore is also used to describe a structure produced by some bacteria, but they are not true spores like the plant spores. The bacterial spore is not produced by meiosis, and they do not grow into reproductive structures. These bacterial spores are a dormant stage from which a bacterium can regenerate itself. They are more accurately called endospores (*endo* means *within*), and they are produced by certain bacteria as a means of surviving during periods of adverse environmental stress. When conditions become unfavorable, a bacterium can enclose itself in a multilayered structure composed of proteins, calcium ions, and various other chemical compounds. Special proteins bind to the DNA and protect it from ultraviolet light. The endospore remains dormant until conditions become favorable, and then the bacterium regenerates itself. Bacterial endospores can be very resistant, and it generally requires high temperatures and high pressures to destroy them.

Plants

Familiar examples belonging to the plant kingdom include the trees, shrubs, flowers, ferns, mosses, and some types of algae. Plants can be generally divided into two groupings. The simplest plants are the different types of algae. The entire algae plant, called a *thallus*, is made of similar tissue so these plants do not have well defined roots, stems, or leaves. Although they are separated into several phyla (or divisions), these simple plants are known collectively as *thallophytes* (after the word thallus).

Thallophytes are usually divided into three phyla, the chlorophytes (green algae), the chromists (brown algae), and the rhodophytes (red algae). The chlorophytes are usually grouped with the land plants, while as we have seen, the chromists are often elevated to the rank of kingdom.

Species of red and brown algae are mostly confined to marine (saltwater) environments, where we commonly call them seaweeds. Giant Kelp is a type of brown algae that often forms dense masses along the western coastline of North and South America, Australia, and South Africa. Species of brown algae belonging to the genus *Sargassum* form the dense mats that cover the Sargasso Sea. Red algae are generally less well known as they tend to be smaller and grow at greater depths than other types of algae.

Many chlorophytes are found in fresh water where they often form green filamentary masses in ponds and streams. Common members of this phylum are species belonging to the genus *Spirogyra*.

From an evolutionary point of view, the green algae are of interest because they possesses chlorophyll a and b, which are the types of chlorophyll found in the land plants. Because of these and other similarities, we believe that the land plants evolved from green algae ancestors. The closest living relatives of the land plants are in the family Characeae (Figure 6.19).

On the other hand, brown algae contain both chlorophyll a and c (chlorophyll c is not found in land plants). Red algae usually contain chlorophyll a along with chlorophyll d (chlorophyll d is not found in any other group of plants).

Thallophytes generally produce simple reproductive cells called gametes that unite to form a simple fertilized egg called a zygote. The more advanced plants are collectively known as *embryophytes* because a baby plant or embryo develops in the ovary of the female plant. Embryos can easily be seen in the seeds of certain plants such as beans. Two groups of embryophytes are the bryophytes (whose best known members are the mosses) and the

Figure 6.19 Stonewort *This freshwater green algae belongs to the family Characeae and the genus Chara. It is a close relative of the green land plants. The thallus is a gametophyte which has structures that look like leaves and stems. It is relatively complex for an algae.* [Christian Fischer]

tracheophytes or vascular plants. Both are commonly called land plants.

Bryophytes are the group of embryophytes that do not have true vascular tissues. There are three groups: the hornworts, liverworts, and mosses (Figure 6.20a). Since they do not have a vascular system, they have a difficult time transferring water and nutrients to different parts of the plant. Also, they do not contain lignin which is a complex chemical compound that strengthens the cell walls of vascular plants. Consequently, all bryophytes are fairly small, seldom growing taller than about 20 centimeters.

The visible liverwort plant is the gametophyte plant. Some liverworts can reproduce asexually by producing structures called gemma cups. These cups contain gemmae, which are cell masses that are dispersed by rain drops and then grow into new plants.

They also produce sexually by growing umbrella shaped structures that produce either eggs or sperms. When an egg is fertilized, it grows into a small sporophyte plant inside the umbrella shaped structure (called an archegonial head) that produces spores.

Figure 6.20a Non-Seed Bearing Plants (from left to right) Field Hornwort (Anthoceros agrestis) [BerndH]; Common or Umbrella Liverwort (Marchantia polymorpha) with gemma cups (lower left corner) and umbrella shaped archegonial heads (center left and right) [Ian Steel]; green gametophyte moss plants with red sporophyte plants [Vaelta]

The spores are dispersed and grow into new gametophyte plants.

Vascular plants, also known as tracheophytes, are named after their tube-like structures (called xylem and phloem) that transport water and nutrients to different parts of the plant. The xylem tubes transport water and dissolved minerals from the roots to other parts of the plant while the phloem tubes transport carbohydrates and amino acids to the various parts of the plant. These tubes are most easily seen as the veins in the leaves of broadleaf plants. Along with the transport of water, xylem tissue (also called wood in many plants) is used to support the plant, an important feature if a plant is to support itself on land. In their aquatic environment, the algae do not need this type of support structure.

The vascular plants are often separated into several phyla or divisions, but the whole classification process is still in a state of flux. Exactly how to separate a group of organisms into classes or phyla is sometimes a matter of opinion. The whole classification system will become clearer with time as more DNA evidence accumulates. We will briefly discuss three of the phyla.

Club mosses (Figure 6.20b) and quillworts are vascular plants that belong to a phylum (or division) commonly called the lycophytes or lycopods. They are the most primitive of the living vascular plants and have appendages or *leaves* that are scalelike or needle-like. Each leaf is supplied by a single vein that has no branching. All the other vascular plants have leaves that are supplied by a network of veins. Although club mosses are not well known today, they were one of the dominant land plants of the late Devonian and Carboniferous times (about 380 to 300 mya). These were the times when many of the great coal deposits were being formed, and lycopods were one of the main coal producing plants.

Another phylum or division is the ferns and fern allies (Pteridophyta). Three of the classes in this phylum are the whisk ferns (Psilotopsida), the horsetails or scouring rushes

Figure 6.20b Non-Seed Bearing Plants (from left to right) Stag's-horn Club Moss (Lycopodium clavatum) with terminal strobili which are cone-like spore-producing structures. [Jason Hollinger]; whisk fern (Psilotum nudum) with cream colored spore-producing structures. [Eric Guinther]; Equisetum telmateia, which is commonly called horsetail because of its shape. [Rror]

(Sphenopsida), and most of the common ferns (Polypodiopsida).

Psilotopsids are interesting from an evolutionary point of view because the oldest fossils of vascular plants look very similar to the whisk ferns (Figure 6.20b). These fossils date back about 410 million years (the late Silurian period). Whisk ferns have no true roots or leaves. They resemble certain types of green algae in many respects (another reason we believe land plants evolved from green algae). They are the simplest (most primitive) of the vascular plants.

Horsetails (Figure 20b) were generally much larger and more common during the Carboniferous coal-forming times than they are now. There are only about 15 different species of horsetails living today, all belonging to the genus *Equisetum*.

The most advanced group of plants is the seed plant group. This group is divided into several phyla (or divisions), although the classification scheme is still not settled. Some people prefer to put them in a superdivision

called spermatophytes and divide it into several phyla or divisions.

The most familiar phylum of spermatophytes is the group of flowering plants or angiosperms. They are distinguished by the fact that they produce seeds inside an ovary. As the seeds develop, the ovary grows into the fruit. The part of an apple that we eat is the ripened ovary. Technically, the fleshy part of tomatoes, squash, beans (the pod), egg plant, and cucumbers are fruits (or mature ovaries) that cover the seeds of these plants.

With the evolution of the flowering plants during the early Cretaceous about 140 mya, the world was quickly transformed. Members of this plant group now dominate the landscape with about 370,000 species divided into 300 to 400 families. The only larger group of organisms is the insect class that probably contains well over a million species.

The angiosperms are divided into about 8 smaller groupings, three of which contain most of the familiar flowering plants. These 8 groupings could be classes. The classification of the flowering plants is an ongoing project and much work is still needed to sort out the relationships of the various groupings.

DNA evidence tells us that the oldest branch of the flowering plant family tree is occupied by a small evergreen plant (*Amborella trichopoda*) found only on the island of New Caledonia (Figure 6.21). New Caledonia was originally a part of Gondwana, but it separated from Australia about 65 million years ago. Today, it is located about 1200 km east of Australia.

Amborella is classified as a basil angiosperm, and it is the sole survivor of an ancient plant lineage. Some plants are male and others are female, a condition known as unisexual. The petals and sepals of the female flowers are indistinguishable. (Sepals are the leaf-like part of a flower that covers the flower bud.) The flowers develop into small berries, each containing one seed.

One of the three largest groups of angiosperms is the magnoliids, which has about 9,000 species. Some of the familiar trees in this group are the magnolias, bay laurel, sassafras, cinnamon, and avocado (Figure 6.21).

Another group of flowering plants is the monocots, named for the fact that the seedlings have one seed-leaf (called a cotyledon). In addition, most monocots have parallel leaf veins, and the vascular bundles in their stems are scattered everywhere. Their flower parts come in threes (like three pedals and three sepals). There are about 60,000 species, and common examples are the grasses (including many important grains like wheat, rice and corn), palms, bananas, onions, lilies, irises, tulips, and the orchids, which is the largest family of monocots with about 25,000 species (Figure 6.21).

By far the largest group of angiosperms is the eudicots, which contains about 175,000 species. The word dicot means *two cotyledons* and refers to the fact that they have two embryonic leaves. If you observe a bean sprouting, you will see these two oval leaves come up first in the young seedling. The prefix *eu-* means true, and is used to differentiate this group from the magnoliids. The magnoliids have pollen grains with one pore out of which the pollen tube can grow, and the eudicots have pollen grains with three or more pores out of which the pollen tube can grow. The three pores are in three groves called colpi and the pollen is called tricolpate pollen. Some botanists prefer to call this group the tricolpates instead of the eudicots.

Most eudicots have broad leaves with branching veins, and the vascular bundles in their stems are arranged in concentric circles. Common examples are shade trees like the oaks, maples, ashes, elms, and aspens, and many flowers like roses and the numerous members of the aster family, the largest eudicot family with over 23,000 species (Figure 6.21). Many food plants are eudicots such as the various kinds of beans, the cabbages, potatoes, squash, and tomatoes.

The rest of the seed plants are generally called the gymnosperms, or naked seed plants which refers to the fact that their seeds are not enclosed in an ovary. The most well known phylum of gymnosperms is Pinophyta, commonly called the conifers. Examples of these are the pines, cedars, redwoods, cypresses, spruces, and junipers. Since many conifers stay green all year, they are commonly called evergreens (Figure 6.21).

The smallest gymnosperm phylum is Ginkgophyta, which includes the single living species *Ginkgo biloba*, a unisexual tree, so a given tree is either male or female (Figure 6.21). It is considered to be a living fossil. The

Figure 6.21 Seed Plants (clockwise from top left) Amborella, a basil angiosperm [Scott Zona]; Ginkgo in autumn foliage, a gymnosperm [JCB]; Common Cattail, a monocot [H.Zell]; American Tuliptree, a magnoliid [JCB]; Canyon Sunflower, a eudicot in the aster family [JCB]; Sequoia (foliage at lower left), a gymnosperm [JCB]

fossil history of the group goes back to the early Jurassic, but after a rise in diversity, they declined during the Cretaceous. It is possible that they are native to eastern China since they grow wild in the Tian Mu Shan Reserve. Recent studies, however, suggest that these trees may have been planted since their DNA is very uniform. Ginkgo trees have been cultivated in

China for over a thousand years, a practice that may have saved them from extinction.

Another phylum of gymnosperms is Cycadophyta (the cycads), which resemble palm trees although they are not closely related (Figure 6.22). Cycad plants are either male or female. The female plants produce seed cones

Figure 6.22 Cycad *Although they look like palm trees, they are not closely related. They were very common trees during the Mesozoic. [Tato Grasso]*

and the male plants produce pollen making cones.

The first cycads appeared in the fossil record in the late Carboniferous or early Permian period about 300 mya. They diversified and became more abundant with time, and by the Triassic and Jurassic they were one of the more abundant plants. From their peak during the Jurassic they have declined, especially after the evolution of the flowering plants. There are about 300 described species of cycads today, but experts think there may be 100 or so that are yet to be discovered.

The earliest gymnosperm fossils are of a group of plants called seed ferns (phylum Pteridospermophyta). They have been found from the late Devonian period (about 350 mya), and they look very much like ferns. The fossil and DNA evidence supports the idea that the gymnosperms and ferns evolved from the same ancestral group.

One of the main differences between ferns and gymnosperms is that ferns produce spores instead of seeds. Both seeds and spores are reproductive structures, but a seed contains its own supply of food while a spore does not. From an evolutionary point of view, seed ferns are a very interesting division of gymnosperms. Seed ferns have leaves that look like ferns but produce seeds instead of spores. Unfortunately, all the members of this group of plants are now extinct, and we know of them only as fossils.

Embryology

The purpose of this book is certainly not to investigate even a small portion of the errors made in the pursuit of scientific knowledge. If it were, this volume would have to be greatly expanded, for no field of study could have enjoyed the success of science without also enduring the humility of making a few wrong turns. However, we expect critical errors will be exposed because of the self-correcting nature of the scientific process. Theoretical predictions are not verified by just one investigator, but are checked and checked again by many independent scientists. In this way, a mistake made by one scientist is corrected by another.

It is important to understand how science differs from a court of law. To win a court case, one must only convince 12 people that you are right at some instant in time, but in science the trial is never over, and the experiments speak for themselves. Scientific debates are not held on a stage, but in the laboratory. If someone makes a claim, they do not defend their view with eloquent speeches, but with carefully performed experiments that can be checked by even their most ardent critics.

A famous misconception about embryology is credited to Ernst Haeckel. He believed that as an embryo developed, it went through the adult stages of its more primitive ancestors. Thus he believed that a mammal embryo went through a stage where it resembled its fish ancestors. The catch phrase that usually accompanies this idea is that *ontogeny recapitulates phylogeny*. Put another way, it was believed that as the embryo went through its different stages of development, it was recreating its own ancestral family tree.

As scientists began to more fully understand genetics and the processes by which organisms evolve, it was realized that the various stages of embryological development are subjected to the same principles of change and selection as other aspects of the organism. It is certainly to be expected that the DNA recipes that guide the early development of an organism might change with time just as does the DNA that controls the final shape of any structure such as the leg of the organism.

However, the study of embryology is not entirely without evolutionary value. The earlier stages of embryonic development are more general in shape than the later stages. For example, arms start out as buds, and develop fingers at a later stage. The early arm bud of a human and horse are similar, and they become more specialized in the later stages of their development. Closely related organisms tend to exhibit strong similarities in their early embryological stages. Therefore, a study of embryology often gives new insight into the evolutionary history of certain organisms.

Embryological evidence is often very important when studying the interrelationships of the different animal groups. Since we expect closely related organisms to show many similarities in their embryonic development, the early embryological stages of a developing bird or mammal should share many common features with a developing reptile embryo.

We expect each new mutation to produce a small change, and these small changes accumulate to create new species, however, we would not generally expect this series of mutations to produce a radically new sequence of embryological changes.

Animals

The most familiar kingdom to most of us is the animal kingdom, the kingdom to which we belong, although many people do not seem to like the idea that humans are animals. Taxonomists of today have divided the animal kingdom into nearly 30 different phyla. Many of these phyla consist of organisms that most of us would lump into the worm category. Of course we will not try to cover all the groupings here, but will present only a brief listing and description of those that are of interest to our discussion.

Sponges belong to the phylum Porifera and are an important group because they represent a multicellular organism that is in some sense just a collection of single-celled organisms living together (Figure 6.23). Indeed, if the cells of certain sponges are separated (by forcing the sponge through a sieve or fine mesh), the individual cells will reassemble themselves into another sponge.

Most of the approximately 9000 species of sponges are marine (saltwater) dwelling creatures, but many are found in fresh water.

Figure 6.23 Porifera (left to right) Yellow Picasso Sponge, of the class Hexactinellida, the glass sponges [NOAA]; Stove-pipe Sponge, of the class Demospongiae, the demosponges [Nick Hobgood]; Clathrina clathrus, of the class Calcarea, the calcareous sponges [Esculapio]

Sponges are exactly what we would expect the first multicellular animals to be like. They generally lack any type of symmetry, and they have no mouth, digestive cavity, organs, or well-defined tissue, although they do have several different types of cells.

A simple model of a sponge is a vase that is full of holes. The inside of the vase is covered by cells that have a whip-like flagellum surrounded by a circle of threadlike protrusions that form a collar. The flagella work in unison to move water into the holes and out the top of the sponge. Particles

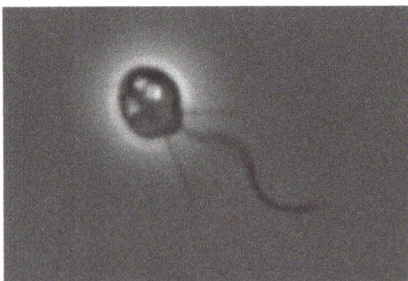

Figure 6.24 Chanoflagellate The photo shows the flagellum and the surrounding collar made of threadlike projections called microvilli. [Stephen Fairclough]

of food in the water are trapped by the various cells of the sponge.

The flagellated collar cells are very similar to members of a group of one-celled flagellated Protists called the choanoflagellates. Genetic studies support the close relationship of the two groups. It appears that all multicellular organisms evolved from these choanoflagellates (Figure 6.24).

As was mentioned earlier, ribosomes (the machines that assemble protein molecules) are themselves composed of RNA and protein molecules. Analysis of the RNA in ribosomes indicates that sponges are at the base of the evolutionary tree of the animals. These ribosomal RNA studies indicate that the fungi are the nearest relatives of the animals, with the plants being more distantly related to both groups.

Simple animals with radial symmetry such as hydras, jellyfish, sea anemones, and corals belong to the phylum Cnidaria, sometimes called Coelenterata (Figure 6.25). Ribosomal RNA studies suggest that they represent the next stage in the evolution of the animals. The oldest fossils of multicellular animals (laid

Figure 6.25 Cnidaria (left) Acropora coral showing several mounds composed of individual polyps [Sonke Johnsen/Duke Univ.]; (center) Zebra-striped Gorgonian Wrapper, an anemone [Nick Hobgood]; (right) jellyfish [JCB]

down more than 700 mya) include representatives of this phylum.

Cnidaria have a simple sack-like body composed of three layers of cells. They have a single opening (a mouth) that leads into a cavity where their food is digested. Undigested waste is expelled back through the mouth.

Except for the hydras and a few freshwater jellyfish, the cnidarians live in marine environments. Corals secrete a hard skeleton of calcium carbonate and are important members of reef communities. Most of the modern corals are colonies of hundreds of individual polyps arranged in a six-fold symmetry.

The next important development in the animal kingdom was the evolution of worm-like creatures with a bilateral symmetry. (Humans and many other animals having a right and left side that are mirror images are bilaterally symmetric.)

Early in the history of the animal kingdom, animals apparently diverged along two different lines. The evidence for this split comes from the study of embryology. As the single fertilized egg cell of a bilaterally symmetric animal begins to divide, a process called *cleavage*, first two cells are formed, then four cells, and then eight. At this point, the two divergent branches of the bilaterally symmetric animals are apparent. Although some variations occur, two general types of cleavage are observed.

In *radial cleavage*, the first four cells form in a plane and the next four

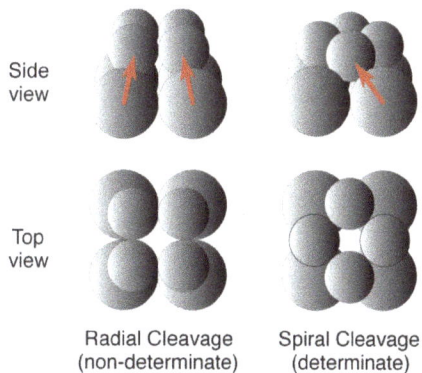

Radial Cleavage (non-determinate)   Spiral Cleavage (determinate)

Figure 6.26 Early Cleavage Deuterostomes (left) have non-determinate radial cleavage and protostomes (right) have determinate spiral cleavage. The fertilized egg divides to form two cells, and these two cells divide to produce four cells in a square. As the four cells divide, the red arrows show where the new cells end up in the two types of cleavage. [JCB]

cells form directly above the first four (Figure 6.26). If a cell is removed early in the development, even at the four cell stage, the organism will develop normally. For this reason, radial cleavage is called non-determinate development.

Radial cleavage is the type of development characteristic of the group of animals known as *deuterostomes*, which includes vertebrates and echinoderms (starfish and sea urchins). Because vertebrates have non-determinate development, identical twins are possible. If the early cell cluster breaks apart, two identical organisms will develop.

The second group of bilaterally symmetric animals is known as the *protostomes*. The pattern of their early cell divisions is called *spiral cleavage*. In spiral cleavage, the first four cells form in a plane, and the second four cells form above the first four, but in the spaces between the first four (Figure 6.26).

Spiral cleavage results in a non-symmetrical development, and every cell is destined to be a particular part of the body. For this reason, removal of a single cell in the early stages of cell division prevents the embryo from developing. This type of development is called determinate development.

In both spiral and radial cleavage, as the cells continue to divide, a hollow sphere of cells called a *blastula* is formed. One side of the blastula then pushes inward, and a cup-shaped structure called a *gastrula* is formed. The outer layer of the gastrula (called the ectoderm) eventually becomes the outer covering of the organism while the inner layer (called the endoderm) becomes the digestive tract. In the space between these two layers, tissue forms, which we call the mesoderm, and it becomes the connective tissue.

At this stage in its development, the organism has only one opening and looks very much like a simple member of the phylum Cnidaria. However, a second opening soon forms and the organism looks more like a fat hollow worm.

In the spiral cleavage group of animals, this second opening becomes the anus while the first opening becomes the mouth. As we saw, this group of organisms is called the protostomes, which means *first mouth*. That is, the first opening becomes the mouth. Members of this group include the Platyhelminthes (flat worms), annelids (segmented worms), mollusks (snails, clams, etc.), and arthropods (insects, crabs, spiders, etc.).

In the deuterostomes (echinoderms and vertebrates), the second opening becomes the mouth, and the first opening becomes the anus. Since the second opening becomes the mouth, this group is called the Deuterostomia, which means *second mouth*.

An important phylum that belongs to the Protostomia is the Mollusca. This phylum contains the clams, oysters, chitons, octopuses, squids, nautiloids, snails, and slugs, (Figure 6.27). Mollusks have a simple body plan consisting of a muscular foot, a body cavity containing the digestive tract and internal organs, and a covering of tissue called the mantle. In most mollusks, the mantle secretes a shell that is easily preserved, making them the most common group of animals in the fossil record.

In many aspects, the primitive mollusk resembles a flatworm that has been covered by a shell. Since the most primitive living mollusks appears to exhibit a segmented pattern, some scientists believe that the mollusks evolved from a segmented ancestor, but the evidence is not overwhelming.

The first of these primitive mollusks to be discovered, whose genus name is *Neopilina*, is considered to be a living fossil because it is similar to fossils of the earliest known members of this phylum. Its shell is a single oval with a blunt point on the front end. In snails, the shell points in the other direction. The 10 or so living members of this class are called monoplacophorans.

The snails, slugs, nudibranchs, and their relatives belong to a class of mollusks called the gastropods, which means *stomach-foot*. It is the largest mollusk class and contains over 40,000 members. Snails have a single coiled shell, and nudibranchs are very colorful marine gastropods whose name refers to the naked gills that protrude from the animals back.

Figure 6.27a Mollusks (clockwise from top left) Greater Blue-ringed Octopus, the Earth's only poisonous octopus. [Cameron Azad]; Giant Clam, found in the Indo-Pacific region (the Indian Ocean and the western and central Pacific Ocean). [Sonke Johnsen/Duke Univ.]; Palau Nautilus, also from the Indo-Pacific. [Manuae]; nudibranch [Chriswan Sungkono]; snail [Yogendra Joshi]; chiton [Kirt L.Onthank]

Figure 6.27b Mullusks The European Common Cuttlefish, the best known species, has an internal chambered cuttlebone used to control its buoyancy. [Hans Hillewaert]

Members of the next largest class of mollusks have two shells or valves and are called bivalves or pelecypods (meaning *axe-foot*). Clams, oysters, scallops, and mussels are common examples of bivalves and there are about 10,000 members in this class.

Squids, octopuses, cuttlefish, and nautiloids belong to a class of mollusks known as the cephalopods (meaning *head-foot*). This group contains the most evolved members of the phylum. The cephalopods have tentacles that evolved from elongated portions of the muscular foot of the mollusk.

Although not closely related to the vertebrates, cephalopods have evolved eyes that look superficially like those of the vertebrates, although they are different structurally. The independent evolution of similar looking features as in this case is known as convergent evolution and will be discussed in a later chapter.

Except for the nautiloids, modern cephalopods either have very small shells or lack shells entirely. However, fossil cephalopods, like nautiloids, belemnites, and ammonites had large shells that were easily preserved.

Another major group of mollusks is the chitons. Chitons have an oval covering made of eight plates or valves. There are over 700 members in this class.

Two animal phyla that are closely related to the mollusks are the annelids (commonly called the segmented worms) and arthropods (the phylum containing the insects). Most of the annelids are marine worms belonging to the polychaete group, but the most well known members of this phylum are the earthworms and leeches. The segmented body, well-developed antennae, and other features of many marine annelids suggest that an annelid-like animal may have given rise to the arthropods Figure 6.28).

The velvet worms (phylum Onychophora) are interesting from an evolutionary perspective because they share similarities with both the arthropods and the segmented worms (Figure 6.28). This small phylum is represented by several hundred species that superficially resemble caterpillars. Their bodies are segmented, but they

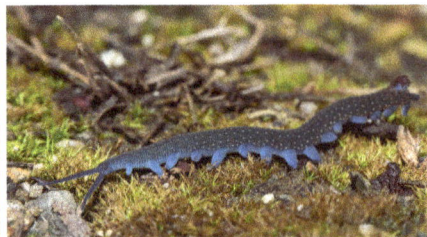

Figure 6.28 Polychaete and Velvet Worms A marine annelid worm (left) is segmented like arthropods and has antennae (not visible in the photo). [Ania Jazdzurka] A velvet worm (right) is also segmented, but has soft legs with claws. [Stephen Zozaya]

have unjointed legs with claws at their tips.

Because they possess both annelid and arthropod features, velvet worms may resemble the ancient organism from which the arthropods evolved. Remember, the velvet worms have also evolved, so we would not expect them to look exactly like the ancestors of the arthropods. However, they appear to have retained some of the features of both the annelids and the arthropods. DNA studies suggest that velvet worms are more closely related to the arthropods. In the next chapter, we will be introduced to an interesting fossil from Cambrian times (530 mya) that appears to be a member this phylum.

As their name implies, the arthropods are animals with jointed feet. This group forms the largest phylum in our classification scheme. Their great numbers are due to the diversity of the class Insecta (the insects), which probably contains well over a million species. Along with the

Figure 6.29 Arthropods (clockwise from top left) Sally Lightfoot Crab [JCB]; reef lobster [Fernando Herranz Martín]; Deathstalker, a scorpion [Ester Inbar]; Golden Silk Orbweaver [JCB]; centipede [Tom Coleman/USDAFS/Bugwood.org]; two-pronged bristletail, a dipluran [Steve Nanz]; symphylan [Sonia Martinez]

insects, this phylum contains the horseshoe crabs, spiders, scorpions, lobsters, shrimp, crabs, centipedes, millipedes, and trilobites, an extinct group familiar to fossil collectors (Figure 6.29). Arthropods have a hard external covering composed of chitin (called an exoskeleton), to which their muscles are attached. It is their segmented body pattern that suggests their kinship to the segmented worms.

The insects are arthropods with six legs and have mouthparts composed of external appendages. By studying the development and anatomy of insects, we can infer much about their evolutionary history. Insects evolved from segmented worm-like animals as did their very close relatives the centipedes. However, the centipedes have retained a pair of legs on each body segment while the insects have lost the legs on their posterior segments.

Symphylans are centipede relatives that are very similar to insects. Newly hatched symphylans have 6 pairs of legs, but they gain an additional pair with each molt until the adult has 12 pairs of legs (Figure 6.29).

The most primitive insects have no wings (Figure 6.31), and although they have only six functioning legs, some still possess hair-like features on their

posterior segments that are the remnants of legs.

Insects belong to the group of arthropods called hexapods. Insects are defined as hexapods with external mouthparts. Some hexapods like the diplurans (Figure 6.29), the springtails, and the proturans have mouthparts that are enclosed in a cavity in the head. They are placed in a separate class, and some may have evolved the six-legged condition independently from the insects. Their relationship with the insects is not yet clear.

Embryological studies indicate that the head of an insect is formed by the fusion of the simple head of the original worm-like creature with six trunk segments, each of which bears a pair of modified legs that form the antennae and external mouthparts. Starting from the front and proceeding toward the rear of the insect, these appendages are called the labrum, antennae, mandibles, maxillae, and labium respectively. Strong evidence to further support the belief that these appendages are modified legs comes from studies of the fruit fly. A single mutation can cause the antennae of a fruit fly to develop into legs (Figure 6.30). Another mutation will cause one of the mouth parts of the fly to become a leg.

When an insect egg is formed, certain genes are activated and messenger RNA molecules are produced that become concentrated in various parts of the egg. Their concentration determines the front and back end of the egg. When the egg is fertilized, ribosomes attach to these mRNA molecules and begin producing proteins that form concentration gradients in

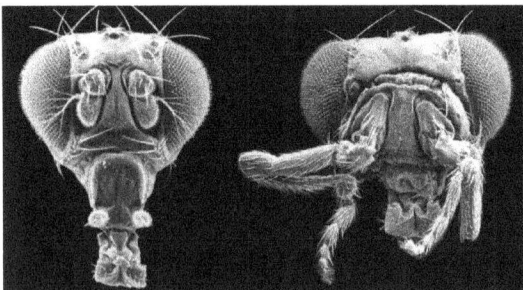

Figure 6.30 Fruit Fly Mutation A normal fruit fly (left) and a fruit fly with the antennapedia mutation that causes legs to grow from the head instead of antennae. [Thom Kaufman; photo, Rudi Turner]

the developing embryo. These gradients determine the front, rear, top, and bottom of the fly.

The embryo is divided into segments by concentration gradients of various proteins and a set of genes called *homeotic genes* are activated in each segment. These homeotic genes produce proteins that coordinate the development of every segment by activating various structural proteins. For example, they regulate the proteins that produce a pair of legs on each segment of the thorax. If one of the homeotic genes is mutated, its target segment could develop differently. The antennapedia mutation (the word is a combination of *antenna* and *pedes*, which means feet) disrupts the function of a homeotic gene in the antenna segment and causes legs to grow instead of antennae (Figure 6.30).

All homeotic genes have a 180 base pair region called the *homeobox*. This region codes for a 60 amino acid sequence called the *homeodomain*. The homeodomain is a helical structure on the protein that binds to a specific gene location on a DNA spiral. This protein regulates the activity of that gene (either activating the gene to produce proteins or repressing it).

Homeotic genes have been found throughout the animal kingdom. They are apparently a very ancient gene family that is highly conserved because of the important role they play in the development of the organism.

The most primitive insects are the wingless silverfish and firebrats (Figure 6.31). Most are about 1 to 2 cm long, and they have chewing mouthparts.

The next major event in the evolutionary history of the insects was the evolution of wings. The most primitive winged-insects are the mayflies, dragonflies, and damselflies that have two pairs of fan-shaped wings. Their wings cannot be folded down and laid flat over the back of the insect as is possible with members of more advanced groups (Figure 6.31).

Figure 6.31 Insects firebrat, a wingless insect [Clemson Univ./Bugwood.org]; mayfly with primitive wings [Richard Bartz]; stonefly has advanced wings that fold down over the back. [Whitney Cranshaw/Colorado SU/Bugwood.org]

Figure 6.32 Crane Fly *The halteres are the ball-on-stick structures just behind the base of the wings. [Janco Tanis/Bugwood.org]*

How or why insects developed wings is not known, but it is interesting that the insects with the most primitive wings are aquatic in their early stages of development. One possibility is that branches on the first two pairs of legs, as are found on many crustaceans,

evolved into wings. Perhaps these structures were aids to swimming and later developed into devices for flying. We do not know at this time.

All the more primitive groups of insects develop slowly into adult forms by a process known as *incomplete metamorphosis*. As the young or nymphs grow, they periodically shed their external covering. With each shedding or molt, the insect becomes more like the adult form. The last molt produces an adult that is capable of reproduction.

Complete metamorphosis is a more recent evolutionary development that occurs in butterflies, beetles, bees, flies, and other more advanced insects. The young are worm-like larvae that grow to a mature size and are then

Figure 6.33 Echinoderms *(clockwise from top left) A crinoid from Indonesia [Alexander Vasenin]; a brittle star [UCSB/Univ.S.Carolina/NOAA/WHOI]; Red Slate Pencil Urchin. [NOAA/David Burdick]; Blue Starfish [NOAA/David Burdick] All but the brittle star are from the Indo-Pacific region (the Indian Ocean and the western and central Pacific Ocean).*

transformed into very different looking adults. Instead of a gradual change as in incomplete metamorphosis, the transformation stages have been compressed into an inactive pupa stage.

An interesting evolutionary development took place in the wings of the flies. This order is known as the Diptera, which means *two wings*. Except for the flies, all other winged insects have two pairs of wings. In flies, the second pair of wings has been modified into balancing organs called *halteres* (Figure 6.32).

Although insects are closely related to centipedes, they are more distantly related to the spiders and their relatives and to the crustaceans (crabs, lobsters, and shrimp). It appears likely that these three arthropod groups developed from different non arthropod ancestors, and therefore, should be divided into separate phyla.

The three phyla of deuterostomes are the echinoderms, hemichordates, and the chordates. The echinoderms or spiny-skinned animals are divided into five classes: the sea stars or starfish, the brittle stars, sea urchins, sea cucumbers, and crinoids (Figure 6.33). These animals live exclusively in salt water, and because of their bottom dwelling habits and hard internal skeleton, they commonly form fossils. The internal skeleton, usually called a test, is a characteristic of the phylum. The test is made of small calcite plates, but it cannot be seen in a live animal because an outer covering encloses it.

Echinoderms are more closely related to the chordates than one might have first imagined. Although the adult forms in this phylum appear to have a five-part radial symmetry, this feature is a secondary characteristic. These radially symmetric organisms evolved from a bilaterally symmetric ancestor. The evidence for this statement comes from studies of echinoderm larvae, which are bilaterally symmetric. If one looks at the tests of many adult echinoderms, their bilateral symmetry is still evident.

The phylum Hemichordata (acorn worms) contains less than a hundred species, but it is important because of its links to the echinoderms and the phylum to which we belong (the chordates). Acorn worms are of interest for two reasons. They have gill slits that serve the same function as the gill slits of fish and the more primitive chordates. The second interesting aspect of hemichordates is that their larval stage very closely resembles the larvae of some echinoderms. These similarities lead us to believe that a common ancestor links the chordates and echinoderms sometime in the distant past.

As we mentioned earlier, animals belonging to the phylum Chordata possess a notochord at some stage in their development. A *notochord* is a flexible rod that runs down the back of

Figure 6.34 Lancelet *This small (about 5 cm long) marine chordate has a dorsal nerve chord just as vertebrates do. Note that the muscles are structured in segments called myomeres. This is the same type of muscle that gives fish its flakey texture. Its head is on the left with its mouth below. [Hans Hillewaert]*

the animal, below the nerve chord. One of the most primitive chordates is an animal called the lancelet or amphioxus (Figure 6.34). There are about 20 species and they belong to a subphylum called Cephalochordata

Another fairly large group of about 2000 marine chordates are the tunicates (subphylum Tunicata). In their early stage of development, tunicates have a notochord and look like a small tadpole. The tadpole is free swimming, but it eventually attaches to some surface and becomes a sedentary filter feeder. The adult is surrounded by a covering called a tunic. The tunic has two opening, one to draw water in and the other to expel the water (Figure 6.35).

In a large group of chordates (the vertebrates), backbones or vertebrae develop around the notochord. The notochord is only present in the embryo stage of a vertebrate's life.

Figure 6.35 Tunicate *This orange and purple tunicate is also called a sea squirt. The water flows in the top hole and exits through the bottom hole. [Nick Hobgood]*

The subphylum Vertebrata (the vertebrates) is typically divided into eight or nine classes, most of which are very familiar to us. The fishes make up four or five of the classes: the agnathans (jawless fish), placoderms (an extinct class of armored fish), the chondrichthyes (cartilaginous fish like sharks and skates), and the osteichthyes (the bony fish). The osteichthyes are divided into the sarcopterygians (lobe-finned fish), and actinopterygians (ray-finned fish). The last two groups are sometimes promoted to the rank of class and osteichthyes is promoted to superclass status. The other four classes of vertebrates are the amphibians, reptiles, mammals, and birds (Figure 6.36).

The first known vertebrate fossils were formed about 500 million years ago (500 mya), well after the first fossils of members of many other phyla. The oldest vertebrate fossils are of a type of fish that did not have jaws. Fossils of fish with jaws did not appear until about 50 million years later.

Today, there are less than 50 species of jawless fish (agnathans), about 1000 species of chondrichthyes, and more than 30,000 species of osteichthyes. There are about 7,200 species of amphibians of which 6,300 or so are frogs. There are over 10,800 reptile species, two-thirds of which are lizards, and most of the rest are snakes. Turtles (order Chelonia) account for about 325 species and crocodiles contribute only 25 species. There are nearly 10,000 bird species and about 5400 species of mammals.

Amphibians evolved from a type of fish with fleshy muscular fins approximately 380 mya. The four fins of these fish evolved into the four limbs of the amphibians. Amphibians and all vertebrates that evolved from them are known as tetrapods (which means *four feet*). Snakes are also called tetrapods since their ancestors had four feet.

Structures that share a common ancestry are said to be *homologous* (Figure 1.19). The common ancestry of many homologous structures is often readily apparent when one studies the details of the structures. For example, the bones in the forelimbs of various tetrapods (all four legged animals)

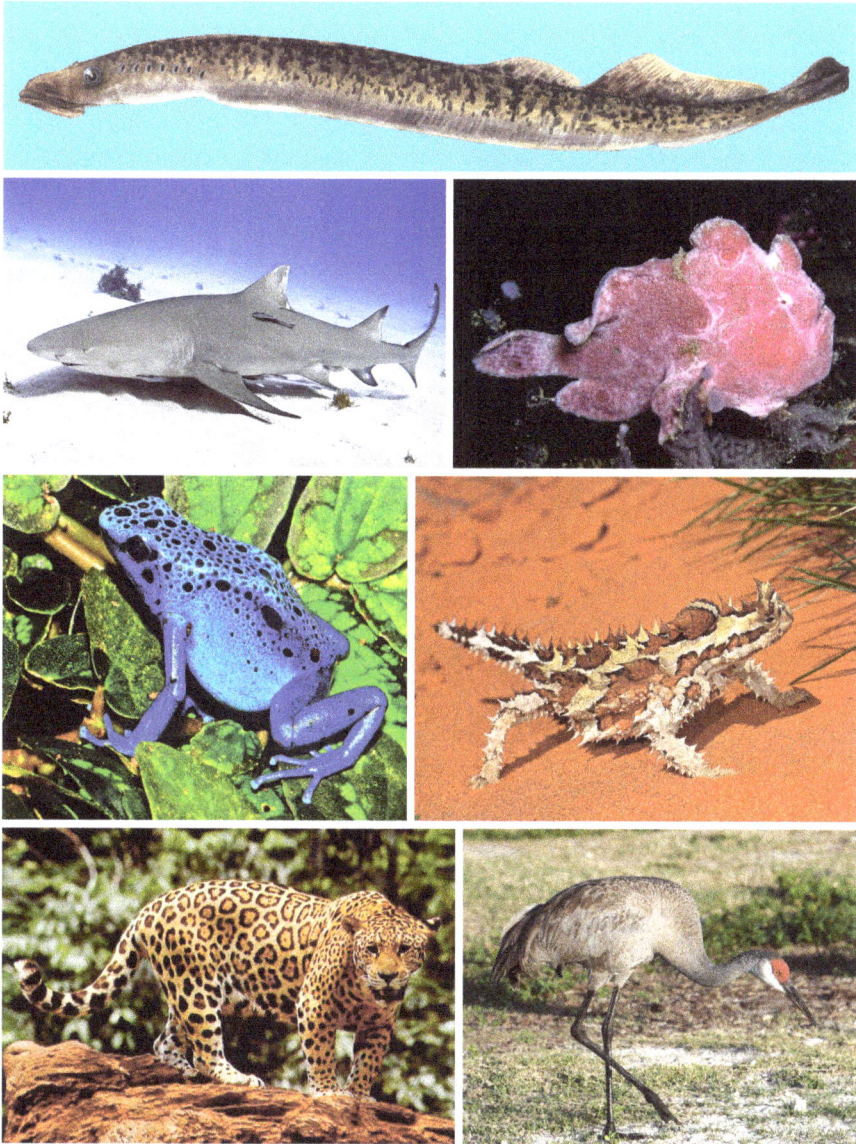

Figure 6.36 Vertebrates (clockwise from top) Sea Lamprey [Ellen Edmonson and Hugh Chrisp]; Ocellated Frogfish, an unusual ray-finned fish [Nick Hobgood]; Thorny Devil, an Australian lizard [Stephen Mahony]; Sandhill Crane [JCB]; Jaguar [USFWS]; Blue Poison Dart Frog [Michael Gäbler]; Lemon Shark [Pterantula]

Figure 6.37 Analogous Structures The wings of a bird like this Sword-billed Hummingbird are analogous to the wings of an insect like this West Coast Lady. The wings did not evolve from the same structure. [JCB]

undoubtedly share a common ancestry. The wing of a bat, the flipper of a dolphin, the leg of a horse, and the arm of a primate have similar bones. Similarities like these tell us to that these animals are closely related, and therefore, must have evolved from a common ancestor. These similarities are why we place all these animals in the same class (mammals).

In contrast, the wing of an insect and the wing of a bird have no structural features in common and did not share a common ancestry. Although both are structures used for flying, they are not homologous and are instead said to be *analogous structures* (Figure 6.37). Analogous structures are structures that perform similar functions, but do not share a common ancestry.

Although the wing of a bird and the wing of a bat are homologous structures as far as their bone structure is concerned, they do not share a common ancestry as functional wings and their wings are said to be analogous structures. Wings have evolved independently on at least four different occasions, in insects, birds, bats, and in an extinct group of flying reptiles known as pterosaurs. The wings of these four different animals are said to be analogous. Although they share a similar function, they do not share a common winged ancestor.

However, the tetrapod limbs of birds, bats, and pterosaurs that evolved into wings are homologous as limbs since their limbs did evolve from a common ancestor with tetrapod limbs.

As we study organisms, we notice many similarities between the different life forms. Indeed, the fact that we can place organisms into groups within groups indicates that all organisms are related. Some are very closely related, while others are related more distantly. This fact alone is strong evidence that the organisms living today evolved from earlier forms. As we have seen, rocks, stars, geometrical figures, and other inanimate objects cannot be classified in this related manner

Now that we have briefly examined the diversity of life, we will look at the fossil record to discover when these various organisms first appeared on the Earth. The observed ordering of the fossil record is interpreted as conclusive evidence that evolution did take place and is responsible for the observed changes in life forms through the ages. No other plausible scientific explanation has been found.

Chapter 7: The Life of Ancient Times

Fossils

Much of the evidence that supports evolution comes from studies of the ancient life forms that have been preserved as fossils. The word fossil means literally to *dig up*, and the word was originally used to describe any object or substance that was dug out of the ground, even ordinary rocks and minerals. As it became evident that some of these objects were the remains of dead organisms, use of the word was restricted to these objects, although it is sometimes used to describe such inanimate things as fossil beaches and fossil riverbeds.

A fossil can be any evidence that has been left by life forms of the past. This evidence can be impressions, tracks, casts, or actual parts of the organism. Bones, shells, teeth, and other durable parts of organisms are more likely to be preserved, so the fossil record is highly biased toward those organisms that possess hard parts. Soft-bodied creatures such as spiders, jellyfish, and various types of worms are much less likely to be preserved as fossils.

Since all organisms are attacked by various decay mechanisms, to be preserved as a fossil, the organism, or part of the organism must be protected from decay. Quick burial by substances such as mud, volcanic ash, tar, resin, or other material is one common way the decay process is suppressed.

Conditions for fossilization are most often met in aquatic environments, therefore, the fossil record is strongly biased toward organisms that live in or around water. Several processes that are responsible for the formation of fossils are reviewed below.

Original preservation or unaltered organisms, preserved with little or no change, can be the most spectacular of all fossils. Extraordinary examples of this type of preservation include mammoths, bison, moose, horses, and rhinoceroses that are sometimes found frozen in the arctic regions (Figure 7.1). These animals are a few tens of thousands of years old, yet the meat is often preserved and even the animal's

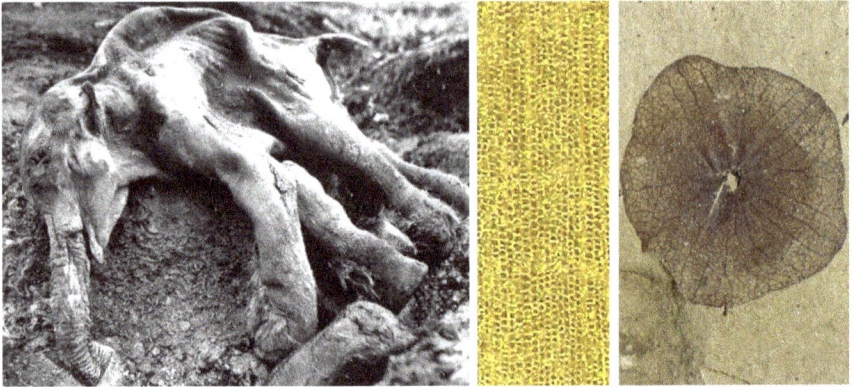

Figure 7.1 Original Preservation, Permineralization, and Carbonized Compression
About 40,000 years ago this seven or eight month old Woolly Mammoth was frozen in the permafrost of northeast Siberia. It was discovered in 1977 and named Dima. [NOAA]; Permineralized wood (middle) from the Devonian Period (about 400 million years old) showing the individual cells of the wood. [Roman Kishkan]; Carbonized compression (right) of a 50 million year old flower showing the vein pattern. [Kevmin]

last meal may be preserved in its stomach.

Bits of hair and large quantities of dung from the extinct Shasta Ground Sloth have been found in several caves in southwestern United States. One of these is Gypsum Cave near Las Vegas, and another is Rampart Cave in northwestern Arizona.

Another unusual example of original preservation was a 39,000 year old Woolly Rhinoceros named Cherskyi discovered frozen in northern Siberia. The entire animal was preserved except for the lower part its right front leg.

A 36,000 year old Steppe Bison named Blue Babe was found in frozen soil near Fairbanks, Alaska and its meat was still edible. Blue Babe was named after Paul Bunyan's blue ox, and because it was partly covered with vivianite, a blue mineral.

The hard parts of various organisms (e.g., shells and bones) are sometimes preserved with very little change, although the crystal structure and chemical composition may be altered to varying degrees. Some of the more easily dissolved chemicals may be carried away, and other chemicals may be deposited in the fossil material. For example, the coloring in shells tends to be bleached out with time, and bones tend to absorb fluorine from the ground water. Many bones of recent origin, however, can be largely unaltered.

Permineralization or petrification occurs when minerals such as silicates, carbonates, and iron compounds penetrate into the cells of an organism and are deposited there. Bones and shells are very porous, and the spaces can be filled with minerals. This type of preservation is often seen in such familiar objects as petrified wood and dinosaur bones. Even the cell structure is sometimes preserved (Figure 7.1). Dinosaur bones and petrified wood are usually not completely permineralized, but contain a substantial amount of the original material. Fossils generally contain large amounts of amino acids (the building blocks of proteins) and sometimes even DNA.

Most people could distinguish a dinosaur bone from a more recent mastodon bone simply by the difference in density. Most recent fossil bones have not been extensively permineralized and are, therefore, less dense. In contrast, dinosaur bones are usually permineralized and are very dense. They have literally been turned to stone. Permineralization preserves not only the general form of the object, but microscopic details are often retained such as the structure of the cells. Because of the many details that are often preserved, permineralized fossils can be an extremely important source of information.

Carbonized compressions and impressions are formed when an organism is compressed and the gas, moisture, and other soluble materials are lost (Figure 7.1). Under pressure, the remaining material, which is mostly carbon, may form a black deposit. Leaves preserved in this way may show intricate details such as the vein pattern and even the microscopic stomata cells that allow air to pass in and out of the leaf. Sometimes the compressed residue is very obscure or has been dissolved away, leaving only impressions in the rock.

Casts or *molds* are types of fossils that preserve just the outer shape of an organism. For example, after the burial of an animal such as a clam, the complete organism may be dissolved away, leaving a mold of the object. If the mold is then filled with another type of material, a cast of the original organism will be formed (Figure 7.2). These casts are usually composed of silicates, but casts of more exotic materials such as iron pyrite are sometimes formed. Since this process only reproduces the object's shape, information about the organism's soft parts is not preserved as it is in some permineralized fossils.

Trace fossils are not the actual remains or casts of organisms, but are traces of the activities of these past organisms. Examples of trace fossils

Figure 7.2 Mold and Cast An ammonite mold (left) and cast (right) from the Jurassic Period. [Mike Viney/Virtual Petrified Wood Museum]

are dinosaur tracks (Figure 7.3), worm trails, and animal burrows. Because of the very indeterminate nature of trace fossils, it is sometimes impossible to be certain of the organism that produced the trace fossil.

However, important information can be obtained from trace fossils. Studies of dinosaur tracks have yielded a great deal of valuable information about the behavior of certain dinosaurs. For example, some dinosaur tracks show young dinosaurs traveling in herds with adults. These tracks suggest that some dinosaurs provided a certain amount of care and protection for their young. Birds and mammals feed and protect their young, but most of today's reptiles lay eggs and leave their young to fend for themselves.

Most fossils are preserved after the organism dies, but the study of trace fossils allows us to glimpse some of the activities carried out by certain organisms when they were alive.

Biogenic substances are fossils of objects made by ancient organisms. Examples are fossilized animal feces called *coprolites* (Figure 7.3), egg shells, and fossilized tree sap. Tree sap or resin can be extruded by the trunk where it forms blobs of various shapes. Insects and other small organisms like spiders and even frogs and lizards are

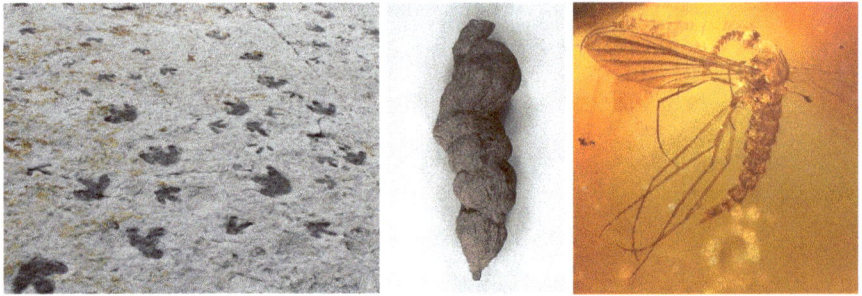

Figure 7.3 Trace Fossils and Biogenic Substances Dinosaur footprints (left) from the Jurassic Period, 150 mya. These prints at Dinosaur Ridge in Morrison, Colorado have been "enhanced" with paint. [JCB]; Next, fossilized dinosaur dung, called a coprolite (8 cm), from the late Cretaceous Period, approximately 70 mya. [North Dakota Geological Survey]; A 20 million year old mosquito (right) preserved in Dominican amber. [Didier Desouens]

sometimes trapped in this resin. Over time the resin will harden into a substance called *amber*, with the organism preserved inside. In some instances the mummified tissue of organisms is preserved in minute detail. Much of our knowledge about prehistoric insects has been obtained by studying insects preserved in amber.

Perhaps the most famous amber deposits are found in the Baltic region and in the Dominican Republic. Baltic amber was formed about 40 mya and Dominican amber is about 20 million year old. Several scientists have reported extracting DNA from insects preserved in amber.

Chemical fossils are traces of the chemical remains of ancient organisms, such as amino acids that are often found in fossils. Since various chemical processes can change molecules, chemical fossils can be hard to identify.

One chemical that can provide evidence of past life is carbon. Since all organisms use carbon in their body chemistry, carbon deposits may represent organic remains. There are two common isotopes of carbon, C-12 and C-13. (Carbon-14 is a very rare isotope of carbon, and because of its

relatively short half-life, it is not present in carbon deposits older than a few hundred thousand years.) About 1% of all natural occurring carbon is C-13 and the rest is C-12.

Since carbon-12 is approximately 8% lighter than C-13, it tends to be selected over C-13 when an organism uses carbon in a chemical reaction. This preferred selection is because at a given temperature, the lighter atoms tend to be moving slightly faster than the heavier atoms, and, therefore, react more easily.

When live plants are analyzed and the carbon content is compared to the natural carbon abundance, we find that carbon-13 is depleted by 2 to 3 percent. On the other hand, carbonate rocks that are not formed by living organisms do not show a depletion of carbon-13. Therefore, a depletion of carbon-13 by 2 to 3 percent relative to carbon-12 would be evidence that the carbon was produced by living organisms.

Although there are many methods by which an organism can be preserved, the vast majority of organisms do not end up as fossils. Most eventually become food for other organisms, as the great cycle of life continues. Darwin once likened the fossil record to a book with many missing pages. However, only rarely is a whole page or even a complete sentence found. Most of the time only a word or phrase is discovered. Because of the incompleteness of the record, it is impossible to know every detail of how life developed, and we must make educated guesses to help fill in the gaps.

All organisms are not equally likely to become fossils. As mentioned above, the fossil record is highly biased toward aquatic animals such as clams and snails that have some type of hard covering, or animals with sturdy bones or teeth that are able to resist decay long enough to be buried and preserved.

Even if evidence of an organism is preserved, it most likely will not be found unless subsequent erosion uncovers the specimen. Even at this point, someone must discover the fossil relatively quickly or the weathering process that uncovered it will soon obliterate it. Of course, particularly rich fossil beds can be excavated and may yield significant discoveries. Nevertheless, the chance of an organism becoming a fossil and then being discovered is quite low.

The fossil record is discontinuous in time and space. At a given location, there are many time gaps in the record. Since fossils are incorporated in sediments, there will be gaps in the record whenever sediment is not being deposited. Because of the way the landmasses are molded and shaped by plate tectonics, continual deposition at any one place is impossible. As plates collide, areas of deposition eventually become uplifted to form mountains, which are then eroded away.

Not only is the fossil record discontinuous in time, it is also discontinuous in space. While fossils may be deposited in one area, the organisms in another area may leave no record. This discontinuity in space can apply to areas as large as continents. For example, the fossil record of Eocene mammals (35 to 55 mya) is fairly good in western North America, but essentially absent in Australia. Considering all the gaps in the fossil record, the amount of important information obtained is quite impressive.

In the rest of this chapter, we will investigate the fossil record, paying particular attention to the appearance and disappearance of various groups of organisms. As we describe the evidence such words as extinction and appear will be used. What is generally meant by extinction is that no fossils of that type have been found in rocks of a younger age. Of course, it is possible that new discoveries will show that the organism lived to a later time. Likewise, the time of the first appearance of a particular fossil in the record would have to be changed if specimens are discovered from an earlier time.

The Earliest Fossils

Exactly when or how life first appeared on planet Earth is not known, but the first direct fossil evidence of life occurs in rocks that are about 3,500 million years old. These rocks from the Warrawoona Group in North Pole Dome, Western Australia, contain permineralized microfossils of what appear to be different forms of bacteria (Figure 7.4). The fossils are so tiny that a microscope is required to view them, hence the name microfossils. One of the green bacteria belonging to the genus *Chloroflexus* looks superficially similar to some of the Warrawoona fossils. *Chloroflexus* is a filamentous organism, some of which can reach 300 microns in length. They are found in hot springs, and although they can tolerate oxygen, they grow best under anaerobic conditions.

These fossils and other extremely old fossils have been found in layered structures called stromatolites that were formed in the ocean by colonies of microscopic organisms. The layers are composed of dolomite [(Ca,Mg)CO$_3$], or chert, which is mostly silica [SiO$_2$]. The permineralized fossils of the organisms that formed the stromatolites are sometimes found imbedded in the structure (Figure 7.4).

Stromatolites formed by living colonies of cyanobacteria are still being made in a few select locations. Today, bacteria are used as food by many other organisms, and they generally get eaten before stromatolites can be formed.

However, in some places such as Shark Bay, Western Australia thick mats of cyanobacteria can still be found (Figure 7.4). The water in the bay is so salty that many organisms, which would normally graze on the bacterial mats and destroy them, are not present. The fossil record tells us that stromatolites were once very common, and many flourished in ocean regions similar to those areas that are occupied by coral reefs today. However, with the evolution of multicellular organisms that used these bacteria for food, the abundance of stromatolites declined greatly.

Microfossils older than about 1 Ga (Ga is the abbreviation for giga annum or 1000 million years) have been found in a dozen or so locations throughout the world, a few of which are, the Tumbiana Formation of the Fortescue Group, Western Australia (about 2.7 Ga), the Torridon Group, NW Scotland (1 Ga), the Kromberg Formation of South Africa (3.4 Ga), the Gunflint Formation, northwestern Ontario, Canada (2.1 Ga), and the Apex Chert of northwestern Western Australia (3.5 Ga).

Chemical analysis of stromatolites, even those in which microfossils are not preserved, indicates that the material is depleted in carbon-13 and is, therefore, of organic origin. The Isua rocks from western Greenland, which are among the oldest rocks that have been found (3.8 Ga.) are depleted in C-13 by about 1 percent. Although the data is not generally considered strong enough to make a definitive statement, this weak indirect evidence

Figure 7.4 Microfossils, Stromatolites and Algae A cyanobacteria-like fossil (left) from the Bitter Springs Formation in central Australia is about 820 million years old. [J.William Schopf]; Next, cross section of a 2.3 billion year old stromatolite. [Virtual Fossil Museum]; Present day stromatolites (third) produced by living cyanobacteria in Shark Bay, Western Australia. [crikey-adventure-tours.com]; A 2 billion year old fossil Grypania spiralis (right), is believed to be a eukaryotic algae. The arcs are a few cm across. [Xvazquez]

suggests the presence of life at this early date.

The early fossil record clearly illustrates the very simple nature of the life forms that existed. From about 3.5 billion years ago until almost 2 billion years ago, the only fossils we find are of single-celled organisms, but they also appear to be the simplest kinds of the single-celled organisms, the prokaryotes. Although they were simple in shape, undoubtedly a great deal of evolution was occurring in the chemical processes employed by these organisms.

Prokaryotes range in size from about 1 to 10 microns, while the cells of eukaryotes are generally 10 to 100 microns in diameter. Although there is some overlap, extreme size differences can determine if microfossils are prokaryotes or eukaryotes.

Measurements of fossil cells suggest that eukaryotes did not evolve until about 2 billion years ago. The earliest known macroscopic fossils are simple small ribbons believed to be a type of eukaryotic algae (Figure 7.4). They were found in the Upper Peninsula of Michigan and have been dated at about 2 billion years old.

So far, the first evidence of multicellular animals in the fossil record is from approximately 580 mya, but they probably evolved earlier. Stromatolites began to decrease in abundance a little more than a billion years ago which may indicate that multicellular animals evolved about that time.

Although all the very earliest fossils that have been found are of prokaryotic organisms, it is reasonable to assume that the first living forms must have been even more primitive. These earliest forms of life either left no clear remains other than a decrease in C-13, or the evidence of their existence has not yet been found. Perhaps we do not know what to look for.

As we saw in Chapter 3, the geologic column is divided into several time intervals, that we call periods. The time sequence for the last 540 million years as shown in Figure 3.8 is repeated in Figure 7.5. Early fossil investigations were carried out in England and Europe where there is a sharp division between the Cambrian and the Precambrian because of a gap in the fossil record (called an unconformity). This unconformity led people to believe that there was a sudden explosion at the beginning of the Cambrian. As more continuous fossil sequences are studied from other parts of the world, it has became apparent that the increase in diversity was not quite as sudden as was once believed.

The Ediacaran Fauna

During the early part of the Cambrian Period, about 540 mya, many animals evolved shells and other hard coverings that were easily fossilized, resulting in a great increase in the number of animals preserved as fossils.

Before the evolution of shelly coverings, the soft-bodied animals that ruled the seas were seldom preserved, and they are quite rare in the fossil record. There are, however, a few important examples of fossilized Precambrian soft-bodied animals such as the Ediacara fauna (Figure 7.6). The Ediacara fauna flourished from approximately 610 mya until near the beginning of the Cambrian. The food source for the multicellular animals must have been photosynthetic bacteria, filamentous algae, and possibly other animals.

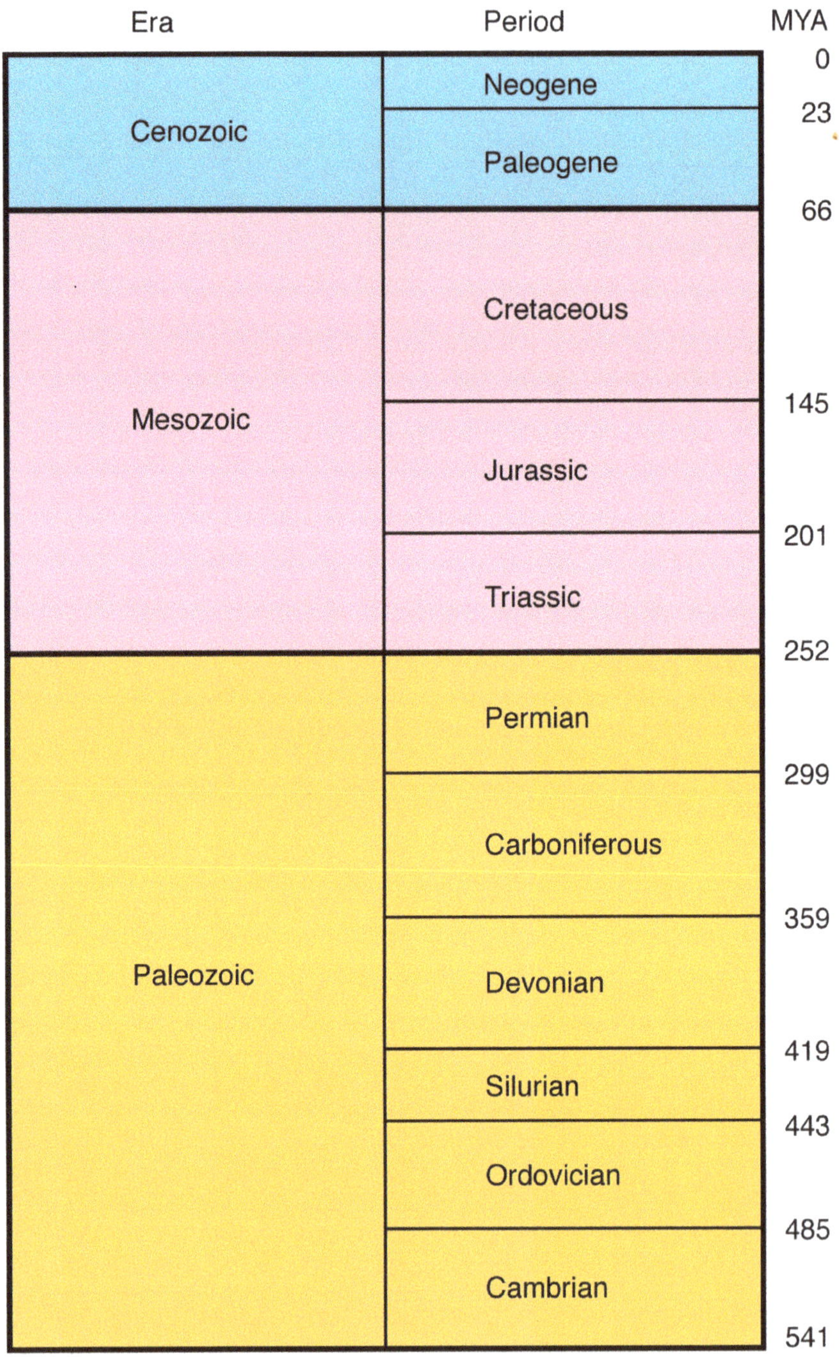

Era	Period	MYA
Cenozoic	Neogene	0
		23
	Paleogene	
		66
Mesozoic	Cretaceous	
		145
	Jurassic	
		201
	Triassic	
		252
Paleozoic	Permian	
		299
	Carboniferous	
		359
	Devonian	
		419
	Silurian	
		443
	Ordovician	
		485
	Cambrian	
		541

Figure 7.5 Eras and Periods of the Phanerozoic Eon *There are three periods in the Cenozoic Era. The last 2.6 million years is known as the Quaternary Period, but it is too short to be shown.. [JCB; data: International Commission on Stratigraphy, 2014]*

Figure 7.6 Ediacaran Fauna How the 850 million year old animals that made these fossil impressions are related to other organisms is not well known. (left to right) Tribrachidium (about 1 cm across) is unusual in that it has three arms. [photo, Arkhangelsk Regional Museum/Aleksey Nagovitsyn]; Spriggina (about 4 cm long) appears to be a segmented animal possibly related to annelid (segmented) worms or arthropods like trilobites. [Smith609]; Next is Dickinsonia. There may be several species ranging in size from a few mm to over a meter in length. This segmented animal appears to have a front (up in the picture) and back end.[Smith609]; Charniodiscus (up to one half meter long) was a stationary animal that was attached to the sea floor by a holdfast (circle at the bottom of the photo). It resembled the sea pens of today which are cnidarians. [Jim Gehling].

All the members of the Ediacara fauna were soft-bodied animals whose imprints were preserved in sandstone. There are several animals that do not appear to belong to any of the known phyla of today. These forms apparently died out without leaving any ancestors. Ediacara-type animals have been found in several locations around the world, but near the end of the Precambrian, the Ediacara fauna disappeared and shelly animals began to appear in the early Cambrian.

Some of the Ediacara fossils appear to be echinoderms, segmented worms (annelids), or arthropods. Sponges, the simplest creatures in the animal kingdom, appeared in the late Precambrian. All these early fossilized animals were sea creatures. There is no evidence that life of any kind lived on the land at this time.

Some of the Ediacaran worm-like animals were fairly large. Specimens such as *Spriggina* grew to 3 cm in length while *Dickinsonia* grew to more than a meter in length. Their large size suggests that predators were absent or uncommon. *Tribrachidium* displays an unusual threefold symmetry and may be an early echinoderm. Many of the echinoderms of today appear to have a fivefold symmetry, although closer inspection reveals a bilateral symmetry. *Tribrachidium* may have been an ancestral form.

The Cambrian

The Cambrian Period began about 541 mya and lasted for approximately 55 million years. It is the first period of the Paleozoic Era, which is the first era of the Phanerozoic Eon.

The first shelly animals appeared near the beginning of the period. They were very small, only a few millimeters long, and were composed of tubes, spines, cones and plates. Many of the

198

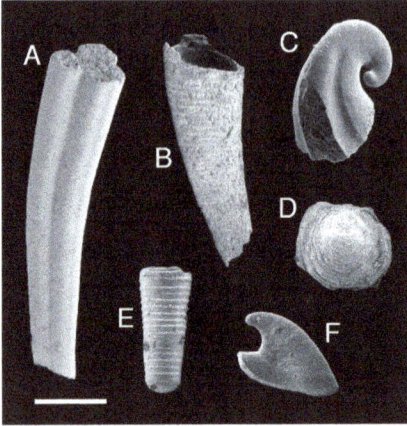

Figure 7.7 Small Shelly Fossils *These fossils are from the early Cambrian, about 540 to 520 mya. The scale bar (lower left) is 0.5 mm long. Anabarites (far left) has an unusual triradial symmetry. [A, Artem Kouchinsky; B,D,E, Susannah Porter; C,F, Michael Vendrasco]*

tube-shaped fossils probably housed worm-like animals (Figure 7.7).

The first half of the period saw an increase in the number and diversity of animals with shells. One example was a group of animals known as the brachiopods. The soft body parts of brachiopods are enclosed in a two-valve shell, and they look superficially like a bivalve clam. However, the two halves of a brachiopod are not

Figure 7.8 Brachiopods and Bivalves *In bivalves (top), the shells are mirror images of each other. Brachiopods (bottom) have asymmetrical shells (the top shell is different from the bottom shell). [JCB]*

symmetrical as are those of most clams (Figure 7.8). The body structure of a brachiopod and a clam is quite different also, indicating that they are not closely related. The brachiopods diversified somewhat during the Cambrian, but did not reach their peak numbers until a much later period.

Bivalves evolved a little later than brachiopods, and did not become diverse until the early Ordovician (the next geologic period). The first brachiopods and bivalves were relatively small in comparison to later forms. Approximately 3,000 genera of brachiopods are known as fossils, but only about 100 genera are alive today.

About 15 million years into the Cambrian, the first trilobites appeared (Figure 7.9). Trilobites are named after the three lengthwise lobes that divide their body. They are primitive arthropods, jointed-legged animals related to crabs. Except for their antennae, all the limbs of a trilobite are built on a common plan. In many modern arthropods, some of their appendages have been modified and are used for functions other than locomotion.

Arthropods are closely related to the segmented worms, and they probably evolved from a segmented worm-like ancestor. The Ediacara animal *Spriggina* may be some type of segmented worm, and it looks like an

Figure 7.9 Cambrian Trilobite *An ancient relative of the insects, lobsters, and horseshoe crabs from the Cambrian Period. This one is called Mesonacis. [S.Gon III]*

animal that could be an ancestor of the trilobites.

Trilobites eventually became the dominant form of animal life in the Cambrian Period. They are one of the most common fossils of the Cambrian, due in part to the fact that they were covered by a cuticle composed largely of calcite that was easily preserved. Over their 300 million year history, about 1500 genera left evidence of their existence in the fossil record. They became extinct at the end of the Permian Period about 250 mya.

Many of the animal phyla apparently originated before or during the Cambrian Period. Unfortunately, since many were soft-bodied, and soft-bodied animals are rarely fossilized, the chances are quite low that many of the evolutionary details of their origin will ever be completely known.

An important assemblage of soft bodied animals has been found in southern China. Known as the 520 million year old Chengjiang biota or fauna, it includes several species that appear to be chordates or close relatives of the chordates. One of these is an animal called *Haikouichthys* which looks very similar to the lancelet or amphioxus of today (Figure 7.10).

Amphioxus is probably the closest living relative of the vertebrates. Amphioxus has a stiff rod, called a notochord, along its back. Notochords give the body stiffness and flexibility, and it provides a structure for muscle attachment. All vertebrate embryos have a notochord that becomes the central portion of the discs between the vertebrae.

Like vertebrates, amphioxus also has a hollow nerve cord located dorsally (toward the back) above the notochord. However, it is not surrounded by bone as it is in vertebrates. As a vertebrate embryo develops, its spinal cord begins as a neural tube. Amphioxus also has

Figure 7.10 Haikouichthys A 520 million year old animal believed to be a primitive chordate (about 3 centimeters long). From the Maotainshan Shales, Chengjiang County, Yunnan Province (southern China). [reconstruction, Nobu Tamura]

muscles arranged in segments called myomeres like vertebrates. These muscle segments are what give fish their flaky texture.

Unusual circumstances near the middle of the Cambrian Period (about 505 mya) resulted in the preservation of about 120 genera of animals, many of which were soft-bodied. This series of fossils was discovered in western Canada and is known as the Burgess Shale deposit.

It is believed that the Burgess fossils were formed in a muddy silt-covered area located in about 150 meters of water. Occasional mudslides carried trapped animals out into the deeper oxygen-poor water where they were protected from scavengers and decay.

If the animals of the Burgess Shale are representative of the time, and it is generally believed that they are, then much of the animal life was of the soft-bodied type. Only about 20% of the animals from the Burgess fauna were composed of animals with hard coverings such as trilobites and brachiopods.

There are several other things that are of interest. First, the animals are generally not advanced in an evolutionary sense. About 10% were members of the sponge group, and about 20% were worms of various types, the most common being the spiny-headed worms. Although spiny-headed worms are not an important component of the animal life of today, they were apparently very common during middle Cambrian times. About 5% of the genera belonged to the annelid worms (the segmented worms).

The first known crinoid (also called a sea lily) is from the Burgess Shale. Crinoids, which are members of the echinoderm phylum, would eventually become much more common. The earliest echinoderms did not resemble the familiar starfish and sea urchins of today.

Almost 40% of the Burgess animals were arthropods, a third of which were trilobites. Although a few of the non trilobite arthropods appear to be primitive crustaceans, most cannot be easily placed in any of the current taxonomic classes (Figure 7.11).

One of the interesting Burgess Shale fossils is an animal named *Aysheaia*,

Figure 7.11 Burgess Shale Fauna (clockwise from top left, maximum sizes are shown in parenthesis) trilobite of the genus Olenoides (10 cm); Marrella, the most abundant fossil in the Burgess Shale, appears to be a crustacean relative (2 cm); Waptia, another crustacean (7 cm); Opabinia, an animal of unknown affinity with a long proboscis on the left (8 cm); Canadia, a polychaete worm (5 cm); Pikaia, believed to be a primitive chordate (4 cm); Aysheaia, similar to the velvet worms living today (5 cm). [Olenoides, Opabinia, and Pikaia by Jean-Bernard Caron; others by Chip Clark/courtesy of Smithsonian Institution]

which was a caterpillar-like worm that resembled modern terrestrial velvet worms (phylum Onychophora). Velvet worms were covered in the last chapter because of their annelid and arthropod features. *Aysheaia* had features that one might expect the ancestors of the arthropod phyla to possess.

Finally, a very important discovery in the Burgess community is *Pikaia*, an animal that appears to be a primitive member of the chordate phylum. *Pikaia*, like the lancelet, was a soft-bodied creature that seems to have a notochord, but it is not a vertebrate. It is of particular interest because it is likely that it, or a creature similar to it, eventually evolved into the fishes.

An interesting development in the plant world was the appearance of multicellular green and red algae. Many types of algae are too fragile to fossilize well, but several groups produce a matrix of calcium carbonate that makes preservation more likely. This rather non homogeneous group is often collectively known as the calcareous algae. Small multicellular green and red calcareous algae of Cambrian age have been found, and it is likely that non calcareous algae were also present. However, there is no evidence that large forms of algae like the Giant Kelp or *Sargassum* of today existed then.

Before the red and green algae appeared in the Cambrian, the largest photosynthetic organisms were strands of single cells such as cyanobacteria and filamentous green algae. Indeed, even today most photosynthesis occurs in the ocean, the work of microscopic organisms that are collectively known as *phytoplankton*.

The Cambrian came to a close about 150 million years after the first appearance of multicellular animals, and we find that representatives from all the major animal phyla were present.

Animals are separated into different phyla according to their general body plan. Since the general design of the body is such a fundamental property, the evolution of different phyla must have been a very early development that we expect occurred when the organisms were still very small and simple. Therefore, it should not come as a great surprise to learn that most of the animal phyla were present a hundred million years or so after the multicellular animals first appeared.

If we could travel back in time to the end of the Cambrian, we would find the continents barren and the sea filled with very unfamiliar creatures. There were no coral reefs, no fish, no whales or other sea mammals, nor any of the great marine reptiles that would inhabit the seas in later periods. The small snails, brachiopods, jellyfish, and some of the other creatures might bear a superficial resemblance to some of the animals of today, but to anyone familiar with today's sea life, the strange echinoderms, trilobites, and most of the other unusual sea creatures of the Cambrian would be totally foreign.

The Ordovician

The division between the Cambrian and the Ordovician is marked by the disappearance of several groups of animals. It appears that some kind of crisis occurred near the end of the Cambrian, as many unspecialized trilobites died out. About two-thirds of the 60 or so families of trilobites became extinct. Although trilobites flourished throughout the Ordovician, they never regained the dominance that they enjoyed during the Cambrian. This may have been due in part to the rise of new predators such as the nautiloids, which are squid-like mollusks that had a snail-like shell. The earliest nautiloids that appeared in late Cambrian times had small straight shells, but larger curved forms evolved during the Ordovician.

Another group of organisms called archaeocyathids (meaning *ancient cups*) disappeared at this time. Archaeocyathids were double-layered, cup-shaped, sponge-like creatures that were important reef-building organisms during the Early Cambrian. Although archaeocyathids do not seem to be an important part of the overall picture of the evolution of life, they illustrate an interesting pattern seen repeatedly in the fossil record. A new organism will appear and flourish for some period of time, only to disappear later without a trace. The process of evolution explains this continually changing pattern in the fossil record. New organisms evolve from ancestral forms, but many of these organisms eventually become extinct for one reason or another.

The Ordovician lasted about 40 million years (from about 485 million to 445 mya) and saw perhaps the greatest increase in the diversity of animal life throughout all of history. Brachiopods diversified greatly as did the more advanced forms of trilobites. Echinoderms diversified as starfish and sea urchins (known as echinoids) appeared, and crinoids became more common. One new addition was a group of echinoderms known as blastoids. Blastoids had the appearance of a flower bud on the end of a stalk (Figure 7.12).

Mollusks became very abundant, especially bivalves, which were rare before middle Ordovician times. Rugose corals first appeared and became an important part of the seafloor fauna (Figure 7.12). Many rugose corals were solitary individuals and are commonly called horn corals because of their shape. Other important reef-builders of Ordovician times were the tabulate corals (Figure 7.12), so named because of the horizontal partitions within the tubular chambers

Figure 7.12a Crinoids, Rugose, and Tabulate Corals Crinoids (left) are relatives of starfish and sea urchins. [Berengi]; Because of their shape (second), solitary rugose corals are commonly called horn corals. A cross section of this 450 million year old fossil is shown at the bottom. The specimen is 3 cm in diameter. [Mark A.Wilson]; Next is the polished cross section of a colonial rugose coral known as a Petoskey Stone from the Devonian Period (415 to 360 mya). [James St.John]; Cross section of a tabulate coral (right) showing the many horizontal partitions (tabulae) in each vertical cell. [R.Weller]

Figure 7.12b Blastoids They reached their peak diversity in the early Carboniferous, but did not survive the great extinction at the end of the Permian. [DanielCD]

that housed the individual polyps. These colonial corals first appeared in the early part of the period.

Stromatolites, the algal mats made by cyanobacteria, are abundant Precambrian fossils as we have already discussed. However, as the sea creatures diversified during the Ordovician, stromatolites declined in abundance. Apparently, with many of the newly evolved sea creatures grazing on the algal mats, large stromatolites were unable to form as they had done in the past.

Of particular interest to us is the appearance of the first primitive fish in the Ordovician (Figure 7.13). These fish were called ostracoderms, which means *bony skin*, because plates of bony material covered them. They belonged to the class of jawless fish called the agnathans, meaning *without jaws*. They were fairly small, and since they lacked jaws, they must have been filter feeders, sucking up food from the ocean floor.

Mostly small scales and fragments of plates have been found in the Ordovician, and a few scales that may have come from a primitive fish have been found in rocks as old as the later part of the Cambrian. There are very few Ordovician fossils that are complete enough to tell what the fish looked like. They seem to have played a fairly minor role in the Ordovician fauna.

The most primitive agnathans had no fins and could only swim by wiggling their tail. Although one group of jawless fish did evolve paired pectoral (front) fins by the later part of the Silurian, agnathans did not ever evolve paired pelvic (rear) fins. The only living members of this group are the snakelike lampreys and the hagfish, which do not have paired fins either. Paired pectoral fins and pelvic (rear) fins are important from an evolutionary point of view because the four limbs of all higher vertebrates evolved from the paired fins of an early type of jawed fish.

A group of arthropods called eurypterids or sea scorpions first appeared in the Middle Ordovician. Eurypterids became more diverse and reached their peak during the Silurian.

Figure 7.13 Agnathans The first jawless fish also had no fins. One of the earliest known forms is shown here. Arandaspis lived 480 mya and was about 15 cm long. [Valdosta State University Virtual Museum of Fossils; reconstruction, Nobu Tamura]

Figure 7.14 Eurypterid *Sea scorpions or eurypterids swam with the aid of two large paddle appendages. Most were 10 to 20 cm long. This fossil is from the later Silurian Period. [Eduard Sola Vázquez]*

One group of eurypterids, grew to more than two meters long, although their average size was 10 to 20 centimeters. They began to decline during the Devonian as fish became more abundant, and they became extinct at the end of the Permian (Figure 7.14).

Eurypterids most likely preyed on the much smaller ostracoderms. All the familiar fish of today have jaws, and many use their jaws effectively to feed on other animals. Lacking jaws, however, the ostracoderms were not predators, but filter feeders. That is, they fed by drawing water into their mouth and filtering out the food particles as the water passed out through their gill slits. It is interesting and humbling to contemplate that the ancestors of all the higher vertebrates were simple filter feeders.

Because of the plate tectonic motion discussed in Chapter 4, the surface of the Earth looked quite different during the Cambrian and Ordovician than it does today (Figure 4.21). South America, Africa, India, Antarctica, Australia, southern Europe, and Arabia formed one large supercontinent, which is called Gondwana. Laurentia (most of North America) was situated very close to the Equator. Although much of Canada was above water, shallow seas covered large parts of the United States. Evidence of these seas comes from the large numbers of marine fossils that are found in many parts of the United States. Northern Europe (called Baltica) was a separate continent situated to the southeast of Laurentia. The body of water separating Baltica and Laurentia has been named the Iapetus Ocean. Siberia was a separate continent, and China formed another separate continent with much of southeast Asia.

The Cambrian and Ordovician Periods lasted for a total of about 100 million years, almost one-fifth of the Phanerozoic Eon (the time span since the beginning of the Cambrian). Although marine fossils from these periods are common, fossils of land-dwelling organisms are completely absent, indicating that neither land plants nor land animals had evolved by this time.

Near the end of the Ordovician, one of the greatest extinctions of all times took place. Perhaps an eighth of all the families of animals disappeared at this time. A possible cause of this crisis was another period of glaciation that lowered the sea level and caused a drop in the average temperature of the world. That part of Gondwana, which is now northern Africa, was located near the South Pole, and evidence of the glaciation that occurred during this time has been found in the Sahara

Desert. Apparently glaciation has played an important part in molding the life forms of Earth. It may be that whenever plate tectonic activity moves a large land mass near the poles, an ice age is potentially possible.

During periods of glaciation, the level of the ocean drops because water is removed to produce the ice that builds near the poles. The dropping ocean level exposes shallow water areas and eliminates the habitats of many marine creatures. A lowering of the water temperature is also associated with periods of glaciation, and this change would have had an adverse effect on some forms of marine life. As some life forms diminish in number, others thrive and new forms eventually evolve to fill the vacant niches.

The Silurian

The third geologic period of the Paleozoic Era, the Silurian, lasted about 25 million years, (from about 445 mya until about 420 mya). During this period, Gondwana moved further south and the African-South American portion would eventually pass over the South Pole.

The Iapetus Ocean, which was located near the Equator, became progressively narrower. As it closed, Laurentia and Baltica were sutured together to form a single land mass known as Laurussia. The collision created the Northern Appalachian Mountains along the coast of northeastern North America and the Caledonian Mountains in Scandinavia northern Great Britain, and Greenland (Figure 7.15).

This mountain building episode is called the Caledonian orogeny. (Caledonia is an old name for Scotland and orogeny means *mountain creation*) The Caledonian orogeny reached its climax near the end of the Silurian. These mountains reached elevations comparable to the Rockies and the Alps of today, but erosion has lowered them over the past 400 million years or so. Laurussia is sometimes called the Old Red Sandstone continent because of

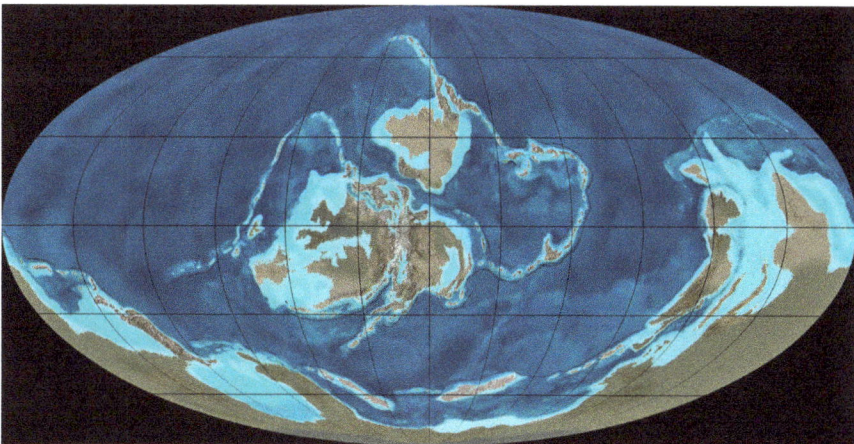

Figure 7.15 Earth Near the Middle of the Silurian On the Equator, Laurentia and Baltica were colliding and the combined land masses will form Laurussia. The Caledonian and Northern Appalachian Mountains were created by the collision. Siberia was to the north of Laurussia and the South America-Africa portion of Gondwana was near the South Pole. [Ron Blakey/Colorado Plateau Geosystems, Inc]

Figure 7.16 Acanthodians Climatius *(meaning "inclined" or "tilted") was a 420 million year old acanthodian. Most acanthodians (possibly the first jawed fish) were about 10 to 20 cm long, and many of their lower fins were paired, each with its own spine. [fossil, Roger Jones collection; reconstruction after Watson (1937), see credits page 415]*

the red sandstone that was formed as the Caledonian Mountains eroded away.

Many new life-forms appeared in the early Silurian as the marine animals recovered from the devastation that occurred at the end of the Ordovician. Brachiopods were probably the most common life-form of the Silurian, but tabulate and rugose corals increased and became important reef-builders as did a type of sponge known as the stromatoporoids.

An important development during the Silurian Period was the evolution of

the earliest jawed fish, possibly the acanthodians. All higher vertebrates possess jaws, and they evolved from these first jawed fish.

Acanthodians (which means *spines*), had many fins that were supported by protective spines on their leading edge (Figure 7.16). Scales and spines believed to be from acanthodians have been found in early Silurian rocks, but more complete specimens are not known before the early part of the next period (the Devonian).

Although acanthodians had bony scales and spines that preserved well, their internal skeleton was mostly cartilage which left a poor fossil record. Since their internal structure is not well known, their relationship to other groups is not well understood. They share features with both the sharks (chondrichthyes) and bony fish (actinopterygians). Perhaps they are ancestral to both groups.

Another type of jawed fish that first appeared near the late Silurian had large plates of armor covering their head and thorax (Figure 7.17). They were called placoderms (meaning *plate skin*). Placoderms are distinctive enough that they are placed into a separate class, one of the four classes into which all fish are placed.

Figure 7.17 Early Jawed Fish Bothriolepis *was a 370 million year old placoderm that was about 30 cm long. The function of the limbs is not well known. [fossil photo, Parc national de Miguasha; reconstruction, Citron]*

While life was making a comeback in the ocean, one of the most important events in the development of life took place on the land. By the middle part of the Silurian, plants had developed the ability to survive on the land. In the sea, organisms are supported by the water and are continually bathed in nutrients. Therefore, they do not need differentiated tissue. Each cell acts like an isolated plant, performing all the necessary life functions independently of the other cells in the plant.

On land, plants must gather water and nutrients from the ground and transport them to the aboveground parts of the plant where food for the entire plant is made by the process of photosynthesis. Some of this food must be carried to the cells below ground.

Most land plants possess a system of vein-like structures for transporting food and water to various parts of the plant. These plants are known as the vascular plants. Because of the different functions that must be performed by a land plant, their cells are generally differentiated into tissues such as roots, stems, and leaves. The most primitive plants do not have roots or leaves, but have an undifferentiated structure called a thallus.

The first and only vascular land plants discovered from the Silurian Period were only a few centimeters high. They are known by the genus name *Cooksonia*. *Cooksonia* had no leaves or roots and was about the most primitive vascular plant that one could imagine (Figure 7.18). It may have been an intermediate form between the vascular and nonvascular plants. This plant had a simple branching thallus with spore-bearing cups at its tips. The establishment of these early plants on the land would lead the way for the animals that would soon follow.

The Devonian

There is not a clear division between the Silurian and the Devonian since no great extinction separated these two periods. The Devonian, the fourth period of the Paleozoic Era, lasted about 60 million years (from about 420 mya to about 360 mya) and is sometimes referred to as the age of fishes. It was during this period that fish diversified and became the undisputed rulers of the ocean.

Of equal importance, however, was the diversification of plant life on the land. The evolution of diverse land

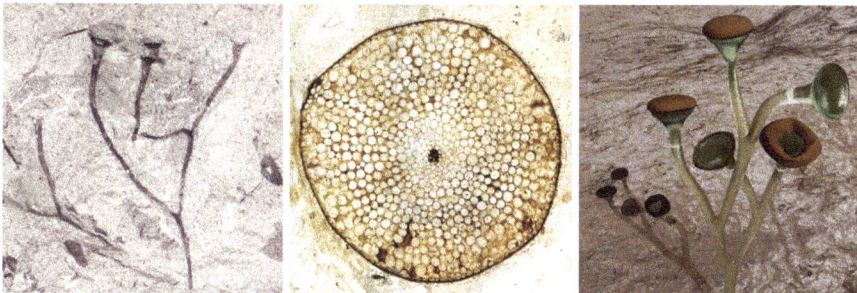

Figure 7.18 Early Vascular Land Plants Cooksonia fossil (left), from the later Silurian, was only a few cm tall. It shows the spore bearing cups. [Hans Steurh]; A cross section (middle) of the permineralized thallus of a related species (Rhynia) from the early Devonian shows the vascular bundle at the center. [Plantsurfer]; A reconstruction of Cooksonia (right) as it probably looked. [Smith609]

plants led to the formation of the great coal forests of the Carboniferous Period that followed. With the greening of the land, the animals would invade the land as well. The land literally did turn green during the Devonian, and the period could accurately be called the age of plants.

Near the beginning of this period, the plants were small. Most were from a few centimeters high to perhaps a meter or two, and they were simple in structure. For example, none possessed primary roots or leaves. They lived in a habitat similar to the marshes of today. These earliest land plants produced one type of spore (by meiosis) and are, therefore, known as *homosporous* plants.

A spore grows into a structure called a gametophyte that produces gametes. Two gametes (generally from different gametophytes), unite to form a zygote, and the zygote then grows into a new plant called a sporophyte. The large plant that you see is generally the sporophyte. (In higher organisms, the gametes are the egg and sperm.)

Much valuable information about the earliest Devonian plants comes from the Rynie chert of Scotland. Chert is silicon dioxide, and the Rynie chert was apparently formed when the area was flooded with silicon-rich water produced by nearby volcanoes. The permineralized fossils that formed in the chert are of such excellent quality that fungi have been found in the cells of the plants. This specimen is the first indisputable evidence of fungi in the fossil record. Fungi often form symbiotic relationships with plants and may even be necessary to help plants absorb moisture and nutrients from the ground.

Along with these early land plants we also find the first spider-like arachnids, millipedes, and the first wingless six-legged arthropods that are close relatives of the insects. As the plants invaded the land, we expect that the animals must have quickly adapted to exploit the readily available supply of plant food.

Near the middle of the Devonian, *heterosporous* plants the size of shrubs or small trees evolved. As the name implies, these plants produce two types of spores (by meiosis), a larger type (*megaspore*) that grows into an egg producing structure and a smaller spore (*microspore*) that grows into a sperm producing structure. Fertilization occurs when a sperm unites with an egg to form a zygote. The zygote then grows into a sporophyte plant that is the spore producing plant. The heterosporous plants were more like seed plants than the less evolved homosporous plants.

The oldest known tall tree with a single trunk, *Wattieza*, is from the middle of the Devonian Period (Figure 7.19). Before this time, plants grew like bushes or small trees. This earliest tree grew to about 10 meters in height with a trunk that was about a third of a meter wide. It had the look of todays tree ferns or palm trees. Its trunk had a pithy core and the foliage sprouted upwards from the top of the trunk. There were no horizontal branches.

A little later in the Devonian we find a heterosporous tree with horizontal branches named *Archaeopteris*, which means *ancient fern* (Figure 7.19). Fossils of these trees are found worldwide. It belonged to a group of plants known as progymnosperms that had wood like a gymnosperm (the kind that produces annual growth rings), but did not produce seeds as gymnosperms do. Their foliage was like that of a fern, a large umbrella of feathery fronds. Some branches had spore bearing capsules instead of leaves. They represent one of those intermediate forms (sometimes called missing links)

Figure 7.19 The First Tall Trees Wattieza (left) was a spore bearing tree that grew about 10 m tall and resembled a tall palm tree. Next is a fossil of the crown and stump of the tree. [drawing and crown fossil courtesy of the New York State Museum, Albany, NY]; Archaeopteris (right) grew about 10 m tall with a trunk over 1.5 m thick. A fossilized frond and trunk are shown to the far right. [reconstruction, Ivy Livingstone/Biodidac; frond photo, Parc national de Miguasha; stump, Brad Holt]

that bridge the gap between two major groups of organisms.

Before the end of the Devonian, seed ferns, that are seed-producing plants with fern-like leaves, had appeared. Seeds are formed after the egg is fertilized. They contain the embryo plant, stored food, and a covering called the integument.

By the close of the period, extensive forests of large trees had evolved, and many species probably lived together in plant communities. The first horsetails (sphenopsids), which are sometimes included in the fern group, appeared in the late Devonian.

Lycopods or lycophytes, whose members were closely related to the club mosses of today, arose and diversified in the middle and late Devonian. Sphenopsids and lycopods would become very common and contribute greatly to the coal beds that formed during the next period.

In the ocean, tabulate corals nearly vanished after the Devonian, but the rugose corals were abundant in the Devonian and in the early part of the Carboniferous Period that followed. Brachiopods reached their peak numbers during the Devonian, and although trilobite diversity was generally on the decline, some very unusual species lived during this period, including the largest trilobites ever found (Figure 7.20).

An important group of mostly coiled cephalopods called ammonites appeared in the early part of the Devonian. Ammonites were similar in appearance to the nautiloids but had a different internal structure. Both nautiloids and ammonites had shells that were divided into chambers. As the animal grew, it secreted material to seal off the last chamber. In nautiloids, the chamber walls (called septa) were smooth, gently curved surfaces that

Figure 7.20 Trilobites and Ammonites Walliserops trifurcatus (left) is a very unusual fossil trilobite from the early and middle Devonian of Morocco. The function of the trident rising from its head is not known. [Moussa Direct Ltd.]; The ammonite fossil (right) has been prepared so the last chamber wall is visible. This specimen is a much later form from the Jurassic (about 180 mya). Its seam of contact with the outer shell is very convoluted. [fossilsdirect.co.uk]

were sutured to the inside edge of the shell. The lines of contact between the chamber walls and the edge of the shell (called the septal sutures) were simple curves.

In ammonites, however, the chamber walls evolved into very convoluted structures, especially near the edges, and the suture lines were folded into very complex curves (Figure 7.20). The convolutions in the chamber wall provided greater strength to the shell.

Both ammonites and nautiloids possessed a tube-shaped organ called a siphuncle that was used to regulate the amount of fluid and gas in the walled-off chambers. The amount of fluid regulated the buoyancy of the animal as it changed depth in the ocean. In ammonites, the siphuncle ran along the outer wall of the coil while in the nautiloids the siphuncle ran down the middle of the coil.

As stated above, the Devonian is often referred to as the age of fishes. While the plants were diversifying on land, the fish were expanding in the oceans, but they were not the kinds of fish that are familiar to us today.

Placoderms, which first appeared in the Silurian, diversified, and some became very large. The largest were

Figure 7.21 Dunkleosteus One of the largest placoderms was about 10 m long. It lived in the late Devonian, about 370 mya. Only the head plates have been found so the reconstruction is based on closely related species. Instead of teeth, it had sharp plates used for biting. [skull photo, Sam Noble Museum; reconstruction, Dmitry Bogdanov]

Figure 7.22 Cladoselache This 360 million year old primitive shark was almost 2 m long. Its mouth opened in front of the head instead of below like modern sharks. [fossil photo, AMNH; reconstruction after Bendix-Almgreen (1975); Zangerl (1981), see credits p. 415]

about 10 meters long (Figure 7.21). A 380 million year old placoderm fossil indicates that some gave birth to live young. That means they evolved the ability to reproduce by internal fertilization. Some male fish had claspers to aid in copulation, and some fossils have been found with embryos still inside the mother. During the late Devonian, placoderms began to decline, and they became extinct at the end of the period. Perhaps the rise of other fish was responsible for their demise.

The Chondrichthyes (sharks, rays, and skates) are the cartilaginous fish. Early forms appeared in the late Silurian and diversified throughout the Devonian. Sharks became an important component in the sea during the Devonian.

Since they have skeletons of cartilage which does not fossilize well, the history of sharks is not well known. Most of our information about sharks comes from scales and teeth. Sharks shed their teeth regularly and they fossilize well. However, many well preserved fossils of an early shark

known as *Cladoselache* have been recovered from the Upper Devonian Cleveland Shale of Ohio (Figure 7.22). Some specimens even show the contents of their stomach. They had a streamlined body with a large tail fin, and their dorsal (back) fins were reinforced with spines along the leading edges.

Most modern sharks reproduce by internal fertilization. The males have fleshy finger-like projections on their underside called claspers which are used to direct the sperm during fertilization. Since no claspers have been found on *Cladoselache*, their method of reproduction is unclear.

The first primitive bony fish (of the class Osteichthyes) also appeared near the beginning of the Devonian and became more common and more diversified as the period progressed. An interesting group of bony fish, the sarcopterygians or lobe-finned fish, may have evolved from acanthodians. They appeared near the start of the Devonian and soon invaded fresh water environments. They had fleshy muscular fins that were probably used like legs to push the fish along the bottom of ponds and lakes (Figure 7.23).

Figure 7.23 Osteolepis This 395 million year old lobe-finned fish (sarcopterygian) from the middle Devonian was about 20 cm long. Most lobe-finned fish lived in fresh water. [fossil, Roger Jones collection; reconstruction, Nobu Tamura;]

Members of a group of lobe-finned fish named rhipidistians are believed to be the ancestors of the amphibians. Remember, amphibians are animals that are intermediate between fish and reptiles. Frogs and toads generally have a larval or tadpole stage that is very fishlike, but the adults are generally terrestrial. In the next chapter, we will investigate the evolution of the amphibians from lobe-finned fish in more detail.

Fossils of lobe-finned fishes are mostly found in freshwater sediments, and near the end of the Devonian they were much more common than the ray-finned fishes that also appeared near the beginning of the Devonian. The lobe-finned fish declined after the Devonian, and most of the bony fish of today are ray-finned fish (called actinopterygians).

There are only eight species of lobe-finned fish living today, two marine coelacanths and six freshwater lungfish. Four species of lungfish are found in Africa, one in South America, and one in Australia. One species of coelacanth is found along the east coast of Africa and the other lives in the waters off Indonesia.

The most primitive surviving members of the ray-finned group (the chondrosteans) include the Reedfish and the bichirs of Africa, the paddlefish of North America and China and the sturgeons (Figure 7.24). The non symmetrical shape of their tail fin is evidence of their primitive nature. The Chinese Paddlefish may be extinct since 2007 was the last time one was seen alive. The bichirs are particularly interesting because they have lobe-like fins and paired lungs.

It was near the end of the Devonian that the first amphibians ventured out of the water for brief periods of time. These first amphibians were called labyrinthodont amphibians because of the labyrinth pattern of their teeth. This pattern was also found in the teeth of a group of lobe-finned fishes (the rhipidistians). As we will see in the next chapter, the rhipidistians had many things in common with the first amphibians, which is why we believe they were the ancestors of the amphibians.

The Devonian was a time of great change for life on Earth. At the beginning of the period, the sea life was still reminiscent of earlier times. Dominant animals were the trilobites, ammonites, and the eurypterids. Brachiopods, bivalves, and crinoids dotted the sea floor. Rugose and tabulate corals built up reefs in many areas. A few strange little jawless fish could be found filtering food from the debris on the ocean bottoms.

On the land, small primitive plants were just venturing out onto the edges of shallow wetlands. They were only stalks a few tens of centimeters high and if they had not been green, many of us would not even have recognized them as plants.

Figure 7.24 Ray-finned Fish Atlantic Sturgeon (top) [Brian Reynolds] and American Paddlefish [JCB], two of the most primitive of the actinopterygians (ray-finned fish) alive today. Note their non symmetrical tails.

In contrast, about 50 million years later, near the end of the period, large areas of land were covered with ferns and bushes, and there were even forests of tall trees. However, most of the plants would have been unfamiliar to us. There were no flowering plants or conifers which account for most of the familiar trees and other plants of today.

Life in the ocean had also changed dramatically over the period. There were many jawed fish, such as the armor plated placoderms, some of which were huge. There were also primitive sharks, primitive ray-finned fish, and lobe-finned fish. Many of the lobe-finned fish lived in fresh water. Although amphibians evolved late in the period, they did not venture far from the water. The land was still void of large animals of any kind.

Before the end of the Devonian, another major extinction occurred. The ostracoderm fish (the group to which the first jawless fish belonged) were common near the beginning of the Devonian but declined in numbers as the period progressed, and they became

extinct by the end of the period. Only a few genera of placoderm fish (the armor plated ones) survived beyond the end of the Devonian, but they too became extinct near the beginning of the next period (the Carboniferous).

Several invertebrate groups, such as selected brachiopods and most of the trilobites, also died out at the end of the Devonian. Only one group of trilobites (belonging to the order Proetida) survived into the Carboniferous. Ammonites, lobe-finned fish, and many other groups suffered losses although they were not wiped out completely. Near the end of the Devonian, many of the corals became extinct, ending the reef building that had flourished during the period.

The Carboniferous

The fifth period of the Paleozoic Era was the Carboniferous, which lasted about 60 million years (from about 360 mya to about 300 mya). It was named after the large coal (carbon) deposits

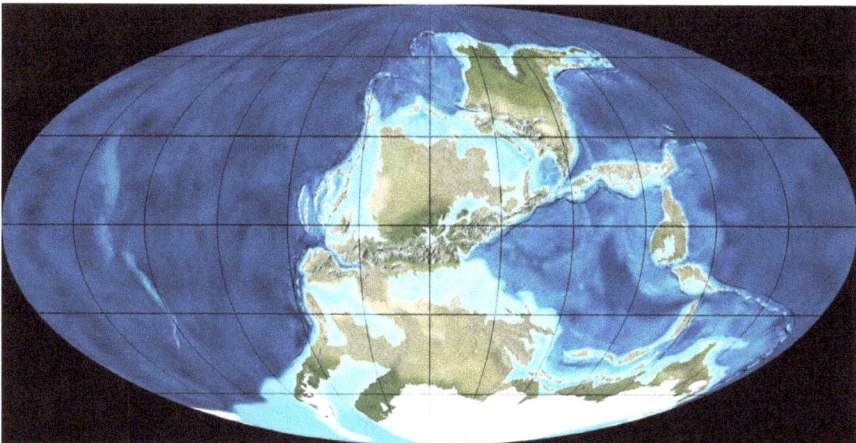

Figure 7.25 Earth near the end of the Carboniferous Laurussia and Gondwana (the northwest African portion) were colliding near the Equator (center of the map), forming the Appalachians. Kazakhstania-Siberia was located further north and fragmented China was to the east, across the Paleo-Tethys Sea. Southern Gondwana was near the South Pole, and those portions were glaciated. [Ron Blakey/Colorado Plateau Geosystems, Inc]

214

that were laid down in parts of eastern United States and Europe during the period. The Carboniferous is divided into two subperiods. The first 35 million years is called the Mississippian and the last 25 million years is the Pennsylvanian.

By the late Carboniferous, Laurussia joined with Gondwana. The collision generated a period of mountain building known as the Hercynian or Variscan orogeny. It created the southern portion of the Appalachians in North America (from New York south to Georgia), the Atlas Mountains in northeastern Africa, and many of the mountains in southern and central Europe. As with any mountain building episode, the formation did not happen quickly, but generally lasted for tens of millions of years, well into the middle of the next period, the Permian (Figure 7.25).

In the sea, the placoderms became extinct early in the period and the other fish, especially sharks-like forms, diversified further and continued to

dominate the oceans. The lobe-finned fishes, which were found mostly in fresh waters, declined during the Carboniferous and were replaced by ray-finned fish which flourished and diversified. These were still the more primitive forms as the advanced ray-finned fish of today had not yet evolved. Crinoids, blastoids, and brachiopods became especially abundant, but a drop in sea level near the middle of the period caused the extinction of many crinoids and ammonites. Trilobites became less common throughout the period.

The Carboniferous was also an important time for the development of the terrestrial vertebrates. It is sometimes known as the age of amphibians, and many of these creatures roamed the vast coal forests (Figure 7.26). Some amphibians probably fed on each other, but others undoubtedly fed on the insects and other arthropods like centipedes that were well established by this time. The first terrestrial scorpions are also from

Figure 7.26 Amphibian Pederpes (left) was a one meter long early Carboniferous amphibian that was better suited for terrestrial life than earlier amphibians. [fossil, Jenny Clack/Hunterian Museum]; Eryops (right), a large amphibian (about 2 meters long) that lived about 300 mya in southern Laurussia (southern North America). [photo, NMNH, Washington/Daderot; both restorations, Dmitry Bogdanov]

near the beginning of the period (remember scorpions are not insects).

The fossil record of insects is generally poor, but the first primitive wingless insect fossils are from about 400 million years ago near the beginning of the Devonian. At least 320 mya, winged insects evolved. There were many small insects, but there were also some very large forms. One of these was a giant called a griffinfly (*Meganeura monyi*), an early ancestor of the dragonflies, that lived near the end of the period (Figure 7.27). It had a wingspan of 65 cm (just over 2 feet).

Also late in the Carboniferous, we find fossils of the first cockroach-like insect, although they were probably present earlier. They belonged to the group that would ultimately evolve into the cockroaches, termites, and mantids. The largest of these insects was from the late Carboniferous. It was about 9 cm long, not quite as big as some tropical cockroaches of today which can be 10 cm long.

Although not an insect, another giant arthropod that appeared in the Carboniferous was *Arthropleura*, a centipede and millipede relative. It grew to about 2.5 meters long, making it a contender for the largest arthropod ever (Figure 7.27). It disappeared during the early part of the Permian. The ancestors of this group first appeared in the late Silurian.

Near the middle Carboniferous, the first reptiles made their appearance (about 50 million years after the first amphibians). The first reptiles, called cotylosaurs or stem reptiles, were the most primitive and show a mixture of amphibian and reptilian characters. It is not always easy to tell the difference between a reptile fossil and an amphibian fossil from this time period. Amphibians slowly blended into reptiles.

Figure 7.27 Arthropods *Griffinfly (top) was a dragonfly-like insect with a wing span of about 65 cm. [photo, MNH Brussels/Hcrepin] The 30 cm wide tracks (bottom) were made by two arthropleuras, relatives of millipedes. Based on the size of the tracks, they must have been about 2.5 meters long. [photo, David Litchfield]*

The word amphibian comes from the Greek word *amphi* meaning *both* and *bios* meaning *lives*. The name refers to the fact that most amphibians spend the first stage of their life as fishlike larvae in water (the tadpole stage for frogs) and the last stage of their lives on land, or at least partially on land. Most amphibians must return to water to lay their eggs which are jellylike. Their eggs would dry out and die without the water. On the other hand, reptiles lay eggs covered by a leathery shell which protects them from drying out.

Since details like those just described are almost never preserved as fossils, they cannot be used to differentiate amphibian fossils from reptile fossils. Reptiles generally have longer fingers with claws, sturdier legs, stronger hips and shoulders, a more upright stance, and a stronger spine. These features allow reptiles to walk with less wiggling from side to side and move faster without dragging their stomach on the ground.

Figure 7.28 Coal Forests *Lepidodendron (left) was a 30 m tall heterosporous relative of the club mosses. Next are fossils of its leaves (top), bark (bottom), and a spore bearing cone (middle). As the leaves fell, they left the regular pattern of scars seen in the bark. Calamites (right), was a giant 30 m tall horsetail (sphenophyte). Fossils (far right) are of its leaves (top) and a cross section of the trunk, which was hollow, but is collapsed in the fossil. [both leaves and cone, Hans Steurh; trunk, Mike Viney/Virtual Petrified Wood Museum; both reconstructions, Sergio Perez]*

However, there is no single skeletal difference that can readily separate a reptile from an amphibian. Therefore, it may not be possible to put your finger on the first reptile. This is true of other transitional forms as well. We will look again at this problem in the next chapter.

During the later part of the period, the uplifting caused by the collision of Gondwana and Laurussia produced a decrease in coastline and an increase of floodplains and deltas in Laurussia. Great swampy forests covered much of eastern United States and Europe. The remains of many of these trees were eventually converted into vast coal deposits.

Giant heterosporous lycopods which were common trees in the coal swamps were related to the inconspicuous club mosses of today. They are most closely related to a little known group of small plants called quillworts, named because of their hollow quill-like leaves. Some lycopod trees like *Lepidodendron*

reached more than 30 meters in height. By comparison, the tallest trees of today are the California Redwoods which reach 115 meters. Lycopods probably contributed most of the plant material to the coal deposits, but their numbers dwindled as the period ended, and they became extinct in the Permian (Figure 7.28).

Horsetails, belong to a group of plants called the sphenopsids. During the later part of the Carboniferous when the coal forests reached their peak, giant horsetails were important members of the coal swamps. Seed ferns, and tree ferns also became important members of these forests (Figure 7.29).

Ferns diversified during the period, and cycads probably arose very late in the Carboniferous. Conifers also appeared near the end of the period, but the familiar flowering plants of today had not yet appeared. The trees of the Carboniferous forests would have

Figure 7.29 More Coal Forest Trees *Medullosa (left), a member of the extinct group of seed ferns, reached heights of 10 meters. It was a dominant tree in the later carboniferous, but went extinct in the early Permian. [reconstruction, Ivy Livingstone/Biodidac]; Next is Psaronius, a homosporus tree fern about 10 to 15 meters tall which, except for the leaves, resembled a palm tree. Its leaves and a cross section through its trunk are shown with the reconstruction. Psaronius was a dominant tree of the late Carboniferous early Permian coal forests. [leaves, Hans Steurh; trunk, Mike Viney/Virtual Fossil Museum; reconstruction after Morgan (1959), credits p. 415]; Sigillaria (far right) was a lycopod almost as tall as Lepidodendron. [Sergio Perez]*

looked very strange to anyone familiar with the trees of today.

As Gondwana was passing over the South Pole during this time, the sea level fluctuated as the ice cap waxed and waned on the southern part of the continent. When the sea rose and flooded the coal forests, the plant material got covered with sediment and was eventually converted to coal. When the sea receded again, the forests grew back and produced a thick layer of vegetation. This cycle of rising and falling sea levels was repeated many times during the period, producing multiple layers of coal.

Another feature that may have contributed to the large accumulation of organic matter that eventually produced the coal deposits is that many animals and fungi that attack and break down dead plants had not evolved yet.

For example, fossil termites are not found until the early Cretaceous, 150 million years after the Carboniferous.

The Permian

Named after Perm, a small town in Russia near the Ural Mountains, the Permian was the sixth and last period of the Paleozoic Era. The Permian began about 300 mya and lasted until about 250 mya, a duration of approximately 50 million years.

During the Permian, the combined landmasses of Kazakhstania and Siberia collided with Laurussia and the Ural Mountains were formed along the suture line. Later they were joined by China, and the formation of a gigantic supercontinent called Pangaea was completed. Recall that the eastern portion of Gondwana had earlier joined with Laurussia, but much of Eurasia was separated from Gondwana by a body of water called the Tethys Sea. The northern portion of Pangaea is usually called Laurasia and the southern portion is called Gondwana. The Tethys Sea separated the eastern portions of these two land masses. A map of Pangaea is in Chapter 4 (Figure 4.20).

Reptiles changed greatly during the Permian, and many became more mammal-like in appearance (Figure 7.30). They developed differentiated teeth (most other reptiles have teeth that are all of one kind), and they also developed limbs that were placed more directly beneath the body, enabling them to move in a more upright manner. This improved stance also enabled the animals to move faster. By the close of the Permian, the mammal-like reptiles had taken over the land and were the dominant life-form.

Dinosaurs did not appear until much later. Large amphibians apparently could not compete with the more advanced reptiles and large amphibians did not survive beyond the Permian.

The Permian Period ended 250 million years ago with what appears to be the greatest period of extinction in the history of the planet. It has been estimated that more than one half of the animal families disappeared at this time. Both land and sea creatures were affected. Many mammal-like reptiles became extinct, but new forms would arise and become the dominant animals of the next period, the Triassic.

The cause of this great extinction is not known, but it appears that there was a climate change associated with it. A great period of glaciation (called the Permo-Carboniferous Glaciation) occurred in the late Carboniferous and early Permian, and it may have had an adverse effect on the life of the times. Evidence of this glaciation can be seen

Figure 7.30 Mammal-like Reptile Lystrosaurus *(left) was common in central Gondwana during the Late Permian and Early Triassic. This meter long animal had two tusks, but no other teeth. [photo, Natural History Museum, Stuttgart/Ghedoghedo; reconstruction, Dmitry Bogdanov];* Cynognathus *(right) was a meter long predator of the Early and Middle Triassic. It had differentiated teeth and is closely related to the mammals. [photo, Museum of Man and Nature, Munich/Ghedoghedo; reconstruction, Nobu Tamura]*

in southern South America, central and southern Africa, Antarctica, Australia, and India. As we have seen, these areas were all connected during this time, and the South Pole was located near the middle of these land masses.

The buildup of ice caused the sea level to drop, and this drop would have produced a decrease in the shallow water habitats available to the sea creatures. Since shallow water supports most of the life, this loss of habitat would have destroyed much of the sea life. It appears that there was also a temperature drop in the oceans at the end of the Permian as evidenced by a reduction in reef growth and limestone deposition.

Since the large land masses were all united at this time, the animals were not isolated from each other, and the increased competition may have caused some of the observed extinctions. It appears that the extinction was not instantaneous, but occurred over a period of several million years. Many of the species that became extinct at the end of the Permian had been on the decline for several million years before their extinction. There are many ideas about the Permian extinction, but we may never know exactly what caused this greatest extinction of all times.

Among the casualties at the end of the Permian were the rugose and tabulate corals, blastoids, all but one group of crinoids, and many families of brachiopods. Trilobites, that had managed to survive for about 300 million years, became extinct as did their close relatives the eurypterids (sea scorpions). Most of the ammonites died out, but one group survived to produce an impressive number of new varieties in the next era, the Mesozoic.

The Permian saw the disappearance or reduction of several groups of fish, including the acanthodians and the rhipidistians. However, the first of the more modern ray-finned fishes (the neopterygians) made their appearance during the period. These fish are distinguished by their modified jaws and associated musculature (the jaws had more freedom of motion). These changes improved feeding, while changes in their scales and fins provided greater maneuverability. The only primitive neopterygians (called holosteans) that are still with us are the seven species of gars which are found in North and Central America, and the Bowfin (*Amia calva*), a single species of eastern United States and southern Canada.

With the end of the Permian, the Paleozoic Era was brought to a close. About 300 million years in length, the Paleozoic Era had witnessed a dramatic rearrangement of the continents and saw many changes in the organisms that lived in the sea and on the land. Starting with the soft-bodied worms and small shelly creatures at the beginning of the Cambrian, the seas now harbored various kinds of fish, clams, cephalopods, and other sea creatures. Still, however, there were no modern fish, and no marine reptiles or mammals.

The land, that had been barren for the first 130 million years of the era, was now largely covered with vast forests of ferns, strange looking trees, and unusual plants. Also living on the land were insects, amphibians, and reptiles. Still absent, however, were the flowering plants, the birds, the true mammals (although some of the reptiles had many mammal-like features), and of course, the great dinosaurs had yet to make their appearance.

220

The Triassic

The middle era of the Phanerozoic Eon is called the Mesozoic. During this era of approximately 185 million years duration, the land and later the air and the sea would be dominated by reptiles.

Today, most of the landmasses are located in the northern hemisphere, but throughout most of the Mesozoic, the landmasses were about equally distributed in the northern and the southern hemispheres. Perhaps because of this distribution of continents, the Mesozoic had a mild climate with no evidence of glaciation.

The Mesozoic era is divided into three periods, the Triassic, the Jurassic, and the Cretaceous. The first of these periods, the Triassic, began about 250 mya and ended about 200 mya, a span of approximately 50 million years. The Triassic, which obviously has

something to do with three, was so named because in Germany, the rocks of this period are divided into three distinct layers.

New to the ocean environment during the Triassic were several groups of aquatic reptiles. The nothosaurs (of the sauropterygian group) were the first to appear in the early Triassic (Figure 7.31). They were aquatic reptiles with sharp teeth, which indicates that they were fish eaters. Although nothosaurs had paddle-like limbs, they probably spent some time on land, much as the seals do today. The placodonts were a group of stocky marine reptiles, some of which were covered with bony armor. They had large block-like teeth that were used to crush shellfish such as clams and mussels.

Near the early part of the Triassic, the first ichthyosaurs appeared. They resembled and generally seemed to

*Figure 7.31 **Aquatic Reptiles** A Nothosaurus (top) from the middle Triassic was about 4 meters long and probably ate fish. [photo, Museum für Naturkunde, Berlin/Elke Wetzig]; Ichthyosaurus (bottom) was about 1 m long, but other members of the group reached 9 m in length. They gave birth to live young. [photo, Holzmaden, Germany/Didier Descouens]*

Figure 7.32 Pterosaur This specimen lived during the late Jurassic and is from the famous Solnhofen limestone deposits of southern Germany. It belongs to the genus Rhamphorhynchus, whose members reached lengths of over 1 meter with wing spans of just under 2 meters. Note the sharp teeth and long tail. It most likely fed on fish. [photo, Royal Ontario Museum/Daderot]

Bony fish were also present in the ocean, but they were still of the more primitive type, having diamond shaped scales and jaw mechanisms that were not the modern type.

Near the end of the Triassic, we find the first evidence of the flying reptiles which are called pterosaurs (Figure 7.32). Although the whole Mesozoic Era is popularly known as the age of dinosaurs, the first dinosaur fossils did not appear until near the middle of the Triassic, and the really huge dinosaurs did not appear until the Jurassic. Many mammal-like reptiles had become extinct at the end of the Permian, but new forms arose near the beginning of the Triassic and dominated the land along with a group of crocodile-like reptiles called thecodonts. They were the ancestors of the dinosaurs.

The ancestors of the modern lizards, had already appeared by the late Permian. As we will see in the next chapter, lizards can be told from other reptiles by the absence of the lower temporal bar in their skull.

The earliest turtle fossils are also from the Triassic. Although these turtles had lost their marginal teeth, they still had teeth on their palate. Modern turtles have a horny beak and are completely toothless. Besides having palate teeth, the early turtles show some differences in the bone structure of the head and shell. Also of

occupy a niche similar to that of the present day porpoises (Figure 7.31). As a group they survived until about 90 million years ago, dying out about 25 million years before the dinosaurs.

Invertebrate life in the ocean gradually became more modern in appearance during the Mesozoic. Bivalves (clams) and gastropods (snails) became increasingly more common. Oysters appeared and clams developed siphons, which enabled them to become burrowing creatures, a feature that may partially account for their great success. As the clams became more abundant, brachiopods became scarce. By the end of the Jurassic, brachiopods had dwindled to the two orders that are still found today. The ammonites, which had suffered great losses at the close of the Permian, diversified again during the Triassic. Modern corals appeared, and sea urchins diversified greatly.

Figure 7.33 Early Mammal One of the earliest mammals was 200 million year old Morganucodon, about 10 cm long not counting the tail. Its skull was 2 or 3 cm long, with a jaw intermediate between reptiles and mammals. [Michael B.H.]

late Triassic-early Jurassic origin, was the first crocodile (genus *Protosuchus*). In comparison to later members of the group, these creatures were relatively small, being about a meter in length.

The first animals that are considered to be mammals appeared in the Late Triassic (Figure 7.33). These animals were small shrew-like creatures that were probably nocturnal. During the Mesozoic, mammals never developed into forms much larger than a badger, apparently because they were unable to compete with the dinosaurs.

Ferns dominated the plant life for much of the Triassic, but cycads, conifers, and ginkgoes were common trees. The uniformity of the plant and animal fossils from many different locations around the world as well as paleomagnetic studies indicate that the major landmasses were still connected during the Triassic.

Near the end of the period, however, Pangaea began to fragment. As we measure the age of rocks on the floor of the North Atlantic Ocean, we find the oldest parts of the ocean bottom (the extreme western portion and the extreme eastern part) are of Jurassic

age. The fracturing of Pangaea in the late Triassic marked the beginning of the formation of the modern continents. The breakup apparently began as a widening of the Tethys Sea and a rifting of Gondwana away from North America and Eurasia. Probably by the end of the Jurassic, North America had separated from Africa and South America (Figure 7.34).

The Triassic ended with one of the five most devastating extinctions of all times. Perhaps 30 per cent of all families became extinct near the end of the period. The labyrinthodont amphibians became extinct, and most of the mammal-like reptiles were lost. The only marine reptiles to survive were the ichthyosaurs and a group known as the plesiosaurs. The plesiosaurs were sea serpent-like reptiles with legs shaped like large paddles. They descended from the nothosaurs near the end of the Triassic. The ammonites suffered another great hit as only one of the 25 or so families survived the great Triassic extinction.

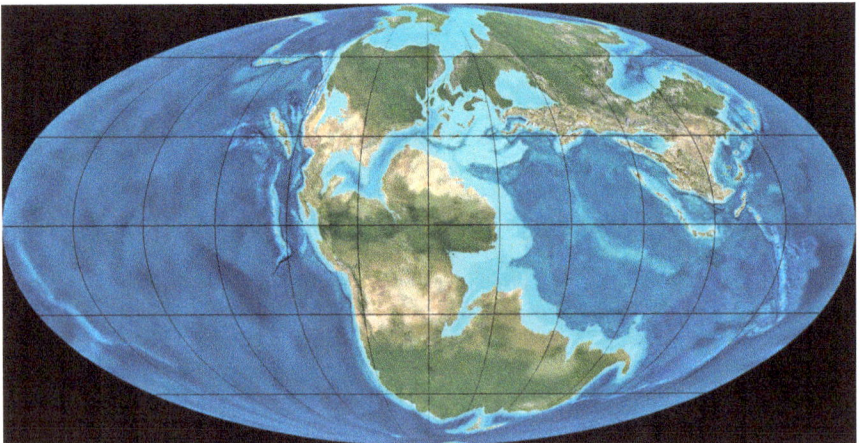

Figure 7.34 Earth During the Jurassic About 170 mya, Gondwana and Laurasia were separating. [Ron Blakey/Colorado Plateau Geosystems, Inc]

The Jurassic

The middle period of the Mesozoic Era was the Jurassic which was named after the Jura Mountains in eastern France and Switzerland. It lasted for about 55 million years, from 200 mya to about 145 mya.

The breakup of Pangaea very likely played a role in the early divergence of the mammals. Like their reptile ancestors, the earliest mammals laid eggs, and the first split was when the egg-laying mammals diverged from the therians (the group that evolved into the marsupials and the placentals). Although the fossil record of the monotremes is poor, based on DNA evidence, this split must have occurred sometime during the early Jurassic.

The metatherians (the marsupials) and the eutherians (the placental mammals) may have split about 160 mya. The oldest known eutherian fossil (*Juramaia sinensis*) is from China and is about 160 million years old. The oldest metatherian fossil is about 125 million years old and is also from China. All mammals continued as a relatively minor component of the world's fauna until the extinction of the dinosaurs.

In the ocean, the Jurassic was a time when the more primitive bony fish (sometimes called the holosteans or

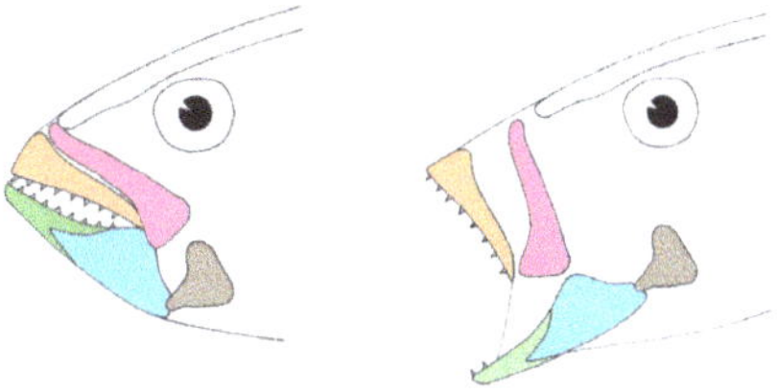

Figure 7.35 The Teleost Jaw The premaxilla (orange), and maxilla (pink) move forward with the bite. The lower jaw is composed of the dentary (green), and articular (blue) which hinges with the quadrate (tan). [JCB]

simply the primitive neopterygians) reached their peak. However, modern bony fish (the teleosts) first appeared in the latter part of the Triassic or the early Jurassic. Their numbers virtually exploded in the Cretaceous and today there are over 20,000 species.

They have a movable upper jaw, which allows the fish to extend their jaws forward while feeding. This action creates reduced pressure behind the jaws, and the food is pushed into the mouth (Figure 7.35).

They also have a tail fin that is symmetrical and produces more efficient locomotion. Their overlapping scales reduce friction and give teleosts more flexibility than the more primitive scales (called ganoid scales) that were thick and did not overlap.

Figure 7.36 Spotted Gar Note the diamond-shaped ganoid scales and the non symmetrical tail of this primitive neopterygian. [Cliff]

Figure 7.37 Great White Shark *This modern shark is able to push its upper jaw forward. The front of its jaw is mobile since it is only connected to the cranium by muscles and tendons. Sharks have five gill slits, but they are not covered by opercular bones as are the gills of bony fish. [Anna Phillips]*

Today, the only actinopterygians (ray-finned fish) that do not belong to the teleost group are the Reedfish, bichirs, sturgeons, paddlefish (Figure 7.24), Bowfin, and gars (Figure 7.36). Gars and bichirs have the more primitive ganoid scales while sturgeons have a few rows of bony plates called scutes. Paddlefish have no scales, and the Bowfin has more modern scales.

An important group of fish, the modern sharks (neoselachians), first appeared in the early Jurassic. Neoselachians have relatively smaller but stronger jaws, and the front of their upper jaw is not connected to the cranium (brain case) like ours is. This allows the shark to push its jaw forward, producing a larger stronger bite, and it allows the shark to catch food more easily and pull it into its mouth (Figure 7.37).

Many modern sharks have sharp triangular teeth so they can tear a chunk of flesh from a larger animal and not have to swallow the whole animal.

They have partially ossified vertebrae (made of bone instead of cartilage), so their swimming muscles have a better attachment, making them faster swimmers. Their pectoral and pelvic girdles are modified and strengthened, and they have large brains, an acute sense of smell, and very good vision.

Also in the ocean, the squid-like belemnites appeared and sea urchins increased in diversity and abundance. Lobsters, crabs, and squids, became common, and the ammonites again diversified greatly. If sea life alone is considered, the Mesozoic might well be called the age of the ammonites.

Aquatic reptiles were represented by the plesiosaurs and the ichthyosaurs, which apparently reached their peak abundance in the early Jurassic. Marine crocodiles also appeared during this period. As the weather became hotter and drier during the Jurassic, conifers and treelike cycads replaced the seed ferns, horsetails, and ferns of the Triassic. Cycads and ginkgoes became very common during the Jurassic.

Although the word dinosaur means *terrible lizard*, dinosaurs are not lizards, but belong to a different group of reptiles. Dinosaurs can be told from other reptiles by the openings in their skull. The number and position of the openings in the skulls of reptiles are important features for the classification

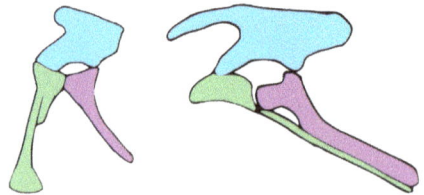

Figure 7.38 Dinosaur Hips *Saurischians (left) and ornithischians are distinguished by the configuration of their hip bones. The head of the animal is to the left. The bones are pubis (green), ilium (blue), and ischium (violet).*

of these animals. The number of fossils discovered from the early to middle Jurassic has been quite limited, but many dinosaur fossils have been found from late Jurassic times.

Dinosaurs are generally divided into two groups, reptile-hipped forms known as the saurischians, and bird-hipped forms called ornithischians (Figure 7.38). The ornithischians were herbivorous (plant eaters), and Jurassic members of this group include the well-known *Stegosaurus*. Some saurischians were plant eaters while others were meat eaters. Common Jurassic examples include *Allosaurus*, a carnivorous member of the group,

and two of the most massive of all the dinosaurs, *Brachiosaurus* and *Diplodocus*. These extremely massive herbivorous dinosaurs form a natural subgroup of saurischians called the sauropods (Figure 7.39).

Near the end of the Jurassic, the first birdlike animal, *Archaeopteryx* appeared. Birds represent the third

Figure 7.39 Late Jurassic Dinosaurs *Stegosaurus (top left), was a plant eater about 7 meters long. [Nobu Tamura]; Allosaurus (top right) was a meat eater about 9 m long. [Emily Willoughby]; Skull is of Allosaurus. [photo, Oklahoma Museum of Natural History/ Bob Ainsworth]; Diplodocus (bottom) was a plant eater estimated to be about 50 m long and weighed about 12 metric tons (12,000 kg). [Dmitry Bogdanov]*

226

independent group of animals to conquer the air, the first being the insects and the second being the pterosaurs. Pterosaurs, which were not closely related to the birds, diversified during the Jurassic and were quite common. Bats would become the fourth group to conquer the air.

Archaeopteryx fossils are quite remarkable, since the preservation of feather imprints suggests that the creature was a bird. But, it had a long lizard-like tail, a reptile skull with teeth, and three separate fingers with claws instead of the fused wing-structure of modern birds. In fact, if it were not for the feather imprints, the skeletons would be classified as small dinosaurs. *Archaeopteryx* is another example of a creature that shares characteristics of two different groups of animals, an intermediate form that bridges the gap between the reptiles and the birds (Figure 7.40).

The *Archaeopteryx* feathers are only one of several examples of the unusual preservation of delicate animal parts in a limestone bed in Germany near Solnhofen. The bed apparently formed in a quiet lagoon located behind a coral reef. The fine-grained limestone has preserved dragonflies, pterosaurs, and even the final tracks of dying animals as they hit the bottom of the lagoon.

Although frog-like amphibians (*Triadobatrachus*) are found in the early Triassic in Madagascar, fossils of modern frogs are not found until the

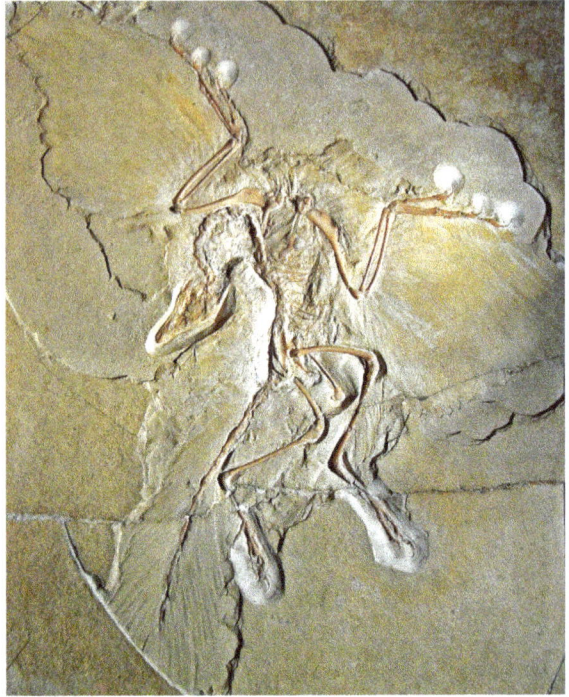

Figure 7.40 *Archaeopteryx* *From the Solnhofen limestone deposits, this earliest known bird had feathers, but also separate clawed fingers, socketed teeth, and a reptilian tail. [photo, Museum für Naturkunde, Berlin/H.Raab]*

Jurassic. Salamanders also first appear in the Jurassic.

Yet another period of extinction brought the Jurassic to a close. Most of the stegosaurids and sauropods (such as *Brachiosaurus* and *Diplodocus*) became extinct, as did the marine crocodiles. Only a few groups of marine reptiles survived the Jurassic extinction.

The Cretaceous

The third and last period of the Mesozoic Era was the Cretaceous, which began about 145 mya and ended about 65 mya (a duration of about 80 million years). The word Cretaceous is derived from a Latin word meaning *chalk*. The period was named because of the large chalk deposits that formed during the period. The most famous of these deposits are the White Cliffs of Dover.

The chalk deposits are composed mainly of disk-shaped plates of calcium carbonate called coccoliths, that cover one-celled phytoplanktonic organisms called coccolithophores. Coccoliths first appeared in the late Triassic (Figure 7.41). They are members of the brown algae group.

Another group of phytoplanktonic organisms, the diatoms (Figure6.15), appeared in the Cretaceous and became very common. As discussed earlier, diatoms are one-celled organisms that are covered by a pillbox-shaped structure of silica, and belong to the brown algae group, the chromists.

During the period, the breakup of Pangaea continued. Africa and South America became separated as the South Atlantic Ocean formed. Near the end of the Jurassic, India separated from Antarctica, and Madagascar moved

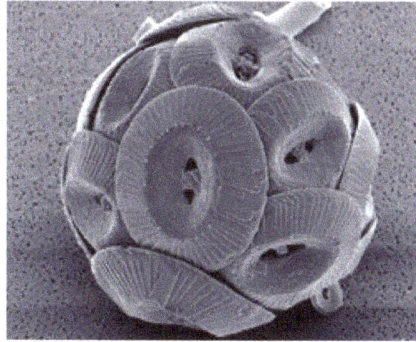

Figure 7.41 Coccolithophore A single-celled algae covered by calcium carbonate plates called coccoliths. [R.Lampett, J.Young/Natural History Museum, London]

away from Africa. Madagascar is still a large island today, and India was an island until relatively recent times, although it may have come very close to Africa as it moved northward.

As the ocean crust was subducted beneath western United States, microcontinents and island arcs were added to the continent. These masses of land, called exotic *terranes*, formed

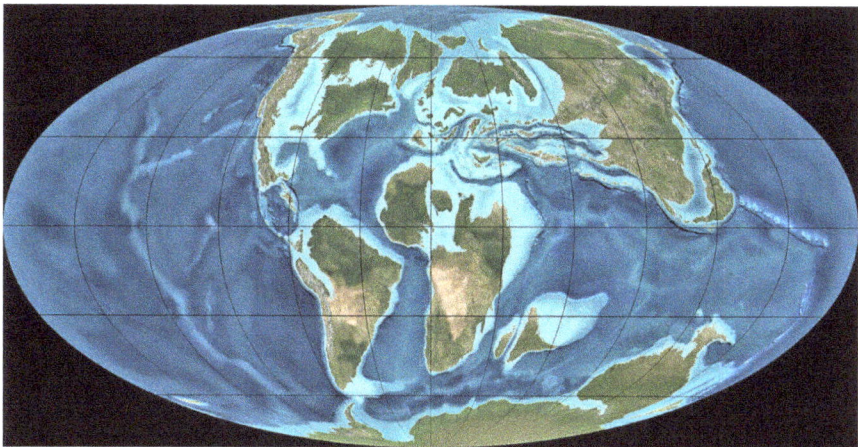

Figure 7.42 Earth During the Late Cretaceous The world as it looked when dinosaurs roamed the Earth some 90 mya. North America was divided by a large sea called the Mid-Continental Seaway. Europe and Asia were separated by the Turgai Sea, and Africa was an isolated continent until it began to collide with Europe and Asia about 40 million years ago. Alaska was connected to Asia allowing the exchange of many plants and animals between Asia and North America. [Ron Blakey/Colorado Plateau Geosystems]

large portions of western United States, Canada, and Alaska. A culmination of this activity was the formation of the Rocky Mountains during the late Cretaceous and early Cenozoic.

During most of the Cretaceous, North America was divided into an eastern and western portion by a large sea known as the Mid-Continental Seaway. This seaway began to dry up near the end of the period. In the later part of the period, the western portion of Alaska became attached to Asia. This connection, called Beringia, allowed an interchange of plant and animal life between Asia and North America (Figure 7.42).

Eastern North America, Greenland, and Europe remained connected throughout most of the period, but Europe and Asia were separated by a marine barrier called the Turgai Sea.

The Cretaceous was an important time for the formation of fossil fuels. The large coal deposits in the western North America and the petroleum reserves in Alaska were formed during the Cretaceous and early Cenozoic. During the later part of the Cretaceous, much of the land surrounding the Mid-Continental Seaway consisted of low marshy areas where coal deposits were formed. The oil in the Middle East was also formed during the Jurassic and Cretaceous by sediments accumulating in the Tethys Sea.

As mentioned above, the numbers and varieties of fish exploded in the Cretaceous due to the tremendous increase in the numbers of modern bony fish (the teleosts).

The reptile population of the Cretaceous changed substantially. The flying reptiles of the Cretaceous had evolved a reduced tail and a more birdlike beak with fewer teeth. Some, like *Pteranodon*, did not have teeth at all (Figure 7.43).

Snakes, which are essentially legless lizards with modified jaws, are first found as fossils from the early Cretaceous. Mosasaurs which are carnivorous marine lizards, made their appearance, and turtles invaded the sea, some of which reached gigantic proportions. Plesiosaurs and a few ichthyosaurs survived the Jurassic extinction, but ichthyosaurs were less common in the Cretaceous and were represented by mainly one genus. The ichthyosaurs become extinct before the end of the period.

Well-known dinosaurs of the Cretaceous include *Alamosaurus* (named after the Ojo Alamo geologic formation), *Ankylosaurus*, *Triceratops*, *Iguanodon*, and *Tyrannosaurus rex* (Figure 7.44 and 7.45).

Figure 7.43 Pterosaur Pteranodon is the genus name for a group of pterosaurs that contains some of the largest flying reptiles. This skeleton replica is of a species that had a wingspan of about 6 meters. The backward pointing head crest may have been for courtship display. [photo, AMNH/ Matt Martyniuk]

Figure 7.44 Triceratops Skull A plant eater of the late Cretaceous in North America. They were about 9 meters long and weighed about 5 metric tons (5,000 kg). [photo, London Natural History Museum/Zachi Evenor]

Near the middle of the Cretaceous, about 140 mya, the first flowering plants appeared. Judging from the enormous success of these plants, the development of the flower was perhaps the most important innovation in the plant world since plants first moved onto the land. Flowering plants (the angiosperms) produce seeds that are enclosed by an ovary. When the ovary matures, it becomes the fruit. Once flowering plants appeared, they diversified very rapidly, and before the end of the period they had become one of the dominant types of land plants.

Undoubtedly, the success of the flowering plants is largely due to the intimate association that many of them

Figure 7.45 Cretaceous Dinosaurs Tyrannosaurus (top left) was a fierce meat eater that could be 12 m long and weighed about 7 metric tons (7,000 kg). [myfavoritedinosaur.com]; Anchiceratops (top right), an early smaller cousin of Triceratops, was a plant eater that weighed about 2 metric tons (2,000 kg) and was about 4 m long. [Mariana Ruiz Villarreal]; Alamosaurus was a gigantic sauropod that was at least 15 m long. [Dmitry Bogdanov]

share with various species of insects. Insects are attracted to the nectar of the flower, and in the process of getting the nectar they aid the plant in fertilization. With the evolution of the flowering plants and deciduous trees, the foliage began to take on a more modern look near the end of the Cretaceous. The rise of the flowering plants resulted in a decrease of the cycads and ferns, and a slight decrease of the conifers.

Most of the mammals of the Cretaceous continued to be small, but the largest discovered so far had a body about half a meter long and weighed about 10 kg. It had the remains of a small dinosaur in its stomach area, indicating that it was carnivorous and fed on small vertebrates.

Although the fossil record of birds is generally sparse, there are several fossils from the Cretaceous. Cretaceous birds were more advanced than Archaeopteryx, but some still had teeth and clawed wings.

The Cretaceous ended with perhaps the most famous extinction of all times. Several extinctions, including the great extinction at the end of the Permian

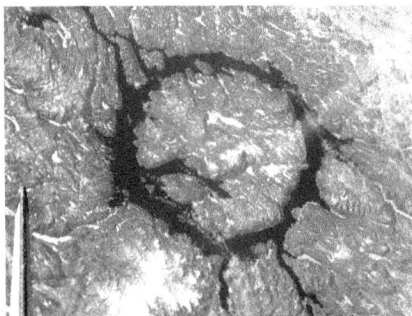

Figure 7.46 Manicouagan Crater This 200 million year old crater in northern Canada is about 70 km across. It was produced by a meteorite about 5 km in diameter. If a meteorite wiped out the dinosaurs, it would have been about twice that size (8 times more massive). The tail fin of the Space Shuttle Columbia can be seen at the lower left. [STS-9 crew/NASA]

and the extinction at the end of the Triassic were more devastating, but the Cretaceous extinction is famous for eliminating the dinosaurs. About half of the plants were also wiped out at this time. There appears to have been a huge die off of many flowering plants and a sudden increase in ferns. The explosion of ferns was short lived, however, as the flowering plants soon returned to the dominant position.

In the sea, plesiosaurs, mosasaurs, what was left of the ichthyosaurs, the large marine turtles, and conical-shaped, reef-building clams known as the rudist bivalves became extinct as did the belemnites. After several near misses, the ammonites disappeared forever. It has been estimated that one quarter of the animal families became extinct at the end of the Cretaceous. Birds and mammals survived, but they did suffer losses. Perhaps one-third of all mammal genera became extinct at the end of the period.

Recent evidence suggests that a large meteorite may have struck the Earth at the end of the Cretaceous, contributing to the demise of many of these plants and animals. The evidence in support of this idea is a layer of iridium-rich clay found in many locations around the world at what appears to be the Cretaceous-Cenozoic boundary. Iridium is a relatively rare element on the Earth, but is much more common in certain meteorites (Figure 7.46). The flowering plant die off followed by the brief explosion of ferns is similar to what is observed today after forest fires.

All the evidence does not support a sudden extinction, however, as fossil evidence suggests that the ammonites and perhaps the dinosaurs had been on the decline for several million years before their extinction at the end of the Cretaceous. Pterosaurs seem to have been on the decline since the end of the

Jurassic. Perhaps because of increased competition with the birds, pterosaurs became extinct near the end of the Cretaceous. Some scientists believe that something other than a meteorite was responsible for the decline of the dinosaurs, perhaps a long-term change in the weather or intense volcanic activity.

Since the dinosaurs were such a large and varied component of the Earth's fauna, it seems doubtful that any single event could have been solely responsible for destroying all of them. Even if a meteorite did strike the Earth, it seems unlikely that such an event would have wiped out all the dinosaurs but left some of the birds, mammals, and such closely related forms as the lizards, alligators, and turtles. More likely, some event or series of events such as a meteorite impact, a series of large volcanic eruptions, and a change in climate, destroyed a large fraction of all the plant and animal life, especially the larger forms.

Whatever happened, the cards of life were reshuffled and the surviving groups of plants and animals were given a fresh start. Those forms that evolved more quickly and efficiently were able to invade the newly opened niches and gain the upper hand. In the midst of this new beginning, the few surviving dinosaurs, if any were present, were unable to compete with the newly evolving forms of birds and mammals.

It is curious that the ammonites, after surviving the great Permian extinction, although just barely, and the Triassic extinction, became extinct at the end of the Cretaceous while their close relatives the nautiloids survived. One idea that attempts to explain this has to do with the differences in the early stages of development of these two groups.

Fossil evidence suggests that newly hatched ammonites were quite small and apparently fed on phytoplankton. With the extinction of the plankton at the end of the Cretaceous, they may have lost their food supply and became extinct. In contrast, nautiloids have larger embryonic shells and are active swimmers at hatching time. This swimming ability enables them to live in deeper water from the start of their life.

The Cenozoic Era

The last 66 million years, a time span known as the Cenozoic Era, is popularly called the age of mammals. Formerly, the Cenozoic Era was once divided into two periods, the Tertiary (from 65 mya to 2.6 mya) and the Quaternary (the last 2.6 million years). This division was of limited value since the intervals were of such unequal lengths, the Tertiary accounting for most of the era

Another more equitable division scheme is now used that divides the Cenozoic into three periods, the Paleogene (from 66 mya to 23 mya) the Neogene (from 23 mya to 2.6 mya), and the Quartenary (the last 2.6 million years).

Each of the periods is divided into *epochs*. Since the Cenozoic interval of time is of such great interest to us and is often better represented in the fossil record, we commonly refer to the seven epochs instead of the three periods (Figure 7.46).

Flowering plants had evolved in the Cretaceous so the foliage on land had a more modern look at the beginning of the Cenozoic. However, the great extinction of the dinosaurs at the end of the Cretaceous left the land void of large animals. With the extinction of the large marine reptiles, the ocean life took on a more modern appearance, although the whales, seals, and other aquatic mammals were still absent.

The vacancies left by the dinosaurs and the marine reptiles allowed the mammals and birds to diversify and fill the niches that had been held by the reptiles. Lest someone gets the idea that the mammals rule the land today because they conquered the dinosaurs, we should point out that the mammals first appeared in the Triassic and lived with the dinosaurs for more than 150 million years as a minor part of the Earth's animal life. It was not until the dinosaurs were eliminated that the mammals could diversify and become

an important component of the animal community. We mammals apparently owe our position of prevalence to the great Cretaceous extinction that, for some unknown reason, spared our ancestors.

Because of their relative youth, fossils formed during the later part of the Cenozoic are often found in softer sediment whereas older sediments tend to be more lithified (turned into rock). For example, most dinosaur bones have been largely turned to stone due to the incorporation of minerals into the bony structure, while mammoth bones will generally have more of the original

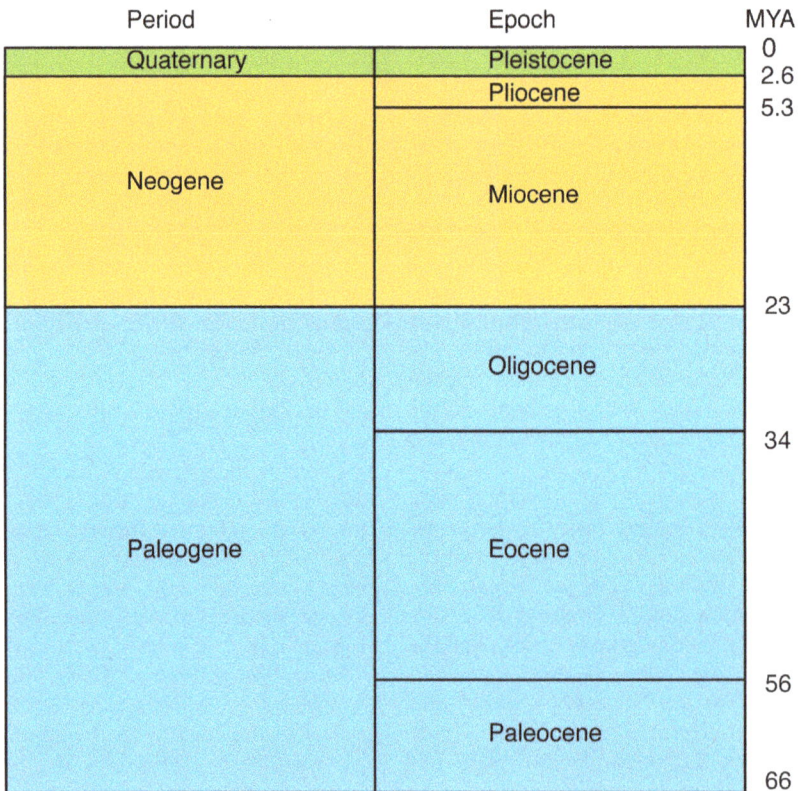

Period	Epoch	MYA
Quaternary	Pleistocene	0
		2.6
	Pliocene	5.3
Neogene	Miocene	
		23
Paleogene	Oligocene	
		34
	Eocene	
		56
	Paleocene	
		66

Figure 7.47 Epochs of the Cenozoic Era The last 12,000 years, known as the Holocene Epoch of the Quaternary Period, is too short to be shown in the figure. [JCB; data: International Commission on Stratigraphy]

bone material with fewer minerals incorporated into the bone matrix.

Since the fossils of the Cenozoic have had less time to be destroyed by geologic activity and weathering, the fossil record is generally more complete than the record of the Mesozoic or the Paleozoic Eras. There are, however, critical gaps in the record during the later Cretaceous and early Paleogene when the mammals were undergoing many very important evolutionary changes.

The Paleogene

The Paleogene Period began 66 mya with the disappearance of the dinosaurs and ended about 23 mya. This 43 million year period witnessed the rise of the mammals to a position formerly held by the dinosaurs in the Mesozoic Era. However, even the largest mammals would probably have been no match for the largest dinosaurs.

As can be seen from Figure 7.47, the Paleogene is divided into three epochs, the Paleocene, the Eocene, and the Oligocene. The middle epoch, the Eocene, lasted twice as long as either of the other two.

The most obvious evolutionary change of the Paleogene was the expansion of the mammals, especially during the Paleocene and Eocene Epochs. During the Eocene, (the dawn of recent forms), we find that mammals had differentiated sufficiently that most of the modern orders could be recognized.

At the beginning of the Cenozoic, most of the mammals were small and resembled insectivores like shrews in outward appearance. In response to the voids left by the dinosaurs, the early mammals diversified very quickly from their primitive stock, and as might be expected, many new groups appear in the fossil record almost simultaneously. DNA evidence tells us that by the late Cretaceous, the placental mammals had split into four separate superorders.

One group, the *Laurasiatheria* superorder, evolved in Laurasia during the later Cretaceous. They diverged into several major groups, namely the even-toed ungulates, whales and dolphins, the odd-toed ungulates, bats, carnivorans (the cat and dog group), pangolins, and a group that included the moles, shrews, hedgehogs, and the two species of solenodons.

Odd-toed ungulates (perissodactyls) include the horses, rhinoceroses, and tapirs. Modern even-toed ungulates (artiodactyls) are the pigs, camels, deer, sheep, goats, hippopotamuses, giraffes, antelope, and cattle.

The Laurasiatheria group included the ancestors of the carnivorans and another group of carnivorous mammals known as the creodonts (Figure 7.48). Creodonts were the dominant carnivore for much of the Paleogene and early

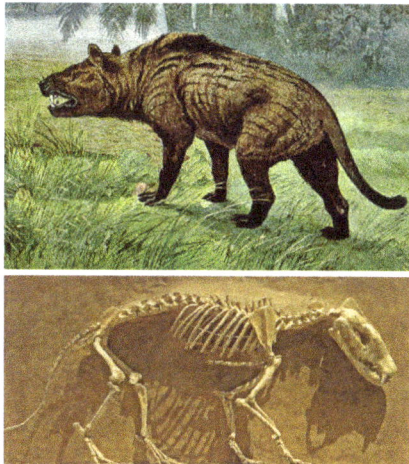

Figure 7.48 Hyaenodon A creodont that was a predator in North America, Eurasia, and Africa from the Late Eocene until the Early Miocene. Various species ranged in size from about 10 kg to 500 kg. [painting, Heinrich Harder; fossil photo, Royal Ontario Museum/Daderot]

Neogene, but were eventually replaced by the modern day carnivorans that include the cats, dogs, civets, and its marine members, the seals, sea lions, and walruses.

Before the Oligocene there were no really large flesh-eating mammals, but by the early Oligocene, the carnivorans had diverged such that the cat and dog families could be distinguished. Bears, which are essentially big, tailless dogs, with slightly modified teeth, and flat feet, first appeared near the middle of the Oligocene.

The first whale fossils are also from the early Eocene. A primitive whale known as *Dorudon*, which lived during the later Eocene, still had tiny hind limbs, reminding us that whales evolved from land mammals with four legs. The early whales had nostrils near the end of their nose instead of on top of the skull as the modern whales do, and they had differentiated teeth (modern toothed whales have peg shaped teeth). Modern toothed and baleen whales did not evolve until the later Oligocene. Baleen whales have a very large head in proportion to the rest of their body, and they do not have

Figure 7.49. Arsinoitherium Although it looks similar to the rhinoceroses, this large member of the Afrotheria superorder was not closely related. It lived during the early Oligocene in northern Africa and was almost 2 meters tall. It weighed about 2,500 kg. [Heinrich Harder]

hind feet although they do have rudimentary pelvic and leg bones.

The first bat fossils are of Eocene age, but little is known of their early evolution.

Based on DNA evidence, we find a second superorder of mammals that are called the *Euarchontoglires*. (From euarchonta meaning *true ancestors,* and Glires which is the rodent and rabbit group.) It probably split from the Laurasiatheria sometime during the Cretaceous. The living members of the group are the rodents, rabbits, colugos, tree shrews, and primates.

Primates, the group to which we belong, have evolved a high degree of intelligence, but some of their features (like their teeth and finger bones) have not changed much. The earliest fossils that may be of primates are from Paleocene times. These animals were very primitive, and some scientists feel they should not be classified as true primates, but early ancestors of the primate group. The first animals that are more like the living primates are from the early Eocene of Europe and North America. The earliest fossil anthropoids, the group to which the New World monkeys, the Old World monkeys, and the apes belong, are from the early Oligocene in Africa.

Again based on molecular DNA evidence, we have identified a third superorder of mammals known as the *Afrotheria* (Figure 7.49). This superorder evolved in Africa, which was isolated from about the middle of the Cretaceous until the end of the Eocene. Living members include the elephants, sea cows, hyraxes, tenrecs, elephant shrews, and the Aardvark.

By the late Eocene, we find the first early elephant fossils. A close relative, *Moeritherium*, was about the size of a pig, and *Palaeomastodon* was larger, weighing about 2,000 kg (Figure 7.50).

Figure 7.50 Moeritherium An early relative of the elephants. [Heinrich Harder]

The fourth mammal group is the superorder *Xenarthra*. The present day members of the group are the sloths, armadillos, and anteaters. The closest relatives of the xenarthrans seem to be the Afrotheria. The two groups likely split in the later Cretaceous when their members were still quite small. Xenarthrans reached South America from Africa when the continents were still close, perhaps by floating on rafts

of vegetation. They diversified in South America, but for some reason, they did not survive in Africa.

The xenarthrans have several traits that are considered primitive such as teeth that do not have enamel, a relatively low rate of metabolism, and a variable body temperature. Their name, which means *strange joint*, comes from their unusual vertebrae that have an additional pair of contacts with the adjacent vertebrae.

The Oligocene mammals were generally larger than those of the Eocene and some were very large. They included forms such as whales, elephants, and many strange creatures that have become extinct (Figure 7.49).

The odd-toed ungulates became more common, and reached their peak in the Oligocene (Figure 7.51). Today, they are on the decline and have been largely replaced by the even-toed ungulates.

An important group of rodent-like mammals (the multituberculates) lived during late Jurassic and Cretaceous times, but became extinct near the early Oligocene, possibly due to the

Figure 7.51 Giant Oligocene Odd-toed Ungulates Indricotherium (left), also known as Pacaceratherium, lived in central Asia during the Oligocene and is the largest known land mammal. It weighed as much as 3 or 4 adult African elephants and stood almost 5 meters tall at the shoulders (giraffes are 5 or 6 meters tall). Megacerops (right) was a gigantic 2.5 meter tall brontothere that lived in North America from the Late Eocene to Early Oligocene. It is also known as Brontotherium. [reconstructions, Dmitry Bogdanov]

expansion of the true rodents. Their name comes from the many cusps (bumps or *tubercles*) on their molars. The multituberculates existed for at least 130 million years, and they provide us with yet another example of a group of animals that appeared in the fossil record, only to disappear at a later time.

As the South Atlantic Ocean widened during the Paleogene, the landmasses of South America, Antarctica, and Australia became island continents. Antarctica is now covered with ice, but it remained warm until it separated from South America, during the Eocene or early Oligocene. This separation allowed water to begin circulating around Antarctica, which effectively isolated it from the warmer ocean currents. This isolation initiated the formation of ice sheets about 35 mya. Today, Antarctica remains isolated by the Antarctic Circumpolar Current which continues to keep it cold.

Marsupials reached Australia before it became fully isolated, but placental mammals did not. Therefore, for most of the Cenozoic, marsupials evolved on Australia free from the competition of placentals. Unfortunately, Australia has

Figure 7.52 Glyptodon *A large xenarthran that evolved during the Miocene. A close relative, Doedicurus, was 1.5 meters tall and had a ball with spikes at the end of its tail. Their were covered by a shell made of bony plates. They became extinct after the last ice age, about the time the first Indians reached South America. [photo, Museum of Natural History Vienna, Austria/Arent]*

a very poor fossil record for the early Cenozoic, and consequently we have little knowledge of this important time interval. By the later part of the Cenozoic, bats and rodents had reached this island continent, and in more recent years man has unfortunately introduced rabbits and several other placental mammals.

South America was apparently isolated when the only mammals present were marsupials, xenarthrans, and a group of hoofed mammals (meridiungulates). Analysis of collagen from bones of *Macrauchenia* and *Toxodon* (Figure 7.56) indicates that the South American ungulates diverged from the odd-toed group about 70 mya. The odd-toed and even-toed ungulates diverged about 80 mya. Because the continent was isolated during most of the Cenozoic, many unusual creatures evolved from these three groups of ancestral stock (Figure 7.52).

The ancestors of the South American rodents and New World monkeys reached this isolated continent near the beginning of the Oligocene, about 35 mya. The earliest South American rodent fossils have teeth similar to the African rodents of the time, and the teeth and skull of early New World monkey fossils suggest that they are related to African primates. Both groups must have come from Africa by rafting on logs or floating vegetation. At this early date, the trip would have been easier since the Atlantic Ocean was less than half as wide as it is today.

The ancestral South American rodents gave rise to a group known as the caviomorph rodents. They were confined to South America until the Isthmus of Panama was formed about 3 million years ago, and then porcupines came to North America. Some of the modern caviomorphs include the chinchillas, capybaras, porcupines, agoutis, and the cavies (also called guinea pigs), which are the namesake of the group.

Large flightless birds evolved in many parts of the world during the Paleocene. They may have been the largest predators of the time, although based on their anatomy, some scientists think they were herbivores (Figure 7.53). Most became extinct in the Eocene, perhaps because they could not compete with the larger carnivorous mammals that began to appear in the Eocene. A few of these large flightless birds apparently survived to a much later date (late Pliocene) in South America. One group, the dromornithids, survived in Australia until humans arrived on the continent about 50 to 60 thousand years ago. One species was about 3 meters tall.

Other Paleogene bird fossils are mostly of large forms such as storks and herons. Penguins, the seals of the bird world, evolved in the Eocene.

A period of extinction occurred between the Eocene and the Oligocene that may have been caused by a cooling of the weather. The weather generally became cooler and dryer in the late Eocene and Oligocene.

Grasses, which evolved in the Paleogene, became more widespread as the dryer climate favored the spread of prairies at the expense of forests.

The new growth of most plants is at the tip of the branches, for example trees grow this way. Grasses, however, grow from the base of the plant. When an animal nibbles off the tips of the grass, the growth is not affected since the new growth is at the bottom of the plant.

The Neogene and Quaternary

The Neogene, a period of about 20 million years, is divided into 2 epochs,

Figure 7.53 Gastornis Popularly known as terror birds, one of these giant birds was over 2 meters tall. Members of this genus lived in North America and Europe during the Paleocene and Eocene. [F. Lucas]

but the Miocene Epoch accounts for all but the last 2.6 million years (Figure 7.47). The Pliocene began 5 mya and ended 2.6 mya. The last 2.6 million years is called the Quaternary Period. The Pleistocene Epoch accounts for all but the last 12,000 years which is called the Holocene or Recent Epoch.

As the whales, dolphins, seals, and sea lions expanded, life in the sea took on a thoroughly modern look. The early whales had disappeared and modern varieties took their place. Invertebrate sea animals changed in less obvious ways, but one Cenozoic addition of interest was the sand dollar, which evolved from sea urchins near the mid Eocene.

Grasses continued to spread during the Miocene and the animal life adapted accordingly. Bovids, which include the sheep, goats, antelopes, and cattle, first appeared in the Miocene. These grazing mammals increased in variety and numbers as the grasses and herbaceous plants became more abundant.

Insects, mice, and songbirds also benefited from the growth of the prairies. Following the increase of mice, insects, and songbirds were increases in predators such as frogs, snakes, predatory birds, and small carnivorous mammals.

Other mammals that first appeared in the Miocene include early forms of the giraffe, modern members of the cat family (Felidae), and late in the epoch, the hyenas. Although deer appeared in the Oligocene, the first deer with antlers appeared during the Miocene.

Figure 7.55 Smilodon This predator of the Pleistocene Epoch was similar in size to a modern Siberian Tiger and was the last of the saber-toothed cats. They invaded South America during the Great American Interchange, but went extinct about 10,000 years ago. [Charles R.Knight/AMNH]

Near the late Miocene (about 7 or 8 mya) there appears to have been some limited interchange between the mammals of North America and South America. Possibly by floating across the water barrier on logs or by island hopping, a few members of the raccoon family reached South America, and large ground sloths entered North America (Figure 7.54).

When the Isthmus of Panama solidly reconnected South America to North America about 3 mya (near the end of the Pliocene), there was a great interchange of mammals between the two continents (known as the Great American Interchange). The mammals of North America were not affected greatly, but many of the unusual South American forms became extinct and were replaced by northern forms (Figure 7.55).

Many of the marsupials failed to compete with the placental mammals from the north, and all the South American ungulates eventually died out, although the last members of the group survived until the Indians reached South America and hunted them about 12,000 years ago (Figure 7.56).

A few native South American mammals, such as the opossum,

Figure 7.54 Giant Ground Sloth Called Megatherium, this 6 meter long giant lived in North America from the late Pliocene until about 10,000 years ago. Smaller species survived in Cuba and Hispaniolia until about 5000 years ago. They died out soon after Indians reached the islands.

Figure 7.56 South American Ungulates (meridiungulates) Mixotoxodon (left), was a rhinoceros-sized member of the Notoungulata order of unique South American ungulates. [karkemish00] Macrauchenia was the last of the Litopterna, another unique order of South American ungulates. It had a trunk-like nose and three toes. They were hunted by the first humans who reached South America, about 12,000 years ago. [Olllga]

armadillo, and porcupine, became firmly established in North America, but they apparently did not displace any of the native mammals.

The last ice age, known as the Late Cenozoic Glacial Age, began about 3 mya. The glaciers waxed and waned (grew and shrank) many times during the Pleistocene before they retreated for the last time about 11,000 years ago. The large ice sheets on Greenland and Antarctica are remnants of the last great expansion that occurred about 18,000 years ago. This expansion is known in North America as the Wisconsin glaciation, and in Europe by the Würm glaciation.

Today, glaciers cover about 10% of the Earth's land (a substantial fraction when you think about it), but about 30% was covered at the maximum expansion of ice in the Pleistocene. In some places the ice was as much as 3 to 4 kilometers thick! The great weight of this ice caused the land to sink, much as a floating board sinks when you push down on it (remember the lithosphere is floating on the more plastic asthenosphere which can flow). When the ice sheets melted, the land

began to rise slowly, and measurements indicate that Scandinavia and Canada are still rebounding in places.

During much of the Cenozoic, North America and Asia were connected by a wide land bridge called Beringia. Many mammals traveled freely between the two continents, and since the climate was mild until the late Miocene, many plants also spread between the two continents (Figure 7.57). Today, we still see the effects of this connection in the similarities of both the animals and plants of these two regions. We find

Figure 7.57 Woolly Mammoths During the later part of the Plistocene, they roamed northern Eurasia and North America. [Charles R. Knight/AMNH]

closely related species of hickory, walnut, beech, oak, elm, maple, and tulip trees in both areas. A close relative of the Sequoia and California Redwood lives in China which should not surprise us since Beringia once connected these two areas.

The Alps and related mountains of Europe and North Africa formed during the Late Cretaceous and early Cenozoic as Africa and several microplates shifted with respect to Europe. Africa joined Asia in the later part of the Miocene.

During the Neogene, uplifts and block faulting produced the Sierra Nevada and the Rocky Mountains. Further south, the subduction of the Nazca Plate beneath South America formed the Andes. The Indian-Australian Plate moved north and India contacted Asia near the middle Eocene, beginning the Himalayan uplift.

By the early Pleistocene, large marsupials and monotremes had also evolved in Australia. The largest monotreme was a giant echidna that weighed about 100 kg. A giant 2 m tall short-faced kangaroo (*Procoptodon*) evolved in the Pleistocene, as did *Diprotodon*, a giant wombat relative that was the largest marsupial known (Figure 7.58). Also present at this time was a carnivorous marsupial about the

Figure 7.58 Diprotodon This ancient relative of the Wombat was 2 meters tall and weighed almost 3000 kg. They died out soon after humans arrived in Australia about 50,000 years ago. [Dmitry Bogdanov]

size of a leopard called *Thylacoleo*. All the large marsupials became extinct soon after humans arrived in Australia about 50,000 years ago.

Even a brief glimpse at the fossil record is enough to reveal many of the dramatic changes that have occurred over the past several billion years. From the simple bacteria of the early Precambrian to the radiation of the mammals in the Cenozoic, a steady progression of new organisms appears in the fossil record. Many periods of extinction seem to dot the past, ending the reign of some organisms while providing new opportunities for others. The victims of these extinctions left niches open that were soon filled by new life forms. The process of change that provides this continual supply of new organisms is evolution.

We have seen that new organisms appear sporadically in the fossil record, but others disappear, often during episodes of extinction. However, the extinct organisms were not merely replaced by new organisms. The fossil record clearly indicates that the total diversity has generally increased with time. If we plot the number of families present in the fossil record as a function of the time, we find that the number of families has generally increased with time. This observation is explained by the evolution of new families as organisms diversify and adapt to new situations.

The fossil record is generally not complete enough to provide us with all the details in the evolutionary sequence of every organism. However, in several instances we have enough evidence to work out the general trends. In the next chapter we will look in some detail at several evolutionary sequences that have been pieced together from the fossil record.

Chapter 8: Missing Links

Introduction

Populations of organisms evolve by incorporating small changes that are passed on from one generation to the next. By accumulating many small changes, a population of fishes evolved into a population of amphibians. If sufficient evidence was available, one should be able to document these changes in the fossil record. Because of the low probability of being preserved as a fossil, however, most organisms are not represented in the fossil record.

Therefore, the likelihood of finding a fossil from the exact population of fishes that eventually evolved into an amphibian is relatively low. However, that particular type of fish might have had a closely related form that was fossilized. For example, we might find a fossil of a fish belonging to the same genus, or at least the same family. Even if we do not find the direct ancestors, we might be able to discover the general trend in the fossil record (Figure 8.1).

Figure 8.1 Family Tree We do not generally expect to find fossils (red dots) of the direct ancestors of a species living today (green dots). We will more likely find fossils of their more distant cousins. [JCB]

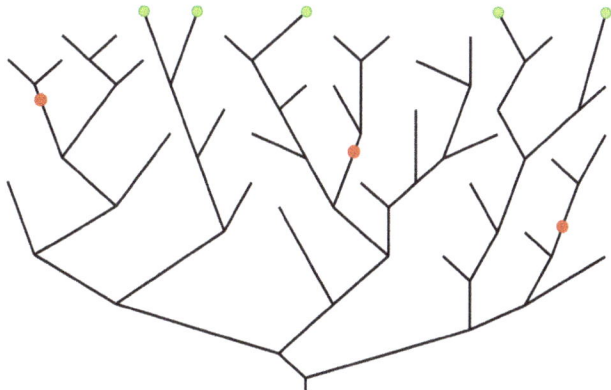

242

The point is this, as we search the fossil record, we do not necessarily expect to find the exact fish ancestor of the amphibians, or the ape-like ancestors of man, or the reptile ancestors of the mammals, but we do hope to find forms that are closely related to the direct ancestors of these animals. There are many side branches on the evolutionary tree, and we are more likely to discover a fish from one of the side branches than we are to find a direct ancestor of the amphibians.

However, if the fossil fish we find are reasonably close relatives of the direct ancestor, it will share many features with the direct ancestor, and we will learn a great deal about the evolutionary process. We will be able to determine the general evolutionary trend of the organism being studied.

It is quite possible that more than one population of fishes evolved amphibian characteristics or that more than one group of reptiles evolved into mammals. If this happened, we would say that amphibians and mammals are *polyphyletic* groups. On the other hand, if mammals evolved from just one group of reptiles, we would say that mammals are *monophyletic*. By relying only on the fossil record, it is sometimes impossible to absolutely differentiate between these two possibilities.

Fish to Amphibians

The first jawed fish evolved about 460 mya and by about 430 mya the four major classes of jawed fish had appeared. They were the placoderms, the extinct class of armor plated fish, the chondrichthyes or cartilaginous fish, the actinopterygians or ray finned fish, and the sarcopterygians or lobe-finned fish.

The fossil evidence indicates that amphibians evolved from a type of lobe-finned fish. In the early Devonian, the lobe-finned fish split into two major groups, the coelacanth group and the rhipidistians. The rhipidistians soon invaded the fresh water streams and rivers. They then split into two groups, the lungfish and the group that contained the ancestors of the tetrapods (amphibians, reptiles, birds, and mammals).

It was once thought that all members of the coelacanth group were extinct until one was discovered in 1938 off the coast of South Africa (Figure 8.2). Most other specimens have been found in the deep waters (around 200 meters) between Madagascar and the African mainland. A second species has been discovered in the waters off Sulawesi (east of Borneo).

Lungfish, which first appeared in the middle Devonian, belong to the subclass Dipnoi. Modern members of this group are found in Africa, South America, and Australia. They live in ponds that dry up periodically, but the lungfish survive by burrowing into the mud and hibernating until the rains return and fill the lake with water.

When we think of the differences between fish and land vertebrates, one of the first things that comes to mind is

Figure 8.2 Coelacanth Latimeria chalumnae is the species found primarily in the waters between southeast Africa and Madagascar. It is almost 2 meters long. Notice the fleshy fins. [Zoo Firma]

that tetrapods (four footed vertebrates) breathe with lungs and fish breathe with gills. However, as their name implies, lungfish possess lungs, as did other bony fish from the Paleozoic.

In the more advanced fish, lungs eventually evolved into the swim bladder, an organ used to control the buoyancy of these fish. Sharks, which branched off earlier, never had lungs and did not evolve a swim bladder.

The early rhipidistian fish were predominantly freshwater animals. During the Devonian, many apparently lived in pools of water that became stagnant during the dry season. The evolution of lungs allowing these fish to survive in these pools of poorly oxygenated water. To evolve into a land creatures, rhipidistian fish did not have to acquire lungs. Their adaptation to an oxygen-poor environment facilitated their move onto the land.

In contrast to fish, amphibians need a means of terrestrial locomotion, and the fish-amphibian transition involved the evolution of legs and strong shoulders and hips. The fossil evidence tells us that legs evolved from fleshy fins that were originally used by the

animals to push themselves over debris on the bottoms of lakes, or perhaps to aid the animals in traveling from one pond to another. If a lake dried up, the fish that could not move to another lake would not survive.

Lobe-finned fish had muscles in their fins that could be used to move the bones that made up the fin. The evolution of a movable limb from such a muscled fin would not have been an unimaginable task.

On the other hand, ray-finned fish do not have muscular fins. Their fins are flat fan-shaped structures, one end of which is imbedded in the body tissue of the fish. Muscles within the fish's body control the movement of the fin, and it is very difficult to imagine this type of fin evolving into a tetrapod limb without many major modifications.

A rhipidistian fish that could have been an ancestor or close relative of the amphibians is *Eusthenopteron*. The bones in the fins of *Eusthenopteron* are similar to the leg bones of early tetrapods (Figure 8.3). In the front fin, the bones that would become the humerus, radius, and ulna can be seen

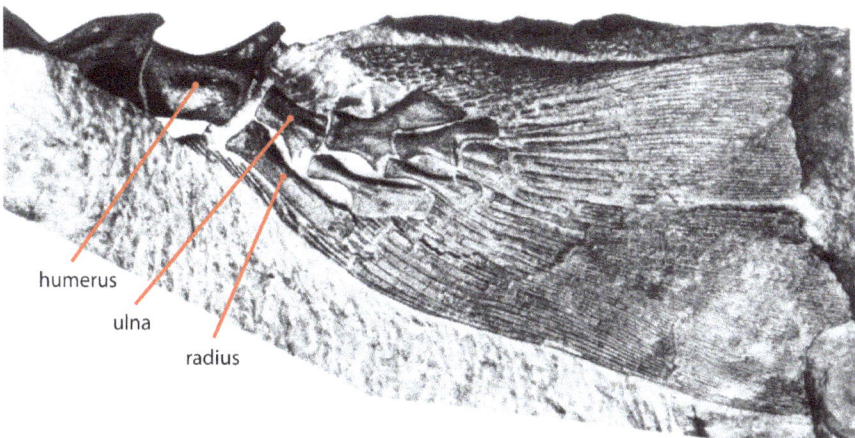

Figure 8.3 Eusthenopteron Pectoral (front) Fin *The bones in this fossilized left fin are similar to the bones in the arm of a tetrapod. [Erik Jarvik (1980), credits p. 415]*

Figure 8.4 Acanthostega *One of the earliest known amplibians. [photo courtesy of Jennifer A.Clack/University Museum of Zoology, Cambridge archives/original fossil, Natural History Museum, Copenhagen]*

clearly. In the rear fin, the similar bones would become the femur, tibia, and fibula of the back leg. In contrast, the bones in the fins of a ray-finned fish are very different from the bones in either the fins of a lobe-finned fish or the legs of an amphibian.

At this point it might be interesting to ask how an amphibian should be defined. If we were to define the first amphibian as a creature with legs, we would have to carefully describe a leg. If we define a leg as an appendage with a humerus, radius and ulna (or a femur, fibula and tibia), then *Eusthenopteron* would have been an amphibian. At what point would we say that it is not a fin, but a leg?

The point of this inquiry is to point out how difficult it is to classify animals that are very similar to one another. Today we place the bony fish in one class and amphibians in another, but during the Devonian, some were very similar, and if we went back in

time to those days, many of those creatures would have been placed in the same family.

Ichthyostega and *Acanthostega* were two of the earliest amphibians and belonged to a group of amphibians, called the labyrinthodonts (Figure 8.4). This group is so named because of the infoldings of their teeth. When viewed in cross section, these infoldings give the teeth the appearance of a labyrinth or maze (Figure 8.5). As might be expected, rhipidistian fish also possessed labyrinthine teeth, but other groups of fish did not, more evidence added to our ever-growing list that demonstrates the close relationship between the rhipidistian fish and the first amphibians.

Another strong piece of evidence that amphibians evolved from a rhipidistian fish is the close correlation of the skull bones of *Eusthenopteron* and *Acanthostega* (Figure 8.6). Except for variations in the relative sizes of the different bones and the loss of a few bones, there is a strong correlation between the skull bones of these two animals. Although *Eusthenopteron* and *Acanthostega* have very similar skull bones, the skull of a typical Devonian ray-finned fish has a very different bone pattern, indicating that they are not closely related to the early amphibians.

Figure 8.5 Labyrinthodont Tooth *This unique type of tooth is found in rhipidistian fish and early amphibians. Cross sectional drawing (right) shows the unique labyrinth pattern. [JCB]*

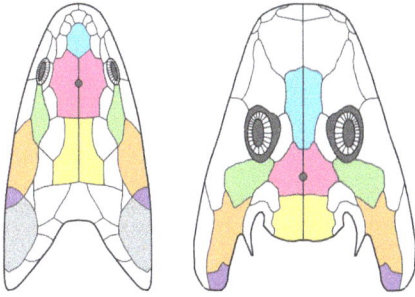

Figure 8.6 Skulls *Eusthenopteron (left) compared to Acanthostega. Bones are frontal (blue), parietal (pink), postparietal (yellow), postorbital (green), squamosal (orange), preopercular (violet), and operculum (gray). The opercular series of bones is the covering that protects the gills in fish. Later amphibians also lost the preopercular bones. [JCB]*

There are many other features common to the two groups, including internal nares and paired nasal passages that open into the throat. Sclerotic plates are bony plates imbedded in the white part of many vertebrate's eyes. Rhipidistian fish and amphibians have about 30 of these plates, but ray-finned fish (the actinopterygians) have larger eyes with only four sclerotic plates.

Since they had well developed legs, *Ichthyostega* and *Acanthostega* are definitely amphibians, but they possessed many interesting fishlike features. For example, *Ichthyostega* and *Acanthostega* had fin-rays in their tail like those found in fish, but we do not find fin-rays in the tails of later amphibians (Figure 8.7).

Ichthyostega and *Acanthostega* retained vestiges of the preopercular bones. In fishes, the opercular bones cover the gills and connect the head to the rest of the body. In amphibians, the loss of the opercular series freed the head from the shoulder and allowed the head more freedom of motion.

The vertebrae of *Eusthenopteron* are similar to the vertebrae of *Ichthyostega* and *Acanthostega*, but significant structural changes occurred in the vertebrae of later amphibians. The bones in the vertebrae of these early amphibians evolved several different configurations (Figure 8.8). The structure of the vertebrae is of interest because to support their weight on land, the first amphibians needed a stronger vertebral column than is found in fish. Eventually, the bones fused to form one unit.

Ichthyostega and *Acanthostega* were early tetrapods that lived at about the same time, and they shared several structures with both fish and tetrapods.

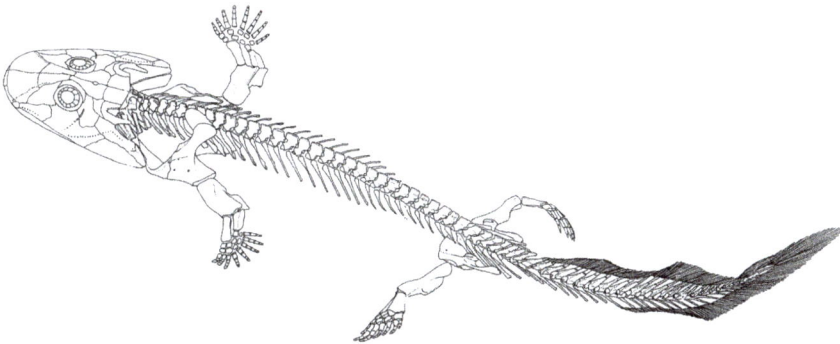

Figure 8.7 Acanthostega *This reconstruction is based on detailed examinations of the fossils. Note the fin-rays and number of digits. Acanthostega, which was about 0.7 meters long, had 8 toes. [Michael I.Coates]*

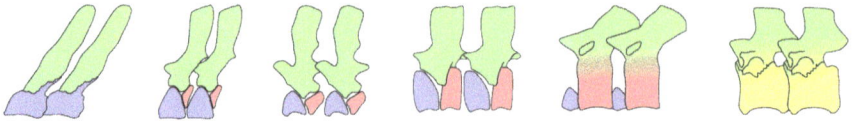

Figure 8.8 Early Vertebrae *(from left to right) Panderichthys (a sarcopterygian fish); Acanthostega (an early amphibian); Eryops (a more advanced amphibian); Archeria (a specialized marine reptile-like amphibian); Seymouria (a labyrinthodont amphibian); Varanops (an early synapsid reptile). Bones are neural arch (green), Intercentrum (purple), pleurocentrum (pink), and centrum (yellow). [Smokeybjb]*

Acanthostega had limbs with eight toes and *Ichthyostega* had seven toes, but they were more like paddles. With no wrists or ankles, they would not have supported the animal on land. *Ichthyostega* probably walked on land, but only with great difficulty, pulling itself along with its front feet. It was still very dependent on water for its survival. The paddle feet were used for swimming and to lift the animal's head out of the water for breathing and capturing food.

Acanthostega had gills, preopercular bones that are lost in most tetrapods, fin rays in the tail, weak hips, and a spine like a fish that would not support the animal's weight. These features tell us that *Acanthostega* was an aquatic animal, and it further suggests that legs evolved before amphibians walked on land.

In retrospect, the evolution of legs by an aquatic animal seems reasonable. Just changing fins to feet would not enable a fish to walk on land unless the fish also had stronger hips and shoulders, and a stronger backbone. *Acanthostega* had feet, but not the stronger hips, shoulders, or backbone needed to walk on land.

On the other hand, *Ichthyostega* had feet and strong shoulders but weaker hips and a fishlike backbone, so it could walk, but with difficulty. It had to wiggle and pull itself along with its front feet and strong shoulders. These fossils seem to be reminding us that the

evolution of walking did not happen in one giant step, but in many small steps.

Was *Acanthostega* an ancestor of *Ichthyostega*? How could it be an ancestor if they lived at about the same time? Placental mammals evolved from egg-laying mammals, but a few egg-laying mammals are still present today. These egg-laying mammals (the Platypus and four species of echidnas) are not the ancestors of present day mammals, but they are more similar to our ancestors in several ways. Since they have features like ancestral mammals, egg-laying mammals are often called living fossils. It is possible that *Acanthostega* was a living fossil living along side *Ichthyostega*. Just because one organism evolves from another, that does not mean that all organisms similar to the ancestor have disappeared.

Although rhipidistians became extinct in the Permian, they were probably the most common freshwater predators during the later part of the Paleozoic. Amphibians did not evolve from a rare type of fish, but one of the most common types of the period.

The amphibians split off from their lobe-finned fish ancestors almost 400 mya. Since then, the fish and the amphibians have been evolving independently, so comparing today's amphibians with today's fish does not reveal a close relationship. As we have seen, modern amphibians are more closely related to birds and mammals than they are to modern fish.

The common fish of today do not even belong to the same group that evolved into the amphibians. Also, there are no amphibians alive today that are similar to the labyrinthodont amphibians that lived in the Devonian. Labyrinthodont amphibians became extinct near the end of the Triassic. Although fish and amphibians of today are not very closely related, 400 mya the two groups were very closely related.

If amphibians evolved from fish, they must have inherited genes that orchestrate the development of the organism as well as the structural genes. Although these developmental genes have been modified by mutations as have some of the structural genes, the general development of amphibians might still be expected to show some similarities to the general development of fish. Most amphibians go through a larval or tadpole stage that is very fishlike. During their early stages of development, amphibians are totally dependent on water and breathe with gills. As they mature, the adults grow legs and become more terrestrial. They are still closely bound to water, since they must lay their soft fishlike eggs in an aquatic environment.

The amphibians of today are divided into three orders, the frogs and toads (order Anura), the salamanders and newts (Urodela), and caecilians or apodans (Gymnophiona). Caecilians are legless and superficially resemble large earthworms. There are about 200 species of caecilians, about 700 species of salamanders, and about 6,500 species of frogs and toads.

Amphibians to Reptiles

Larval amphibians can be easily mistaken for fish, but adult amphibians with legs are easily separated from fish with fins. The difference between an amphibian and a reptile, however, is not as obvious. One significant difference between the two groups is their method of reproduction. Like fish, amphibians lay jellylike eggs that quickly dry out and die if not kept moist.

In contrast, reptiles lay eggs that are surrounded by a watery substance called amniotic fluid, and enclosed in a leathery shell. Reptiles have evolved a way of surrounding their eggs with an artificial pond of their own. The amniotic egg allows reptiles to be completely liberated from an aquatic environment. They are free to populate areas that are far from water, and they do not need to return to water to reproduce.

By examining the fossilized skeleton of an animal, it is not possible to determine which type of egg it laid. Therefore, we need additional criteria. No one feature is used to distinguish an amphibian from a reptile, so generally, the presence of several features is used to make the determination. There is no clear division between the two groups.

Reptiles have 12 cranial nerves instead of 10 as amphibians do. Cranial nerves are the nerve bundles that emerge directly from the brain, and they carry visual signals from the retina, control the motion of the eyes, transmit hearing and balance, transmit smell signals, taste signals, and several other functions of the animal.

Reptiles also have an ankle bone or talus bone which was formed by the fusion of three bones in the amphibian foot.

Modifications in the limbs and hips of reptiles made them more agile and led to improved locomotion. The limbs of amphibians extend out horizontally from the body, causing the animal to move in a slow waddling fashion. Reptiles evolved limbs that were

Figure 8.9 Earliest Reptile A reconstruction of Hylonomus, a 310 million year old reptile. This 20 cm long animal is currently the oldest known reptile. [N. Tamura]

situated further under the body, allowing the animal to move more quickly. Being quicker and more agile would have been a great advantage to these early reptiles since the structure of their teeth and jaw musculature suggests that they fed on insects.

The differences between reptiles and amphibians are relatively subtle and the changes did not happen at the same time. Many Carboniferous amphibians were very reptile-like in appearance. Because they are so similar, trying to classify tetrapod fossils from this time period as either an amphibian or a reptile can be quite difficult.

Currently, the earliest animal widely accepted as a reptile is *Hylonomus*, which means *forest dweller* (Figure 8.9). It lived about 310 mya and looked very much like a modern lizard, although they are not closely related. It was about 20 cm long and its sharp teeth indicate that it most likely fed on insects.

As we saw, some of the early amphibians had more than five digits on their front and back feet, but the later amphibians had just five. The bones in the digits (fingers and toes) are called the phalanges. The pattern of bones in the digits (phalanges) of early reptiles was identical to the pattern found in a group of amphibians called anthracosaurs, the probable ancestors of the reptiles. In early reptiles, there were 2 bones in the first digit (thumb), 3 in the second digit, 4 in the third, and 5 in the fourth. The fifth digit had 3 bones in the forelimb (hand) and 4 bones in the hindlimb.

Very early in the development of reptiles, they differentiated along several distinct lines. These lines of development can be traced by studying the structure of the skull. In several groups of early reptiles, openings between the bones of the skull developed, which provided a stronger attachment for the muscles of the jaw.

These openings, called *fenestrae*, were of several very distinct types, and the number and arrangement of these openings has been used to classify the major groups of reptiles. The earliest and most primitive condition had no openings and is referred to as the anapsid condition (Figure 8.10). The earliest reptiles were anapsids.

Near the middle Carboniferous, soon after the appearance of the first reptiles (the anapsids), a group of reptiles appeared with a single opening in the skull below the postorbital bone and the squamosal bone (Figure 8.10). This condition is called the synapsid

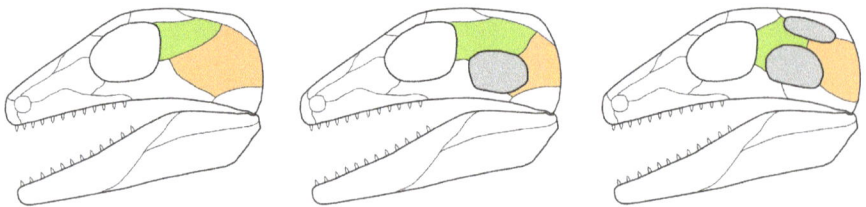

Figure 8.10 Reptile Skulls *(left to right) Anapsid, synapsid, and diapsid. The colored bones are postorbital (green) and squamosal (orange). The fenestrae are shaded gray. [JCB]*

Figure 8.11 Dimetrodon *Representing about 20 species, these early synapsid reptiles of North America and Europe were some of the largest predators of the Early Permian. The largest was about 3 meters long. Since the coloration is not known, artists are free to exercise their imagination. [Dmitry Bogdanov]*

condition, and it is found in the group of reptiles that eventually evolved into the mammals. Because of this evolutionary sequence, the synapsid reptiles are commonly called the mammal-like reptiles. The efficient attachment of the jaw muscles of the synapsids eventually led to the development of the very strong jaws that are found in mammals.

A group of mammal-like reptiles (synapsids) called pelycosaurs became the dominant land animals during the late Carboniferous and first half of the Permian. A well-known pelycosaur was *Dimetrodon*, which is famous for its sail-like back fin (Figure 8.11). Many scientists have speculated on the

purpose of the fin, and one proposal suggests that the fin functioned as a heat absorber and helped to regulate the body temperature of the animal. Pelycosaurs were carnivorous animals, and they evolved slightly differentiated teeth. One characteristic that separates mammals from reptiles is the differentiated teeth that mammals possess.

Near the later part of the Permian, the descendants of the pelycosaurs (called the therapsid reptiles) replaced the pelycosaurs as the dominant land animals (Figure 8.12). The therapsids were very common during the Triassic and they became more mammal-like as the period progressed. The therapsids

Figure 8.12 Therapsid *This therapsid was a mammal-like reptile of the gorgonopsid group that lived during the late Permian. Known as Sauroctonus, it was about 2 meters long. Like other therapsids, it had differentiated teeth. [photo, Natural History Museum Stuttgart, Germany/ H.Zell; reconstruction, Dmitry Bogdanov]*

*Figure 8.13 **Tyrannosaurus** Like typical dinosaurs, it had an opening in front of the eye socket (green). [photo, AMNH]*

were the immediate ancestors of the first mammals.

During the later part of the Carboniferous, we find the first diapsid reptiles (Figure 8.10). As their name implies, the diapsids have two openings on each side of the skull. Except for possibly turtles, all the present day reptiles are diapsids. Dinosaurs where diapsids and birds evolved from diapsid dinosaurs. Turtles have no skull openings, but molecular studies suggest that they may not be true anapsids, but diapsids that reverted to the anapsid condition.

Dinosaurs, which first appeared in the late Triassic, were not giant lizards. They were true diapsids possessing two complete temporal openings, unlike lizards, which lost their lower temporal bar. Another distinctive characteristic of dinosaurs is the presence of an opening in front of the eye socket (Figure 8.13).

Another group of diapsid reptiles called sphenodontids first appeared in the Triassic. All but one of its members eventually became extinct. The lone survivor of the group still lives on a few of the small islands of New Zealand where it is popularly known as the Tuatara. The Tuatara looks superficially like a lizard, but the lower temporal bar is complete so there are two complete openings on either side of its skull (Figure 8.14). The Tuatara belongs to an ancient group of reptiles that appeared before the dinosaurs, making them among of the most interesting reptiles in the world.

Technically, lizards have only one opening on each side of the skull, but the opening is located higher up on the skull than in the synapsids. This condition was derived from the diapsid condition by the loss of the lower temporal bar, the bone below the lower opening (Figure 8.14).

Lizards are first found in the later part of the Permian, but snakes, which are modified legless lizards, did not appear until the Cretaceous. Along with the loss of their legs, the jaws of snakes became more flexible than are the jaws of lizards. This increased flexibility was accomplished in part by the loss of the upper temporal bar (Figure 8.14). Along with the skull's increased flexibility, the lower jaw is very weakly attached and can be dislocated easily, allowing the snake to swallow very large prey.

*Figure 8.14 **Tuatara Skull** The lower temporal bar (ltb) was lost in lizards, and the upper temporal bar (utb) was also lost in snakes. The Tuatara has retained two openings (blue), the original diapsid condition. [N. Tamura]*

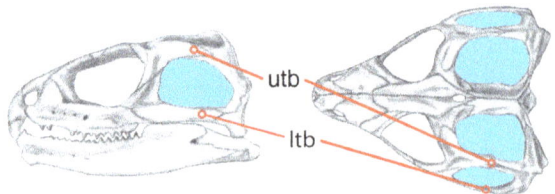

Reptiles to Mammals

The mammals of today differ from reptiles in several important ways, several of which are related to reproduction. Mammals nourish their young with milk that is produced by mammary glands. Mammary glands are highly developed oil glands, and their evolution may have been closely associated with the acquisition of hair, and the advantage of keeping the hair lubricated.

Except for egg-laying monotremes (which include the Platypus and the echidnas or spiny anteaters), mammals give birth to live young that develop to some extent inside the mother. In the case of marsupials, the young are born in a very early embryonic stage. In most placentals (all mammals except the monotremes and marsupials), however, the young have an extended period of internal gestation and are born in a more developed state. Several mammals like some deer, antelope, and horses, are able to run shortly after birth.

Since hair and mammary glands are not often preserved in fossils, skeletal differences are used to separate reptiles and mammals. The bones of reptiles and mammals differ in several important ways. One of the more obvious changes involves the lumbar region. Mammals have reduced lumbar ribs, and their lumbar vertebrae have been modified to provide a more rigid structure. These changes allow mammals to twist their back and lay on their side, a flexibility that reptiles lack (Figure 8.15).

The knees and elbows of reptiles generally point outward, causing the reptile to wiggle as it moves. In mammals, the elbows are pointed backward while the knees point toward the front of the animal, allowing many mammals to run faster than their reptile

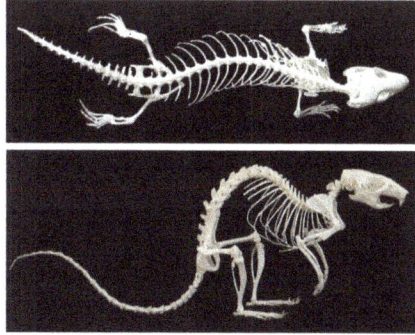

Figure 8.15 Reptile and Mammal The lizard (top) has lumbar ribs, but the mammal (a Norway Rat) lacks these, giving it more flexability. Also, their modified hips allow the legs of a mammal to be positioned more under the body, allowing them to run faster. [rat, Jean-Christophe Theil]

ancestors. Mammal hipbones, shoulder bones, and the connection between the head and the vertebrae allow mammals more flexibility.

As discussed above, the number of bones in the fingers and toes (phalanges) of the earliest reptiles were in the 2,3,4,5,(3,4) pattern, but this original pattern was modified in the different reptile groups that evolved. The group of reptiles from which the mammals evolved, the therapsids, had two phalangeal bones in the thumb and big toe, and 3 bones in each of the other four digits. This 2,3,3,3,3 pattern is found in the early mammals and many modern mammals. The phalanges are modified in the highly evolved feet of horses and deer, but certain other groups such as rodents, bats, and primates (including humans) have retained this primitive bone pattern in their fingers and toes.

The most distinctive features that separate mammal skeletons from reptile skeletons are the bones of the jaw and the type of teeth. Mammals have complex cheek teeth while most reptiles have simple conical shaped teeth.

In reptiles, the two bones that form the jaw hinge are the quadrate and the articular, but in mammals, the two bones forming the hinge are the dentary and the squamosal. The quadrate and articular have been incorporated into the mammalian ear as the incus (commonly called the anvil) and malleus (the hammer) respectively.

The development of the mammalian jaw joint and the incorporation of the reptilian jaw joint bones into the mammalian ear must have occurred in an orchestrated way. Many people have wondered how a jaw joint could have changed in such a dramatic way and how the transition could have been accomplished without losing the use of

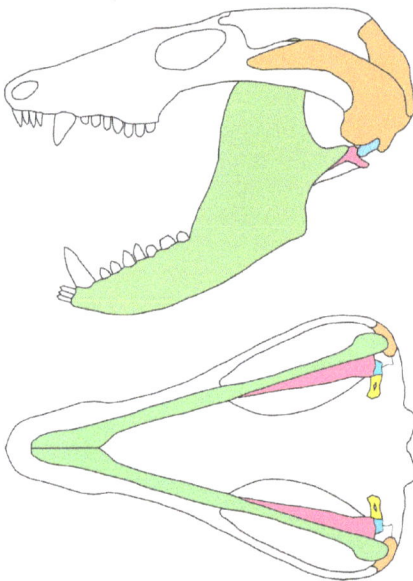

Figure 8.16 Jaws *Probainognathus (top), an advanced cynodont reptile, has a jaw joint formed by the quadrate (blue) and articular (pink). The dentary (green) and squasomal (orange) did not touch. [after Romer] In Morganucodon (viewed from below), the dentary has expanded and made contact with the squasomal. It has two jaw joints, the reptile joint and the mammal joint. The stapes are shown in yellow. [JCB; after Kermack. et al.; Romer, credits p. 415]*

the jaw. Fortunately, the fossil record is fairly good during the time when this transition occurred. Fossils of several mammal-like reptiles show us that some of the transitional animals had two jaw joints, one like the reptile joint formed by the quadrate and articular bones, and another next to it, formed by the dentary and squamosal bones (Figure 8.16).

In reptiles, the stapes is a bone that transmits sound waves directly from the eardrum to the inner ear (the sound sensing area). As the jaw of the mammal-like reptiles evolved, the stapes migrated very close to the jaw joint, eventually contacting the quadrate and articular bones. These bones became incorporated in the ear mechanism after the squamosal and dentary bones formed the mammalian jaw.

Why did these changes take place? The fossils of course can only tell us how, not why. We must use a certain amount of detective work to come up with reasons for the observed changes. The mammal-like reptiles were mainly carnivorous creatures, and many of the changes that led to the mammals can be understood by considering their mode of feeding. Early carnivorous reptiles killed their prey by biting them with their sharp teeth. The prey was then swallowed whole or at least large chunks were torn off and swallowed without chewing.

Chewing the food before swallowing would allow the digestive process to proceed faster and, therefore, provide a quicker supply of energy. The mammal-like reptiles that developed differentiated teeth were better suited for this task and prospered. They evolved teeth for cutting, teeth for stabbing, and teeth for grinding or chewing.

The diversification of the teeth was accompanied by several related changes. The lower teeth form in the dentary bone, and as the rear teeth became more important for chewing, the dentary expanded accordingly (Figure 8.16). Eventually, the dentary expanded to such an extent that it contacted the squamosal bone of the upper jaw. As we see in the fossils, some of these animals possessed a double jaw joint.

Again we should remind ourselves that during the later part of the Triassic when the first mammals appeared, the difference between a therapsid reptile and a mammal was very subtle. The fossils show a gradual transition from the reptile jaw to the mammal jaw. The exact point at which we begin calling these creatures mammals is somewhat arbitrary. If we could examine the live animals that existed then, we would classify both the reptiles and mammals as being members of the same family, or in some cases even the same genus. During the late Triassic and early Jurassic, the differences between mammals and some reptiles were very minor.

In the process of forming the dentary-squamosal joint, the stapes contacted the older jaw joint (quadrate-articular). This contact improved the hearing of the animal and those animals with the best hearing were the ones that survived to reproduce and pass any improvements in hearing to their offspring.

Since the quadrate-articular joint was no longer needed to articulate the jaw (open it and close it), the quadrate and articular bones eventually became completely separated from the jaw and were incorporated into the bones of the middle ear (the malleus, incus, along with the stapes). The addition of these two new bones to the middle ear created a greater mechanical advantage that amplified the sound vibrations of the eardrum. This change resulted in an improvement in the hearing of these animals.

As the embryos of marsupials and monotremes develop, the jaw changes from the reptilian form to the mammalian form. Young marsupials and monotremes have a reptilian jaw formed with the quadrate and articular bones, but as the animal develops, the dentary-squamosal jaw joint is formed and the quadrate and articular bones migrate to the middle ear where they become the malleus and incus.

The lifestyle of the early mammals would help us to understand why improved hearing was so important. All the early mammals were very small and probably would have been easy prey for some of the smaller carnivorous dinosaurs. These early dinosaurs were bipedal and were able to walk and run very effectively on their long birdlike hind legs. In contrast, the early mammals moved more slowly on all fours.

One of the earliest mammals was a shrew-like creature, *Morganucodon*, that is well known from very complete fossils of the early Jurassic Period (Figure 7.33 and 8.16). *Morganucodon* had a body that was about 10 centimeters long.

It is generally believed that the early mammals managed to survive by adopting a nocturnal way of life. Most mammals are nocturnal to this day. While avoiding the reptiles that ruled the day, mammals adapted to living in a world of darkness where the senses of hearing and smell were extremely important. It is this way of life that apparently molded the evolution of early mammals.

There are other changes in the structure of mammals that can be related to their way of living. For example, the nasal openings of reptiles enter directly into the mouth. While eating, reptiles hold their breath until they swallow. As chewing became important to the early mammal-like reptiles, the bony palate (the hard bone on the roof of the mouth) expanded and eventually separated the mouth from the nasal passage. This separation enabled the mammals to breathe while chewing their food. Of course, the bony palate did not form immediately, and its development can be traced in the fossils of the mammal-like reptiles.

As perhaps an extra bonus, the bony palate is also used with the tongue to help manipulate food while chewing. Possibly related to these changes, the separate nose openings in early forms eventually merged to form a single opening into the skull. Reptiles have two separate nose openings into the skull but mammals have only one.

We should keep in mind that these evolutionary changes did not occur because the animals wanted the changes. However, since enlarged palates were advantageous, those animals that happened to have enlarged palates had a better chance of surviving to the reproductive age and passing these traits to their offspring.

The development of hair cannot be traced in the fossil record, but hair is a good insulator and would obviously be advantageous to small nocturnal animals. Being able to maintain a constant body temperature (a condition known as warm-bloodedness) would also be extremely important to small animals that become active as night falls. Small reptiles must warm up in the morning sunlight to become active. If an animal is to remain active during the cool nights, it must be able to maintain an elevated body temperature. A covering of hair would help to conserve the animal's body heat.

Hair does not appear to be a modified scale because it develops in a different manner. It may have evolved as an outgrowth between the scales and may have been present on some of our reptile ancestors. Closely associated with each hair follicle is a sebaceous gland that secretes fat which lubricates and prevents drying of the skin and hair.

It is speculated that the young of mammals gained some nourishment by licking oil off their mother's fur. Eventually some of these glands evolved into mammary glands. It is interesting that monotremes do not have nipples, but the young drink milk that oozes out of mammary glands on the mother's stomach. Their mammary glands are situated at the base of certain mammary hairs.

Monotremes have more similarities with their reptile ancestors than do any of the other mammals. They have no external pinnae (ears) and are the only mammals that lay leathery reptile-like eggs. While forming in the egg, many reptiles develop a special egg tooth that is used for escaping the leathery shell. Monotremes also develop an egg tooth, identical to that of a typical reptile.

The word monotreme means *one hole* and refers to the fact that monotremes have a single opening that passes eggs, urine, and fecal material to the outside. Reptiles and birds also have a single opening, however, in most female placental mammals, there are two or three openings, the anus, the urinary opening, and the genital opening. Marsupials have two openings, the anus and a urogenital opening.

The monotremes have also retained several reptile-like bones in their skeleton. Both coracoids are present as are the interclavicle and epicoracoid. (These bones are in the shoulder area.) Fossil therapsids also have these bones, but they have been lost in other living mammals. The modifications of the shoulder bones allowed the mammals more freedom of motion and enabled them to become faster runners than the reptiles from which they evolved. In contrast to other mammals, the limbs of monotremes stick out to the sides, more like the therapsid reptiles.

Modern monotremes are highly specialized to their present way of life as anteaters (the echidnas) and an aquatic predator of worms and other invertebrates (the Platypus). The echidnas have no teeth nor does the adult Platypus although their young do have six pairs of teeth that are lost before the Platypus matures. Since teeth are so important in the study of mammals, the loss of teeth in monotremes somewhat limits the study of their relationship to other mammals, both living and extinct.

Monotremes have so many distinctive features that some scientists believe they may have evolved independently from a different group of reptile ancestors than the other mammals did. To be sure, the monotremes, which are confined to the Australian region, are one of the most interesting members of the animal kingdom. The preservation of these unique mammals should be of prime concern to all of us.

Although mammals evolved from therapsid reptiles in the late Triassic or early Jurassic more than 200 mya, they represented a relatively minor component of the world's fauna until the extinction of the dinosaurs about 65 mya. With the demise of the dinosaurs, the mammals evolved quickly to fill some of the niches left by the dinosaurs. Today there are almost 4,300 species of mammals of which about 40 percent are rodents and 20 percent are bats. The mammals are grouped into approximately 21 orders and 130 families.

Unfortunately, many mammals and other life forms are in danger of becoming extinct because of one member of the group, the humans. As the human population grows, there is increased pressure to destroy the habitats that are vital to the survival of other mammals. To save many of the more unusual members of the mammal class such as the Tiger, Giant Panda, rhinoceroses, and elephants, their habitats will have to be protected somehow from the increasing pressures of the rapidly exploding human population.

Reptiles to Birds

Most birds are fairly delicate creatures and, therefore, do not fossilize well. The fossil record of birds is understandably very sparse. However, *Archaeopteryx* fossils from the 150 million year old Solnhofen limestone deposits in southern Germany are some of the most spectacularly preserved fossils that have been discovered (Figure 8.17).

Archaeopteryx was a crow-sized animal that lived near the end of the Jurassic. It was the first fossil discovered that possessed feathers. Because of the feathers, it is considered to be a bird, but its skeletal features are virtually identical to those of several small coelurosaurian dinosaurs, a group belonging to the saurischian order of dinosaurs. *Archaeopteryx* skeletons are so much like these extinct reptiles that a few specimens with obscure feather imprints were originally classified as coelurosaurs.

Since *Archaeopteryx* is essentially a feathered reptile and a classic transitional form, it is instructive to compare its skeleton with that of a modern bird (Figure 8.18). We find that *Archaeopteryx* differs structurally in many significant ways. The long tail and pelvis of *Archaeopteryx* are significantly different from the expanded pelvis and shortened tail structure (called the pygostyle) of modern birds. As an aid to navigation, the long floppy tail of *Archaeopteryx* must have been difficult to control. In modern birds, the tail feathers are attached to the compacted pygostyle, and they form a fan of feathers that can be easily controlled by the bird in flight. It is interesting that the early forms of a group of flying reptiles (pterosaurs) had long tails, but the later Cretaceous varieties evolved shorter tails.

Another improvement that has evolved in modern birds is the enlarged sternum and attached ribs. There has been considerable debate among scientists as to whether *Archaeopteryx* was a strong flyer or was merely a

Figure 8.17 Archaeopteryx The Berlin specimen at the Museum fur Naturkunde. [H.Raab]

gliding bird similar to the pheasants of today. The large sternum of modern birds is the attaching structure for the enlarged flight muscles of modern flying birds. Since *Archaeopteryx* did not have a large sternum, some contend that it may have been a poor flyer. Perhaps it could only glide for short distances. Other scientists point out, however, that bats do not have a large sternum but are still very strong flyers.

The forelimbs of *Archaeopteryx* are interesting because, although large feathers were attached to them, each forelimb was composed of three fingers with attached claws (Figure 8.19). Except for their longer length, the forelimbs of *Archaeopteryx* were identical to those of many small dinosaurs. In modern birds, the fingers and some wrist bones have been fused into a rigid structure. For powered flight, there is a distinct advantage to such fusion as it produces a more stable attachment for the flight feathers and allows a more controlled stroke when flapping the wing.

The arrangement and number of the flight feathers of *Archaeopteryx* are the same as is found in modern birds. The flight feathers are non symmetrical so they can form an airfoil, an indication that *Archaeopteryx* used its wings for lift. The shape of the flight feathers is often sighted as evidence that it could fly, although the airfoil shape would also have been of value for gliding.

Archaeopteryx's skull is essentially that of a reptile. Its jaws contained socketed reptile-like teeth, but all modern birds have a horny beak (Figure 8.19). Teeth are relatively

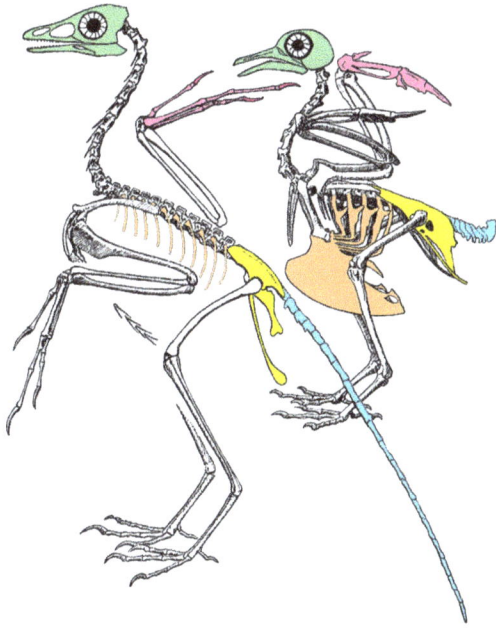

Figure 8.18 Archaeopteryx and Pigeon Compared *The differences are highlighted with various colors. [after Gerhard Heilmann]*

heavy, and one advantage of a beak is its lighter weight. Along with the loss of teeth, the general shape of the skull has also been modified in modern

Fugure 8.19 Archaeopteryx Head (top) showing the teeth. Wing tip (bottom) showing the three clawed fingers. [H.Raab]

birds. Again, it is interesting that some of the later pterosaurs did not have teeth, but evolved a beak-like structure instead.

The bones of modern birds are very light because of the presence of air ducts, and the bones of *Archaeopteryx* also contained air ducts as are found in modern birds. Again it is interesting that pterosaurs also evolved bones with air-filled cavities, a valuable adaptation for a flying animal where weight is so important.

The evolution of feathers was an important series of events that eventually gave birds the ability to fly. Scales and feathers are made of two types of proteins called alpha and beta keratin. The very early development of a feather is identical to that of a reptile scale, which begins with the thickening of keratin-producing cells in the epidermis. In feathers, these cells form an elongated tube. Down feathers are

Figure 8.20 Caudipteryx A small (1 meter long) 125 million year old dinosaur with tail and arm feathers. [photo, Houston Museum of Natural Science/Daderot]

formed when the tube divides into several strands, which are connected at the base like the individual blades in a clump of grass.

In flight and contour feathers, the barbs branch off a spine (called the rachis) that forms on one side of the tube. Birds are not completely covered with feathers, but have scales on their legs and feet. Although we do not know all the details, the evidence suggests that feathers evolved from scales.

The early evolution of feathers was certainly for some reason other than flight. The first feathers were probably just elongated tubes or downy feathers and would have been useless for flight. Many scientists have suggested that the original function of feathers was for insulation, either from the cold or perhaps even the heat. Insulation is still the primary function of the down feathers.

Another function may have been for sexual displays, which is an important function of feathers for many of the birds of today.

Recent fossil discoveries from the early Cretaceous in China show that several dinosaurs also had spiny or downy-feathers. Examples include *Sinosauropteryx*, a turkey sized dinosaur with down-like feathers, and *Protarchaeopteryx* with down-like feathers on its body and vaned feathers on its tail. *Caudipteryx* was also a turkey sized dinosaur, but with a fan of vaned feathers at the end of its tail and symmetrical feathers on its short forelimbs (Figure 8.20). Its forelimbs are too short to be useful for flight.

Many dinosaurs had feathers, either for warmth or for some sort of mating display. Currently, the largest dinosaur known to have had feathers is a 125 million year old 10 meter long species named *Yutyrannus* which means beautiful feathered tyrant. Its feathers were simple filaments about 15 to 20 cm long, like the fuzz on a baby chicken only longer. *Yutyrannus* was a theropod dinosaur like its later cousin *Tyrannosaurus rex*.

Although there is some question as to the age of a few of the feathered dinosaur fossils from China, some are several million years younger than *Archaeopteryx*. If so, they could not be the ancestors of *Archaeopteryx*, but we do not expect that all the dinosaurs closely related to the ancestors of the birds died out as soon as the birds evolved. Feathered dinosaurs may have lived until they were wiped out at the end of the Cretaceous.

Again we see that during the later part of the Jurassic, when birds first evolved from small dinosaurs, the distinction between a dinosaur and a bird was not clear-cut. Many of these transitional creatures would have been very difficult to classify as either birds or dinosaurs, and we probably would have lumped them together in the same family. All the dinosaurs are gone as are the early primitive birds, so as we compare modern birds with lizards we find very large differences. All the transitional forms have become extinct,

accentuating the differences between modern birds and modern reptiles.

As mentioned above, bird fossils are not very abundant, but there are a few specimens from Cretaceous times (an 80 million year time span) that give us clues to their evolutionary trends. A sparrow-sized bird known as *Sinornis* lived in China about fifteen million years after *Archaeopteryx*. This bird fossil gives us important clues to the evolution of birds. *Sinornis* had a toothed snout and stomach ribs like *Archaeopteryx*, but its unfused wing bones were intermediate between those of modern birds and *Archaeopteryx*. In addition, *Sinornis* had other advanced features such as a broad sternum and tailbones fused into a pygostyle.

So far, the oldest bird with a beak instead of teeth is *Confuciusornis,* a 120 million year old specimen from China. Several of the *Confuciusornis* fossils show that some (presumably the males) had two long tail feathers (Figure 8.21). Since they would have been of little use for flying, they must have been used for mating displays.

Although *Confuciusornis* and some other birds had beaks, others retained teeth until near the end of the Cretaceous. Among the examples of

Figure 8.21 Confuciusornis *This 120 million year old fossil from China of the oldest known bird with a beak. Because of the long tail feathers, it is probably a male. Specimans with short tails are believed to be females. [photo, Museum of Natural History Vienna, Austria/Gyik Toma]*

later Cretaceous birds with teeth was *Hesperornis* and related forms that were large diving birds superficially resembling the loons of today (Figure 8.22). They were highly specialized birds that lacked a keel, and had wings that were greatly reduced in size, indicating that they could not fly. However, *Hesperornis* had a shortened tail and was more modern than was *Archaeopteryx*.

Because of its special adaptations as a diving bird, *Hesperornis* probably represents a dead end as far as bird evolution is concerned. However, we do see from *Hesperornis* that some birds still retained teeth throughout the Cretaceous period, but their tail had evolved toward the modern form. Its bill was composed of several plates similar to the albatrosses of today.

Another late Cretaceous bird was *Ichthyornis*, a gull-like or tern-like bird about the size of a pigeon. *Ichthyornis* lived between 85 and 90 mya. It had a large keel and must have been a strong flyer. Although there may be some argument about whether *Archaeopteryx* was a strong flyer, studying *Ichthyornis* shows us that many birds did become strong flyers during the Cretaceous. *Ichthyornis* also possessed teeth and the compressed tailbones of modern birds.

Figure 8.22 Hesperornis Flightless members of this genus lived 80 mya. It was a large (2 meters long) marine bird that probably had a lifestyle similar to the penguins of today. [photo, Smithsonian Natural History Museum]

Perhaps partly because of the loss of the flying reptiles (pterosaurs) at the end of the Cretaceous, birds became very abundant and diverse during the early Cenozoic Era. Today there are almost 9,000 species of birds organized into about 160 families and 27 orders. Their contribution to the control of insects cannot be overstated. Because they are one of the most visible groups of animals, birds add a great deal of grace and beauty to the natural world. They make a major contribution to the aesthetic quality of our lives.

Dogs, Bears, and Raccoons

As we look at a bear, a dog, or a cat, we see mammals that appear quite different and distinct. However, if we investigate the fossil record, we can more easily see how these creatures are related. The further back in time one goes, the more similar the various members of this group become, making it more difficult to classify the fossils than the living forms. We will use this group to illustrate the evolutionary divergence of a typical order of mammals during the past 60 million years or so. A similar picture would emerge if we were to investigate any of the other mammal groups.

Carnivorans are familiar mammals belonging the order Carnivora whose living members are divided into about sixteen families. The more well known families are the cats, dogs, bears, skunks, raccoons, weasels, hyenas, and civets. Although they are occasionally placed in a separate order, the seals, sea lions, and walruses are highly evolved members of the Carnivora order that have invaded the sea. A few other carnivorans such as the Sea Otter and the river otters have also adapted to an aquatic habitat. While these families appear to be quite distinct, as we will see, the divisions are not so clear when one studies the fossil ancestors of various members of these families.

Fossil remains of the condylarth ancestors of the modern carnivorans are found from early Cenozoic times, as are the remains of another group of flesh-eating mammals known as the creodonts (Figure 8.23).

The tooth formula of the earliest placental mammals was 3 incisors, 1 canine, 4 premolars, and 3 molars on each side of the upper and lower jaws (a total of 44 teeth). Incisors, canines and premolars are replaced as the animal grows (they are often called baby teeth), but molars are generally not replaced.

Early carnivorans are distinguished by the presence of two teeth on each side of the jaw that are used for shearing flesh. These modified teeth, the last upper premolar and first lower molar, are known as *carnassials* (Figure 8.24).

There were two groups of creodonts that may not have been closely related, but they also had carnassial teeth. Their carnassials, however, were formed by different teeth than those of the carnivorans. One group of creodonts, the oxyaenids, had a pair of carnassials formed by the first upper and second lower molars. The other creodont group, the hyaenodontids, generally had carnassials, formed by the second upper and third lower molars.

The largest known creodont was a hyaenodontid called *Megistotherium* that lived about 25 mya in northern Africa. It was a huge 500 kilogram animal that was about 4 meters long and fed on mastodons and other large mammals.

These two groups of flesh eating mammals (the carnivorans and the creodonts) also differed in a few skeletal areas, but similarities suggest that they probably evolved from a

Figure 8.23 Arctocyon This condylarh is a member of the ancestral group that gave rise to the carnivorans and creodonts. [Dmitry Bogdanov]

common ancestor without carnassial teeth in the late Cretaceous. Creatures that may have been the common ancestor of these two groups of flesh-eating mammals have been found from late Cretaceous times.

However, many other groups of mammals began to diversify during these times, and unscrambling the very early family tree of the mammals is quite difficult. Most of the early mammals were very similar. Even the early ancestors of the grazing animals such as the deer and the horse were

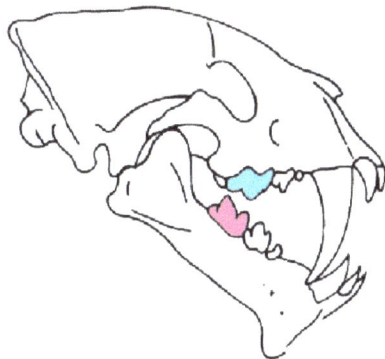

Figure 8.24 Carnassials The last upper premolar (blue) slides to the outside of the first lower molar (pink) producing the slicing action. The drawing is of a sabertooth cat (Smilodon). [Steven M.Carr]

very similar to the early ancestors of the carnivorans.

Near the end of the Eocene, most of the creodonts became extinct, and the true carnivorans began to diversify. At first, the carnivorans were small mammals, about the size of a small cat. The modern civets, although slightly larger, are very similar in form to the ancestral carnivorans.

The fossil record near the end of the Eocene is somewhat poor, but the two major groups of modern carnivorans had begun to diverge by this time. One of these groups, which we might call the cat group, includes the cats, civets, and hyenas. The other group, which we will call the dog group, contains the dogs, bears, raccoons, weasels, and the marine carnivorans (seals, sea lions, and walruses). The marine carnivorans are more closely related to the dogs than the dogs are to the cats.

In the late Eocene, to differentiate between these two major groups, one has to get technical. The differences involve the internal cartoid artery and the bone structure of the auditory bulla, a vase-shaped bone that forms the middle and inner ear cavity. These are relatively minor differences, and had we been around in the early Oligocene, it would have been difficult to separate the two groups.

As time passed, the cat group and the dog group further diverged. Dogs and bears have lost a pair of upper molars so they have 42 teeth. Cats have the most highly developed carnassials and the molars behind the carnassials have been lost or reduced to vestiges. Cats have only 28 or 30 teeth.

To catch their prey, the early carnivorans adapted a running way of life and the bones of the feet changed as those best suited for running survived to reproduce and pass on their advantageous traits. Dog's front legs have 5 toes, but the *pollax* (thumb) is

high on the foot. Their back legs have only 4 toes. In cats, the pollax is also high (it does not touch the ground), but not as high or as reduced as it is in dogs. Cats also have 4 toes on their back legs. Perhaps the most obvious difference between the feet is the fact that cats have retractable claws, but dogs do not.

In early Oligocene times, the ancestors of the raccoons diverged from the main dog line. Raccoons evolved teeth with more rounded cusps for crushing as they generally had a more omnivorous diet than dogs. These two families shared a common ancestor approximately 30 mya.

The Red Panda (Figure 9.48b), with a raccoon-like tail and mask, diverged very soon after the raccoon group split off the main dog line. It is the sole survivor of its family.

The first bears occur in the middle Oligocene. One difference between dogs and bears (with the exception for Polar Bears) is that the carnassial teeth of bears have lost some of their shearing function. The teeth of bears are generally better suited for crushing, an adaptation to their more varied diet. Also, dogs walk on their toes, but bears walk on their feet like we do (called plantigrade locomotion).

Today, no one would mistake a dog for a bear, but during Miocene times creatures existed that were a mixture of the two. These bear-dogs had more of the general body shape of a bear, but the teeth of a dog (Figure 8.25). Since none of these intermediate creatures have survived, the differences between bears and dogs has been accentuated.

The family tree of the dog group has been pieced together from the fossil record, but DNA sequencing of living forms has been used to more accurately refine the results.

One of the more unusual members of the dog group is the Giant Panda,

Figure 8.25 A Bear Dog *Amphicyon was a genus belonging to the dog group of carnivorans. They lived in temperate North America and Eurasia from the Middle Paleogene until about 2 mya. During that time period, they evolved from dog-like to more bear-like mammals. The largest species was about 2.5 meters long and weighed 600 kg. [reconstruction ©2014 Lew Delport]*

which also has a raccoon-like tail, although a very short one (p. 376). It is classified as a bear, but it represents an early offshoot from the main bear lineage. This wonderful creature is one of our rarest mammals and leads an extremely fragile existence in the forested mountains of central China. Wild Giant Pandas living free may soon be only a memory unless their habitat is protected and carefully maintained, an important undertaking for the most populated country in the world.

Horses

Horses walk on a single toenail so their evolution is very interesting, Like all early mammals, their ancestors had five toes. The horse is an odd-toed ungulate like its close relatives the zebras and donkeys, and their more distant relatives, the rhinoceroses and the tapirs.

The rhinoceroses have three toes and the tapirs have three toes on the back feet, but four toes on the front feet. However, the outside toes on the front

feet are reduced in size and the middle of the three remaining toes is larger than the other two (Figure 8.26).

The feet of the horse are more evolved in that its toes have been reduced to one. The evolution of the reduction of the toes in the horse and its closest relatives is preserved in the fossil record.

By the early Eocene, the even-toed ungulates and odd-toed ungulates had split into separate groups. While there are many fossils that are relevant to the ancestry of the horse, we will only mention a few. As we said in the beginning of the chapter, we do not expect every fossil to be the direct ancestor of the horse. They may be just distant cousins, but we expect the general trend to be obvious.

By the early Eocene of Europe and North America, about 50 mya, we find fossils of an animal given the name *Hyracotherium*, which means *hyrax-*

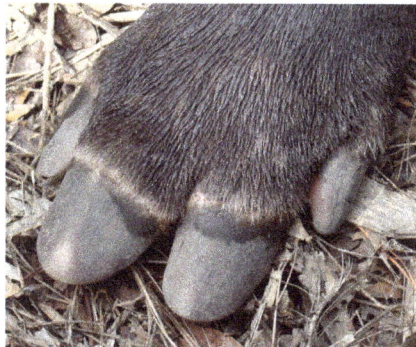

Figure 8.26 Tapir's Foot *The left front foot of a Baird's Tapir has 4 toes, but the outside toe (far right) is smaller than the other 3, and the middle toe of the remaining three is the largest. [Bjørn Christian Tørrissen]*

Figure 8.27 Eohippus *This early relative of the horse had 4 toes on its front feet and 3 on its back feet. [Charles R. Knight/AMNH]*

like beast. It was so named because its discoverer thought it resembled a hyrax. Hyraxes, however are members of the Afrotheria group, which are not very closely related to the odd-toed ungulates. Later a closely related fossil was discovered and named *Eohippus*, meaning *dawn horse* (Figure 8.27).

These early horse ancestors were small animals about 20 kilograms and 20 cm high at the shoulders. They had 4 toes on its front feet and 3 on its back feet with small hoofs on each toe. The outer toes were smaller and higher up, similar to what we see on tapirs today.

Odd-toes ungulates were a very successful group of ungulates throughout the Eocene and Oligocene. They were browsers that ate softer plants and probably lived in more wooded areas. They had low crowned teeth (the crown is the part of the tooth above the gum line). However, as the climate became dryer during the Miocene and grasslands became more widespread, the even-toed ungulates became more abundant and the odd-toed ungulates declined.

Except for the pig group, camel group, and hippos, even-toed ungulates have a more complex digestive system than their odd-toed relatives. Even-toed ungulates with this complex digestive system are known as ruminants. Familiar members of this group are the deer, antelope, giraffes, cattle, sheep, and goats.

Ruminants have a four-chambered stomach. In the first and second compartments, the food is partly digested with the help of bacteria and then lumps of this mixture, called cuds are regurgitated and chewed again. Because bacteria are able to break down cellulose, this process allows ruminants to eat coarser plant matter, an advantage as grasses increased during the Miocene.

By the Oligocene, the climate was becoming dryer and the grassland areas were expanding. In this environment we find fossils in North America of the genus *Mesohippus*, which means *middle horse*. *Mesohippus* was larger than *Hyracotherium*, being about 0.6 m tall (about the size of a sheep), and it had just 3 toes on each foot. The middle toe was much larger than the outer toes Figure 8.28).

Mesohippus was a browser like *Hyracotherium*, and it fed on tender leaves and fruit. It had chewing teeth which were still low crowned, but there was a space between them and the front teeth. The chewing surfaces of the back teeth had ridges of enamel that acted as cutting edges to grind the plant food.

By the middle Miocene a larger ancestor of the horse appeared called *Merychippus* which means *ruminant horse*. The members of this genus varied in height, but typical species were about 0.9 m tall at the shoulders and weighed about 100 kilograms. They had three toes on each foot, but the middle hoofed toe was much larger and obviously supported most of the weight (Figure 8.28). The side toes did not touch the ground except perhaps when the animal was running.

Merychippus looked more like a modern horse than *Mesohippus*. Its chewing teeth became higher crowned as it adapted to a diet of coarse grasses. To accommodate the longer roots of its teeth, *Merychippus* had a deeper, stronger jaw, more like that of a modern horse.

A relative of the *Merychippus* genus evolved into the genus *Pliohippus* in the late Miocene. In *Pliohippus*, the outer toes were lost and the chewing teeth evolved even higher crowns (Figure 8.28). It was about 1.2 m tall at the shoulders, about the size of a modern pony.

Pliohippus may not have been one of the direct ancestors of the modern horse, but it was clearly a close relative.

Pliohippus means *Pliocene horse*, although its fossils are mainly from the late Miocene.

The genus *Equus*, in which modern horses, asses, and zebras belong, made its appearance about 3.5 mya in North America. They soon spread to Eurasia, Africa, and then to South America during the Great American Interchange. They became extinct in

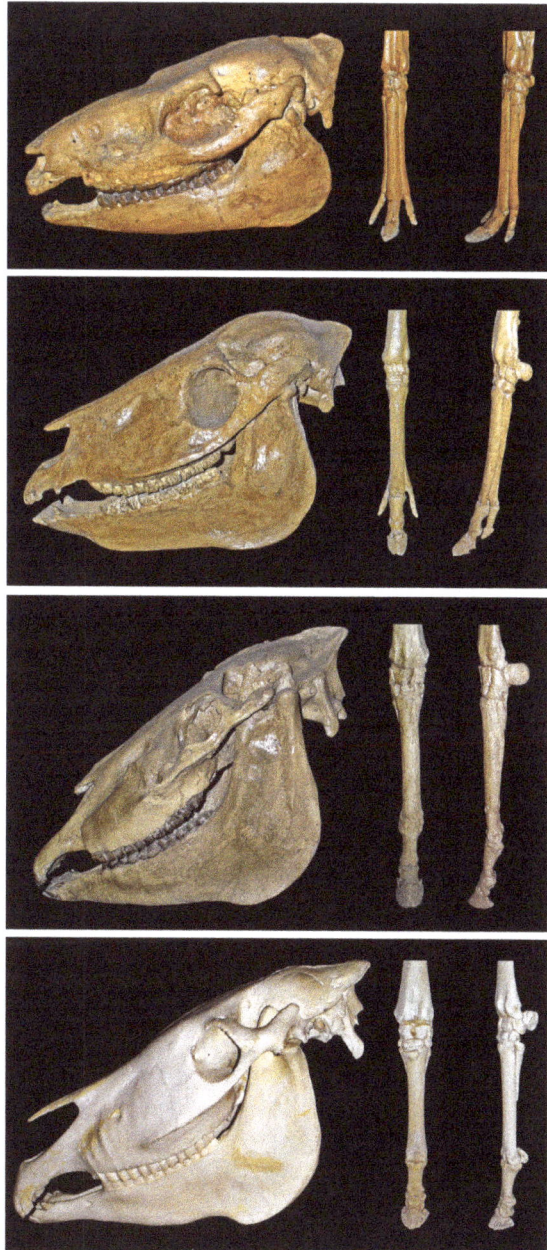

Figure 8.28 Horse Evolution (top to bottom) Mesohippus from the Oligocene, Merychippus from the Middle Miocene, Pliohippus from the late Miocene, and Equus from 3.5 mya. This series shows the loss of the outer toes and the deepening of the jaw as the chewing teeth increased in length. [photo, Natural History Museum Stuttgart, Germany/H.Zell]

the New World about 12,000 years ago, about the time that big game-hunting humans arrived in the New World.

Modern horses are generally larger than *Phiohippus*. They are taller than 1.5 meters at the shoulders, and they have a longer face than *Phiohippus* (Figure 8.28).

Whales

The sequencing of a few genes suggests that hippopotamuses are the living land animals that are most closely related to the whales, dolphins, and porpoises (the cetaceans). However, the relationship of whales to other closely related groups is not currently settled. It appears that about 60 mya, the ancestors of the hippos and whales split off the rest of the even-toed ungulates. The hippo group split off about 5 million years later. By the Oligocene, about 30 mya, we find a relatively large early hippopotamus relative called *Anthracotherium* (Figure 8.29).

Although there are many fossils of early whale ancestors, we will look at just a few to see how whales evolved from early even-toed ungulates.

Near the beginning of the Eocene, about 55 mya, we find fossils of an animal believed to be near the base of cetacean family tree. *Pakicetus* was

Figure 8.29 Anthracotherium An early hippo relative that lived during the Oligocene in Eurasia and North America. It weighed about 250 kg and was about two meters long. [Dmitry Bogdanov]

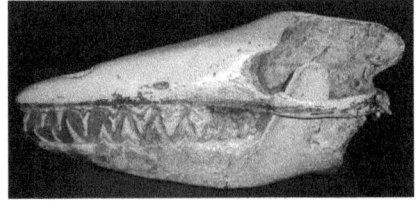

Figure 8.30 Pakicetus Identified as a whale ancestor because of its distinctive teeth and elongated skull. [photo, Natural History Museum, London/Nordelch]

about the size of a wolf and was the earliest member of the group known as the archaeocetes, which means *ancient whale* (Figure 8.30).

As the name implies, *Pakicetus* fossils have been found in Pakistan. We might remember that the Indian subcontinent, which included Pakistan, was an island in the Indian Ocean during much of the Mesozoic and about the first half of the Cenozoic.

Pakicetus fossils are found in areas that were coastal regions at the time of their formation, suggesting that these animals lived near water. Additionally, the bones are thickened and very dense, which would have allowed the animals to stand in the water and be almost submerged.

Pakicetus had a distinctive elongated skull like we find in later whales. Such skulls are often associated with the eating of fish. They had differentiated teeth, but their back teeth were triangular with serrations similar to shark teeth which is another indication of a fish diet. Modern odontocetes (toothed whales) have simple peg-like teeth since they do not chew their food.

Pakicetus had a thickened auditory bulla, characteristic of underwater hearing. The auditory bulla is a hollow bony capsule that encloses the middle and inner ear. The middle ear is where the hammer, anvil, and stapes are found, and the inner ear houses the

Figure 8.31 Ambulocetus A whale ancestor from the Early Eocene. [photo, Certosa di Pisa Museum/Notafly; reconstruction, Nobu Tamura]

sound detecting cochlea and a balance organ called the semicircular canals.

About 5 million years after the appearance of *Pakicetus*, from the early Eocene, we find fossils of 3 meter long *Ambulocetus* in Pakistan. It had functional legs so it could still walk on land in a sea lion-like fashion, but not efficiently because their upper leg bones (femurs) did not have large attaching points for the muscles (Figure 8.31).

Since mammals have a spine that is more flexible in the up and down direction than the sideways direction, it is not surprising that whales evolved a mode of swimming that involves up and down motion. Based on the structure of its spine, *Ambulocetus* swam by moving its back up and down as modern whales do.

Whales have a fluke (tail fin) that is horizontal, and it creates thrust as their tail moves up and down. Other aquatic mammals like seals, otters, and manatees also swim by moving their rear up and down. On the other hand, fish have a vertical tail fin that they move sideways to create the thrust that propels them forward.

The teeth and skull of *Ambulocetus* were very similar to later archaeocetes.

The structure of its ears allowed it to hear under water, and it is likely that *Ambulocetus* lived much like the crocodiles of today.

Oxygen isotope analysis of its bones and teeth indicate that *Ambulocetus* drank fresh water and some salt water. Salt water contains a higher abundance of oxygen-18 than fresh water. A water molecule of O-18 is about 11% heavier than a water molecule made of O-16. As ocean water evaporates, the lighter molecules evaporate more easily, and fresh water, which comes from rain, contains less O-18. Therefore, animals that drink salt water have a higher O-18 abundance in their bones and teeth.

About 46 mya, near the middle of the Eocene, we find fossils of an animal called *Rodhocetus* which was even more whale-like (Figure 8.32). It had weaker limbs and large tail muscles for swimming. It had shorter femurs than *Ambulocetus* which means it probably could not have walked as well on land, and it had a longer skull and less differentiated teeth than earlier archaeocetes. Its nostrils were further back on the head and were situated approximately above the canine teeth.

Dorudon lived during the later Eocene from about 40 to 35 mya. It was about 4 to 5 meters long and had tiny hind limbs and a pelvic girdle that

Figure 8.32 Rodhocetus An ancestral whale that lived 1 or 2 million years after Ambulocetus appeared. [Pavel Riha]

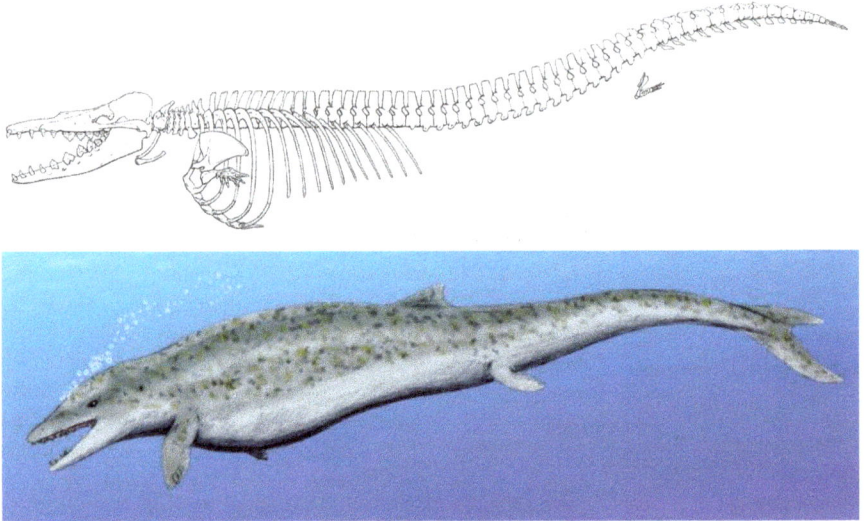

Figure 8.33 Dorudon *This fully aquatic whale was about 5 meters long and lived during the Late Eocene, about 40 to 35 mya. Its tiny hind limbs may have extended outside the body. [skeketon reconstruction, Doug Boyer, Philip D.Gingerich et al., see credits page 415; reconstruction (bottom), Nobu Tamura]*

was not attached to the vertebral column. These legs may have stuck out of the body, but were too small for effective propulsion and were useless for terrestrial movement. Tail flukes do not contain bones so they do not fossilize well, but its vertebral column suggests that it may have had a tail fluke (Figure 8.33).

Its skull lacked the *melon* which is a mass of fat and wax in the forehead of toothed whales that is used for echolocation. The melon focuses the sound waves the whale makes so they are sent out in a beam. The whale hears any returning waves that are reflected off nearby objects, allowing the whale to know the location of the object.

Dorudon fossils are found in marine deposits all over the world indicating that it was entirely free of land. They had a large single nostril that had migrated back above the middle of its teeth.

Basilosaurus (which means *king lizard*) lived at about the same time as

Dorudon. It was a 15 meter long serpent-like animal that looked like a sea serpent, for which it was named. Richard Owen, who lived in the 1800s tried to rename it *Zeuglodon* (meaning *yoked tooth*), but the rules of naming species gives the earlier name priority.

Its long length is due to the fact that its vertebrae are longer than those of other whales of the time. Because of its serpent-like body, it is believed that it is not likely the direct ancestor of modern whales. *Basilosaurus,* like *Dorudon*, had tiny hind limbs and a pelvic girdle that was not attached to the spine. We see in *Basilosaurus* and *Dorudon* whales in the process of loosing their hind legs.

Basilosaurus had a foot that is symmetric between the third and fourth digits which is characteristic of artiodactyls (even-toed ungulates) and is rarely found in other groups.

A close relative of *Dorudon* and *Basilosaurus* was *Artiocetus* (the name is a combination of *artiodactyl* and *cetus*). Its ankle bones are preserved and they have the double-pulleys which are found in artiodactyls. This modified ankle bone is what makes many artiodactyls fast runners. The presence of this ankle bone in early whales is the main piece of evidence that tells us that whales and even-toed ungulates evolved from the same group.

Molecular studies of various proteins also show that whales and hippos are more closely related to artiodactyls than they are to any other living mammal group.

Figure 8.34 Baleen *Surfacing with its mouth open and full of fish (some can be seen jumping out), this Humpback Whale exposes the hair-like masses of baleen on either side of the roof of its mouth. Yes, that is a Brown Pelican in its mouth, and yes it got away in time. [Brad Schram]*

By the later part of the Oligocene, we find the two modern groups of whales. Odontocetes are the modern whales with teeth and the mysticetes are modern whales without teeth.

Mysticetes or baleen whales do not have teeth, but they have rows of brush-like structures called *baleen* growing on either side of the roof of their mouth. The baleen hangs down like brushes from the upper jaw of the whale.

The baleen is used to strain out the krill and small fish that the whale eats. Krill are small shrimp-like crustaceans that is the primary food of baleen whales. The whale takes in large quantities of krill and water and then presses its tongue up against the roof of its mouth, squeezing the water out the sides of its mouth through the baleen, leaving the food behind (Figure 8.34).

Baleen is made of keratin, as is hair and fingernails so it does not fossilize well. However, it grows from the roof of the whale's mouth (palate) and is nourished by blood. The palates of living baleen whales have numerous groves that hold the branching arteries and nerves that serve the baleen. These groves are not present in toothed whales. Fossils of a 25 million year old whale *Aetiocetus* show these groves, but *Aetiocetus* also had teeth similar to those of the archaeocetes.

We should not be surprised that the ancestors of baleen whales had teeth since the fossil record clearly shows that the first whales had teeth. Also as we saw in Chapter 1, fetal baleen whales like the Fin Whale develop teeth that are reabsorbed before birth. Recent DNA studies have shown that baleen whales have the genes for tooth formation, but several of these genes are disabled by mutations.

The Mammal Brain

With the extinction of the dinosaurs at the end of the Cretaceous, the mammals were free to occupy the many newly vacated habitats. The earliest mammal fossils indicate that these creatures were very similar in size and appearance to modern day shrews. As is true of many mammals of today, they were probably nocturnal.

As mentioned above, because of their nocturnal way of life, the senses of smell and hearing were more important than the sense of sight. Indeed, most of today's mammals have very keen senses of smell and hearing, but many have relatively poor eyesight, especially in comparison to most birds.

To see and capture prey while flying, many hawks have developed unusually keen eyesight. (Birds with keen eyesight are more likely to survive and pass their improved eyesight genes to the next generation.) In contrast to birds, mammals often detect the presence of other animals by their sense of smell before they see the visitor.

As the sense of smell was developed in early nocturnal mammals, the smell part of the brain also expanded. The development of certain parts of the brain must generally accompany the development of some improved ability. For example, if a primitive creature developed a complicated eye, it would be of little use if the creature did not have enough brainpower to process the information. This is one reason why the insect eye is so different from the vertebrate eye. Insects primarily detect motion and color. They do not recognize individuals as a dog can do. Even more primitive creatures such as certain worms have much simpler eyes that are only able to detect light and dark. Their brain is so simple that they would be unable to process more complex information.

As birds developed the ability to fly, the part of the brain that controls coordination became enlarged. Flight muscles would be of little use if the bird did not have the coordination to use them. In addition, the part of the brain that controls sight is highly

developed in birds. To be able to fly into a tree and land on a small branch requires a great deal of coordination and keen eyes that can be focused quickly.

Smells generally are more difficult to interpret than are visual images. With visual signals, a great deal of information is collected by the eye and sent to the brain. The animal need not use much thought to differentiate between friend and foe. Studies have shown that certain birds react primarily to shapes and colors.

With smells, however, a simpler signal is sent to the brain and a certain amount of thought must go into deciphering the signal. For this reason, the part of the brain controlling smell and the part that controls reasoning ability became highly developed in mammals. This development can be traced in the fossil record since the size of different parts of the brain can be determined from the shape of a skull. It appears that the nocturnal way of life that the early mammals adopted, and their increased reliance of the sense of smell is partly responsible for their increased intelligence. All of this because of their need to avoid the marauding dinosaurs that ruled the daylight hours.

Again we should point out that these changes were not caused by desire. Traits that increase the mammal's chances of surviving to the reproductive age will be passed to their offspring.

The early primates evolved for a life in the trees, and with the disappearance of the dinosaurs, many became creatures of the daylight hours. This new life in the daylight led to a greater reliance on sight. Compared to other mammals, primates have developed good vision, although not as good as the sight of many birds. Breezes usually do not carry scents from the ground up into a tree, so for a tree dwelling creature, being able to see an intruder approaching the tree is very important.

Stereoscopic vision, which helps in depth perception, is an important feature for judging nearby distances. When judging a jump from one limb to another, well-developed stereoscopic vision could mean the difference between life and death. Stereoscopic vision is produced when both eyes look at the same object, and primate eyes have evolved in this direction.

Primates have eyes that are located more towards the front of the face in contrast to many other mammals that have eyes located more to the sides of their head. Although, eyes located on the sides of the head are not as good for stereoscopic vision, they permit the animal to see more of its surroundings, a definite advantage if you are a deer grazing in an open meadow, and a lion is trying to sneak up on you.

Along with the development of better vision came an even larger brain to interpret what was being seen. Panic at the sight of an enemy that is not in your tree many not be the correct response to every situation. A little careful thought could save a life. We find in the primates, a well-developed

sense of sight and the largest brains in the mammal class.

Early Primates to Humans

Before the disappearance of the dinosaurs at the close of the Mesozoic Era, mammals lived in the shadows. It appears that most of these early mammals were nocturnal hunters of insects.

Near the beginning of the Cenozoic, the primates were just differentiating as a group. As with other groups of mammals, they diversified quickly after the dinosaurs became extinct. The earliest mammals that were the direct ancestors of the primates were small insect eating mammals, superficially resembling squirrels. Today, a group of mammals called tree shrews are good examples of what these early primate ancestors might have looked like (Figure 8.35). There has been some debate whether the tree shrews should be included in the primate order, however, they do represent a creature that closely resembles the earliest mammals that evolved into the primates.

As the placement of their eyes moved more towards the front of the face to improve their stereoscopic

Figure 8.35 Earliest Primates *The Common Tree Shrew (left) is probably the best example of how the animals that evolved into the primates looked. [Alan Hill] The Gray Slender Loris (right) is likely a good example of a very early primate. [Kalyan Varma]*

vision, the faces of primates became flatter. In addition, the early primates evolved hands and feet for grasping the tree limbs instead of using their claws to cling to trees as squirrels do. The grasping action of primates was accompanied by the evolution of flat nails and the development of sensitive pads on the ends of their fingers and toes. The living primates that most closely resemble these early primates belong to a group known as the prosimians, a name that means *before monkeys*. Members of the prosimian group include the tarsiers of Southeast Asia, the lemurs of Madagascar (Figure 9.19) and the lorises of Asia and Africa (Figure 8.35).

Perhaps a more natural classification scheme puts the tarsiers, lemurs, lorises, Bosman's Potto, and the galagos (also called bushbabies), in a suborder called the strepsirhines, often referred to as the wet nosed primates. The rest of the primates are then put into a group called the haplorhines, or the dry-nosed primates.

The Old and New World monkeys, and the apes belong to an infraorder called the anthropoides. Today, there are about 15 species of apes that belong

Figure 8.36 Lesser Ape White-handed or Lar Gibbons are found from northern Burma south through the Malay Pennsula and in northern Sumatra. All but one of the 17 gibbon species are endangered due to the loss of suitable habitat. [JJ Harrison]

to a superfamily called hominoids. They are divided into two groups, the smaller or lesser apes and the great apes. The lesser apes are the gibbons, one of which is named the Siamang (Figure 8.36). The larger hominoids or great apes are called hominids (family Hominidae). They include the gorillas, orangutans, chimpanzees, and humans.

The first placental mammals had 3 incisor teeth, 1 canine, 4 premolars, and 3 molars on each side of their upper and their lower jaws. This dental formula has been reduced somewhat in living primates, as the maximum formula is 2,1,3,3 on each half of the upper and lower jaw. In humans the number of teeth has been further reduced to 2,1,2,3 on each side of the upper and lower jaws. Most humans probably would have been happier if our molars had been reduced to two since the third molar generally causes nothing but trouble for most of us.

Currently, the oldest known primate is a 55 million years old fossil from central China that has been placed with the group of dry-nosed haplorhines. At about the same time a group called the adapiform primates were a common component of the primate communities in Europe, Asia, and North America. They are placed in the strepsirhines group and most closely resembled the lemurs (Figure 8.37) and lorises of today.

The first anthropoids evolved about 40 mya, near the end of the Eocene. These early monkeys were herbivorous leaf eaters. South America was already an island continent at this time, but somehow a group of early monkeys reached the continent about 35 mya, perhaps by rafting from Africa.

These New World monkeys (Figure 9.38a) evolved independently from the Old World monkeys. Some New World monkeys can be told from their Old World cousins by the presence of a prehensile tail. These New World monkeys use this tail as a third hand for climbing and grasping.

Apes first evolved in the early Oligocene about 30 mya. Hominoids, especially the early forms, were primarily fruit eaters. The molar teeth of apes have a chewing surface with five-points or cusps and groves between the cusps shaped like the letter Y (often called the Y-5 pattern).

Apes differ from monkeys in that they have no tail and developed long arms that could be fully extended with flexible elbow and shoulder joints, broad chests, and mobile hips. These adaptations allowed them to hang from tree branches and swing from limb to limb like the gibbons of today. Later forms that moved about on the ground used their long arms and knuckles as supports for walking. Monkeys generally do not use their arms to hang from trees, but scamper over the tops of the tree limbs.

During the Miocene (25 to 5 mya), many new species of apes and monkeys evolved. The Miocene was a time of great tropical forests in Europe and Africa that were well suited for these arboreal animals. Some have even called the Miocene the age of the apes.

The earliest known likely ancestor of the apes is a cat sized animal called *Aegyptopithecus*. It had characteristics of both apes and monkeys and lived about 35 mya in what is now Northern Africa, the Fayum region of Egypt to be more exact. Today, this area consists of large areas of desert, but during earlier times it was heavily forested. Radar images show us that great rivers once flowed through this desolate area. The Fayum has yielded numerous primate fossils of prosimians and early anthropoids. Unfortunately, it is one of the few fossil sights of the entire Oligocene epoch.

Figure 8.37 Early Primate This amazing fossil is of a 47 million year old adapiform primate named Darwinius masillae. It is from the Messel Pit near Frankfurt, Germany. In life it looked similar to a lemur with a head and body of about 24 cm long. An X-ray of the fossil is shown below. [Jens L. Franzen, et al., credit p. 415]

By the early Miocene, about 24 mya, we find larger species with larger brains and skeletons that are more similar to the living apes. The earliest fossil ape was discovered in 1909 in Kenya and was of a gibbon sized ape called *Proconsul* (Figure 8.38). It may have been named after a pair of

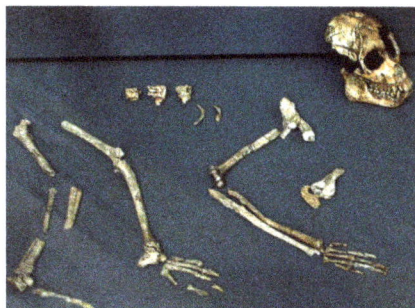

Figure 8.38 Proconsul The earliest known ape fossil from Kenya. [photo, National Museum of Natural History, Paris (MNHN)/FunkMonk]

Millions of
years ago

0 –
　Pan troglodytes　　　　| Homo sapiens
　Pan paniscus　　　　　　 (1300 cc)
　　(380 cc)
1 –　　　　　　　　　　　Homo erectus
　　　　　　　　　　　　　(950 cc)
　　A. boisei　　　　　　　(Acheulian tools)
　 | A. robustus
2 –　　(540 cc)　　| A. sediba　　Homo habilis
　　　　　　　　　　(430 cc)　　(620 cc)
　　A. africanus　　　　　　　(Oldowan tools)
　　(470 cc)　 | A. aethiopicus
3 –　　　　　　　　(410 cc)
　　　　　　　　　　　　　　(Bipedal footprints)
　　Australopithecus afarensis
4 –　　　(410 cc)

5 –　　| Ardipithecus ramidus
　　　　　(330 cc)

6 –- - - - - -　　　　　　　(Chimpanzee human split)
　　　　　　Sahelanthropus tchadensis
　　　　　　(350 cc)
7 –

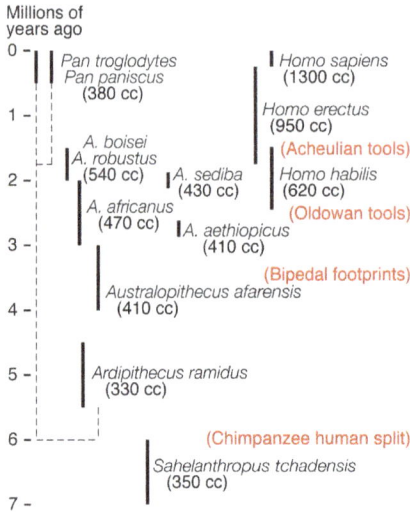

Figure 8.39 Human Ancestory Timeline
Pan is the genus for chimpanzees. The
brain volumes (in cubic centimeters) are
approximate averages for each species.
[JCB]

chimpanzees that performed in Europe during the early 1900s, both of whom were called Consul or it may have been named after a chimp that lived in the London Zoo. There appears to be differing opinions on how *Proconsul* got its name.

Numerous other fossil apes have been found in Africa representing more than a dozen genera. They ranged in size from about 4 to 70 kg, and lived in many different environments. Their teeth suggest that most were fruit eaters, but some may have also eaten leaves. The best-preserved cranium

indicates that the brain was slightly larger than that of a present day Old World monkey. One or more of these groups moved out of Africa into Europe and Asia. Fossil jaws and teeth have been discovered in Southern Europe and China from between 13 and 10 mya. They are given the genus name of *Dryopithecus*.

In India and Pakistan, several fossil jaws, a partial cranium that includes the face, and many bones from below the neck have been found. They are from an animal between 30 and 70 kg known as *Sivapithecus* that lived between 13 and 8 mya. Since the face resembles the modern orangutan, it is believed to be its ancestor. As the climate became drier and many of the forests disappeared near the end of the Miocene Epoch, the apes dwindled in numbers and died out in Europe.

The most significant differences between humans and the other great apes is the volume of the brain and the fully upright stance of humans. Although the volume of the brain generally increases with body size, the 380 cc (cubic centimeters) brain of a Bonobo (Pygmy Chimpanzee) is considerably smaller than that of a modern human, which is about 1350 cc. In the fossil record, however, we find creatures that had intermediate sized brains.

By considering DNA evidence, we have determined that the chimpanzee-human split occurred about 5 to 8 mya.

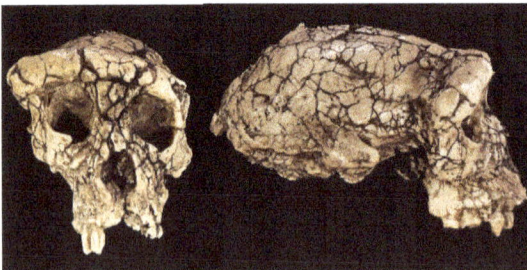

Figure 8.40 Sahelanthropus tchadensis This early hominid lived between 6 and 7 mya. It could be a common ancestor of both humans and chimps. [photo, Laboratoire d'Anthropologie Moleculaire et Imagerie de Synthese, Toulouse/ Didier Descouens]

In the time interval since the chimpanzee-human split and today, the fossils of many human ancestors have been discovered. A summary of most of our current knowledge about the family tree of human relatives over the past 7 million years is shown in Figure 8.39.

A fossil from near the time of the chimpanzee-human split is a hominid called *Sahelanthropus tchadensis*. This species is known from an almost complete cranium (Figure 8.40), a number of lower jaw fragments, and teeth, but no bones from the neck down have been discovered. It had small canine teeth, which are characteristic of later human ancestors, and a small brain of approximately 350 cc. Since this animal lived between 6.7 and 6.3 mya (close to the chimpanzee-human split), it may be close to the common ancestor of humans and chimpanzees. More complete fossils would be of great interest.

Ardipithecus ramidus was an apelike creature that lived approximately 4.4 mya. Nicknamed "Ardi" it had a brain that was a little smaller (about 300 to 350 cc) than that of a modern chimpanzee. A little less than half of the total skeleton was found including most of the skull, teeth, pelvis, hands and feet (Figure 8.41). The skeleton is more complete than the famous Lucy skeleton. Ardi was a 50 kg female whose height was about 1.2 meters. Its pelvis and limbs suggest that it used all fours when moving about in a tree, but was bipedal on land. Their feet were better suited for walking than those of a chimpanzee, but were more primitive than those of later hominids, so Ardi could not travel long distances on the ground.

Between 3.9 and 3.0 mya, an animal called *Australopithecus afarensis* left abundant fossils (Figure 8.42). This species displayed a rather large range

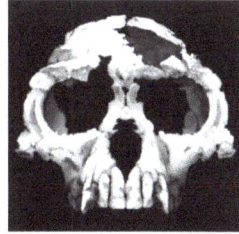

Figure 8.41 Ardi This 4.4 million year old hominid skull, is from the most complete skeleton of any early hominid. [photo, T.Michael Keesey]

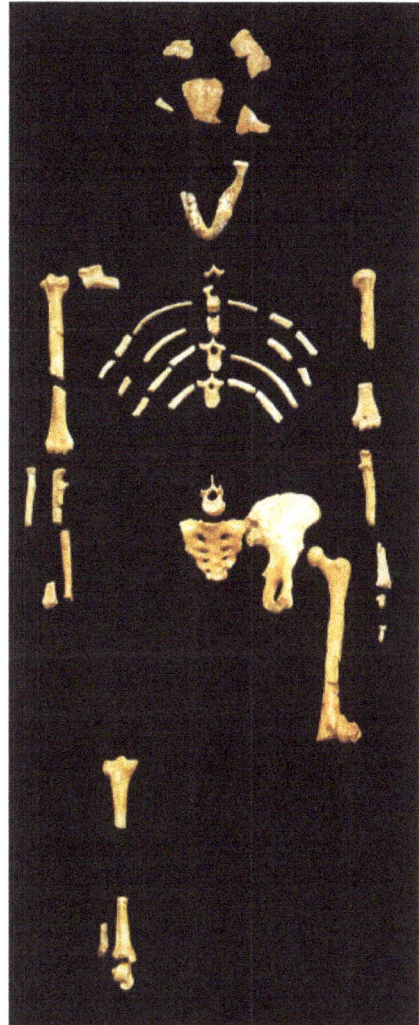

Figure 8.42 Lucy Before Ardi, this 3.2 million year old fossil of *Australopithecus afrensis* was the most complete early hominid known. About 40 percent of the skeleton was found. [photo, MNHN/120]

Figure 8.43 Laetoli Footprints *These 3.6 million year old footprints were made in volcanic ash, probably by two members of the species Australopithecus afarensis. They were made by either a male and female or an adult and child. [photo, National Museum of Nature and Science, Tokyo/Momotarou2012]*

in size, varying in height between 1.0 and 1.5 meters and weighing between 25 and 45 kilograms. This difference was due to sexual dimorphism, the males being larger than the females. The famous skeleton called "Lucy" belongs to this species.

A. *afarensis* walked upright and had a brain with a volume between 380 and 450 cc. The head is ape-like, with a low forehead, a face projecting forward, and a brow ridge, but the rest of the body was more human in form, demonstrating that A. *afarensis* walked upright. Apparently, the bipedal method of walking evolved before the brain capacity increased.

Further evidence that *afarensis* walked upright was provided by the

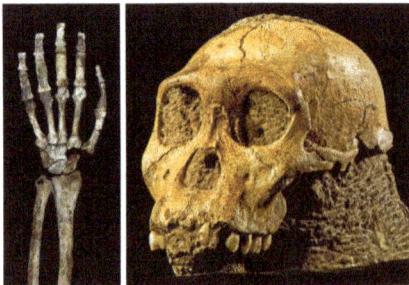

Figure 8.44 Australopithecus sediba *A relatively small brained hominid with relatively modern hands (right hand shown). [photo, Brett Eloff, courtesy of Lee R. Berger/Wits University]*

discovery of 3.6 million year old footprints at the Laetoli fossil site in Tanzania (Figure 8.43). Although it is not possible to say absolutely that *afarensis* made the footprints, the timing strongly suggests that it did. There is no evidence that any of these animals lived outside Africa.

Several other australopithecines followed *afarensis*. Between 3 and 2 mya we find fossils of A. *africanus*, and about 2.8 mya we find *Australopithecus aethiopicus*, which is best known by a skull from East Africa.

Two million years ago we find two larger forms called A. *boisei* and A. *robustus*. As their name implies, A. *robustus* had a large skull. It also had a sagittal crest (a bony ridge running along the top of the skull), a massive face, and a somewhat larger brain of about 550 cc. Its very large grinding teeth suggest a course diet like nuts and roots that required much chewing. A. *boisei* had a brain size of about 530 cc, but an even more massive face and teeth than *robustus*. Because of their heavy features, these two species are not believed to be direct human ancestors, but evolutionary dead ends.

Australopithecus aethiopicus had a brain of about 410 cc and a very large sagittal crest located on the rear of the head. It appears to be an intermediate between A. *afarensis* and A. *boisei*. On the other hand, A. *africanus* had a brain between 450 to 500 cc and could also be an ancestor of the two later australopithecines. More fossils could help clarify the situation.

A. *sediba* is represented by a series of 2 million year old fossils discovered in 2008 that had several interesting characteristics. The fossils include partial skeletons of a female and young male as well as material from several other individuals. Although it had a relatively small brain (430 cc), it had relatively modern hands and a more modern pelvis (Figure 8.44). Exactly where it fits in the family tree remains to be determined.

Contemporaneous with the larger australopithecines were animals that are placed in the genus *Homo. Homo habilis*, which means *handy man*, lived in Africa between 2.4 and 1.5 mya and had brain volumes ranging between 500 and 750 cc. Except for its larger brain, this animal was superficially similar to *Australopithecus africanus*, being about 1.3 meters tall and weighing about 40 kilograms. Electron microscope examinations of the tooth wear patterns of *Homo habilis* indicate that they were primarily fruit eaters. Because of the wide variations in this species, some of the specimens have been separated into other species. One of these species with a particularly large brain (750 cc) may be a separate species called *Homo rudolfensis*.

The earliest primitive stone tools, known as Oldowan tools, first appeared about 2.5 mya and were probably made by *Australopithecus africanus* or perhaps *Homo habilis*. These tools appear to be crude scrapers and choppers that were formed with minimal effort (Figure 8.45). They are named after the famous archaeological site at Olduvai Gorge in Tanzania.

Between 1.9 mya and the present, many fossils have been found and enough variation exists that some scientists divide them into several species. The general trend is from a shorter robust form to a taller lighter body with a larger brain. The early African forms are placed in the species *Homo ergaster*. As their descendants migrated out of Africa into Asia and later Europe about 1.7 mya, some scientists classify them as the species *Homo erectus*.

Early specimens of *H. erectus* were given popular names based on the location of their discovery such as

Figure 8.45 Early Tools Oldowan tools (top), were formed with minimum effort and first appeared about 2.5 mya. [Melka Kunture, Ethiopia]; More carefully formed tools like this hand axe (bottom) are called *Acheulean tools. They appeared* approximately 1.6 mya. [photo, Museum of Toulouse/Didier Descouens]

Peking man and Java man. Some scientists consider *H. ergaster* to be the probable ancestor of modern humans while others believe the differences between the two are not great enough to justify separate species status. The early forms had a brain that measured about 800 cc and *H. ergaster* had a thinner skull than *H. erectus*.

An almost complete skeleton of an early *Homo ergaster* was discovered in 1984 near Lake Turkana, Kenya. This 1.6 million year old skeleton, popularly known as Turkana Boy or Nariokotome Boy (named after the dig site) is

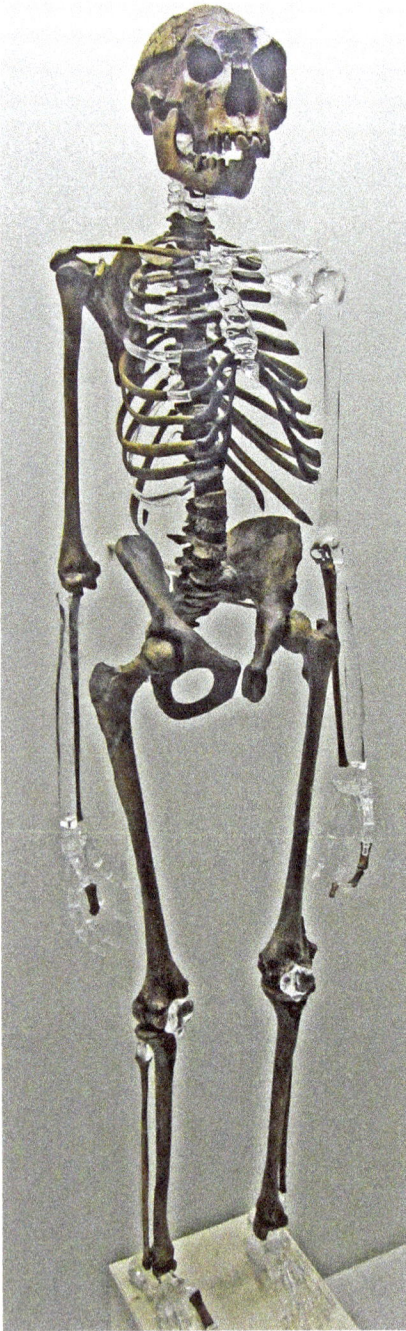

Figure 8.46 Homo ergaster Nariokotome Boy with missing bones in plastic. [photo, World Museum, Liverpool/ReptOn1x]

estimated to be about 12 years old when he died (Figure 8.46).

Tooth wear patterns of *Homo ergaster* indicate that they ate other animals, and many sharp flakes that were used to butcher animals have been found in association with these creatures. These tools are of the Oldowan type.

The more advanced Acheulean tools begin to appear about 1.6 mya. These Acheulean tools are larger implements that are carefully shaped on both sides and include teardrop shaped hand axes (Figure 8.45). Charred bones indicate that fire was used, and similar evidence in Europe suggests that early *H. erectus* may have been able to control fire when they moved out of Africa (Figure 8.47).

As we progress toward modern humans, the brain continues to increase in size. By 500,000 years ago the average brain size was about 1200 cc, almost the size of a modern human brain. These and later fossils are generally considered to belong to the species *Homo sapiens*. However, since they still were not quite modern, some have placed them in a new species called *Homo heidelbergensis*. The name comes from a fossil discovered near Heidelberg, Germany. These forms display a few primitive features and are sometimes referred to as archaic *Homo sapiens*.

As evidenced by fossils from several hundred individuals, one group of archaic humans, called Neanderthals, inhabited Europe and parts of western Asia between about 250,000 and 40,000 years ago (Figure 8.48). They had very dense bones, with shorter limbs, were very muscular, had a protruding brow ridge, a big nose, and their brain capacity was about 150 cc larger than the brain of a modern human which has a volume of about 1350 cc.

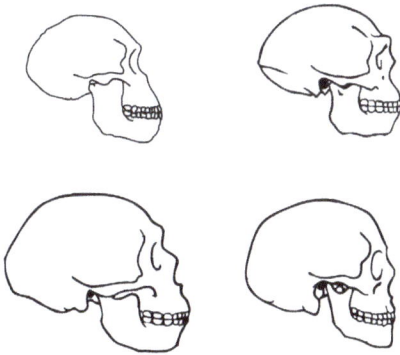

*Figure 8.47 Skulls (clockwise from top left)
Homo habilis; H. erectus; H. sapiens; H.
neanderthalensis. [V.P.Volkov]*

Neanderthals used tools, buried their dead, and hunted large game like the Woolly Rhinoceros.

Modern humans first appeared in the fossil record about 150,000 years ago. They had an average brain size of about 1350 cc and had lighter bones than the earlier forms.

About 45,000 years ago in Europe we find evidence of a new culture of modern humans that had sophisticated tools, some made of antlers and bone, and they produced various forms of artwork, which included carvings and cave paintings. These modern humans lived in the same areas as the Neanderthals for 10,000 years or so before the Neanderthals disappeared for unknown reasons about 28,000 years ago. DNA from Neanderthal bones has been compared to modern humans and it is found that between 1% and 4% of European and Asian DNA came from the Neanderthals, indicating that inbreeding occurred.

Evidence of another subspecies of *Homo sapiens* that lived about 40,000 years ago was discovered in Denisova Cave in the Altai Mountains of Siberia. These hominids, named Denisovians, left fossils of two teeth and a finger bone. However, it was possible to analyze the DNA from the finger bone, and it was found that the DNA was more similar to the Neanderthals than to modern humans. It was discovered that Melanesians and Aboriginal Australians have 3% to 5% of their DNA in common with the Denisovians.

Some researchers believe that the Neanderthals represented a separate species (*Homo neanderthalensis*), a belief that is supported by studies of DNA from mitochondria (mtDNA).

The mtDNA of humans has about 16,570 base pairs, and an analysis of mtDNA from several Neanderthals showed 202 base pair differences between Neanderthals and modern humans. The Denisovians differed from modern humans by 385 base pairs. However, when we compare mtDNA from two random humans, we find about 30 base pair differences, and mtDNA from modern humans and chimpanzees differs by about 1460 base pairs.

Analysis of mtDNA also indicates that our last common ancestor with the Neanderthals lived between 300,000 and 600,000 years ago, and that the

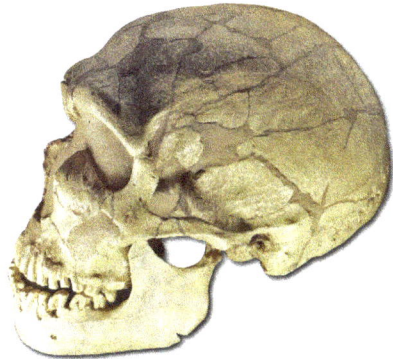

Figure 8.48 Neanderthal This 70,000 year old skull discovered in La Ferrassie, France had a low vaulted cranium, large browridge, and more pointed rear of the skull. [photo, Museum of Man, Paris/120]

Neanderthals and Denisovians diverged about 200,000 years later.

Another DNA study of humans from around the world, known as the search for Mitochondrial Eve, suggests that all living humans may have descended from a population that lived in Africa about 150,000 years ago. This conclusion is based on studies of the DNA in mitochondria.

Mitochondria are not present in sperm cells so all mitochondria are derived from those present in the egg at the time of fertilization. (Recall that mitochondria reproduce by binary fission. They cannot be manufactured in any other way by the cell.) In other words, mitochondria are passed to the offspring from the mother, but not the father. Scientists think the data provides supportive evidence that modern humans arose in Africa about 150,000 years ago and spread out from there, replacing the more archaic humans that were living in other parts of the world.

A similar study called the search for Y-chromosomal Adam is based on the fact that the Y chromosome is inherited only from the father. This study gave similar results.

An interesting finding is that about 30,000 years ago the number of modern humans living long enough to be grandparents rose sharply. Some feel this was an important evolutionary change because older people would have had time to pass on their accumulated knowledge to the next generation.

Most plants and animals have well defined geographic distributions. You will not find a Tiger in Africa or a Giraffe in China. Lemurs are only found in Madagascar, kangaroos only in Australia, and kiwis only live in New Zealand. The Giant Panda is found in China, penguins live around the Antarctic, and if you want to see a Polar Bear you must travel to the Arctic.

How do we explain these very specific distributions? The answer is found in the study of plate tectonics in conjunction with evolution. The breakup of Pangaea which began about 200 mya resulted in the isolation of several continents for very long periods of time. Then, 65 mya, the dinosaurs became extinct. Isolated and given a fresh start, many plants and animals diverged and evolved into new forms. The result of this isolated evolution is apparent in the geographic distribution of plants and animals. The study of this distribution is called biogeography, and it is the topic of the next chapter.

Chapter 9: Strange Creatures in Distant Lands

Islands and Isolation

Isolation plays an important role in evolution. If a population of organisms is somehow separated into two groups, the groups will evolve independently and in time will become distinct species. Since the two groups will, often, be subjected to different environmental conditions, they will tend to evolve in somewhat divergent directions. Given enough time, even random mutations will lead to the evolution of distinct species.

There are many ways a population can become fragmented or divided. A rising mountain range, or a desert region might divide a population distributed over a large area, or a river could be large enough to divide a population of animals that do not swim well.

The fragmentation of a population might occur slowly as in the formation of a mountain range, or it could happen very quickly. For example, seeds and small animals can be transported to new locations by larger animals, by the wind, or on rafts of vegetation (if water is the obstacle). Land birds can be blown off course by storms and end up isolated in some remote location.

An island is rather loosely defined as a body of land smaller than a continent (whatever that means) surrounded by water. An island can isolate many organisms, and studies of the unique organisms present on islands provide strong evidence of the evolutionary process. Such studies can also lead to estimates of the rate at which evolution takes place.

Plants and animals found nowhere else in the world populate virtually every island that is sufficiently separated from other landmasses. We say these unique organisms are *endemic* to the region. Why are so many endemic organisms found on islands? There are two possible answers to this question. Before we consider the answer, we need to investigate the two types of islands.

Islands can be classified broadly into two categories, continental islands and oceanic islands. Continental islands are islands that were once connected to a continent, but have been separated from it at some time in the past. Examples include the British Isles, Japan, Newfoundland, Madagascar, New Guinea, Cuba, Hispaniola, New Zealand, Tasmania, Borneo, and the Philippines.

Oceanic islands are volcanic islands that have grown from beneath the surface of the water and have never been in contact with a continent or other large mass of land. Examples of oceanic islands are the Hawaiian Islands, Galapagos Islands, St. Helena (in the South Atlantic Ocean), Easter Island, and many of the Pacific islands of volcanic origin.

Without the help of humans, it would be extremely difficult for certain types of animals to reach an oceanic island. For example, freshwater fish that are unable to survive in salt water, would not be expected on an oceanic island, but might be found on a continental island. Similarly, most amphibians, lizards, snakes, terrestrial worms, and mammals (except for bats) would not be expected to easily reach oceanic islands if the travel distance is too great.

Small mammals, snakes, snails, and other land animals, could reach an island by rafting. Occasionally, during times of flooding, rafts of vegetation float down rivers and are carried out to sea. Small animals are often trapped on this floating vegetation. If one of these rafts were to carry a pregnant female or a mating pair of animals to an island, a new population could be established.

Birds, of course, can fly to islands, but small land birds generally do not fly over great expanses of open-ocean and only rarely reach remote islands. Small birds, however, are sometimes blown great distances by storms and occasionally reach isolated islands.

Because it is difficult to reach an island, we generally find a smaller variety of organisms on islands than on continents. The few organisms that do colonize an island generally experience less intense competition than their continental relatives. Because of this reduced competition, island species tend to evolve in a direction that makes them more vulnerable to predators that reach the island at a later time. For example, birds can loose the ability to fly and might be wiped out by the first predators that become established on the island. Humans or the predators that humans have introduced to remote islands have exterminated many flightless birds. The list includes the famous Dodo of Mauritius Island, the moas of New Zealand, and numerous flightless rails from various islands.

Plants and animals established on continental islands sometimes survive long after their continental relatives have become extinct. Such plants and animals are called *relicts*. Examples include the Tuatara of New Zealand, the lemurs of Madagascar, the lungfish and monotremes of Australia, and the solenodons (primitive insectivorous mammals) of Cuba and Hispaniola. Relicts are a great natural treasure as they often allow us unique glimpses of ancient life forms.

So why do we find so many unique plants and animals on islands? Some endemic species are relicts, but most endemic species have evolved from *founder species*, which are individuals (either a small group or a pregnant female) that were transported to the island and established or *founded* a new population. If a population establishes itself on a small group of islands, over time a different species could evolve on each island, producing many closely related species. As these species diverge and adapt to a wide range of habitats, we describe the process as *adaptive radiation*. The honeycreepers of Hawaii, Darwin's finches of the Galapagos, the lemurs of Madagascar, and the marsupials of Australia are examples of adaptive radiation.

Although technically an island is a body of land surrounded by water, from an evolutionary point of view, any isolated habitat can be thought of as an island. A lake is an island to the fish and other aquatic creatures that live in it. An isolated mountain peak is an island to many of the plants and animals living on its slopes, if the land around the mountain is not suitable for those particular plants and animals. Many endemic insect species and several endemic plants are found on the slopes of Kilimanjaro, the highest mountain in Africa.

Darwin visited the Galapagos Islands as he was formulating his ideas on evolution, and our first investigation will be of life forms living on the Galapagos Islands.

The Galapagos Islands

The Galapagos Islands are a group of volcanic islands composed of 13 major islands located about 1000 kilometers west of Ecuador, South America. These islands were formed from lava generated by a hot spot beneath the lithosphere (the outer 50 kilometers or so of the Earth's surface). Remember the lithosphere is broken into several plates, one of which is called the Nazca Plate. As the Nazca Plate moves over the hot spot, lava periodically flows up through the plate, forming volcanic peaks. This hot spot happens to be near the spreading ridge of the Cocos-Nazca Plate boundary. Several of the volcanic peaks formed by the hot spot protrude above the surface of the ocean, forming the Galapagos Islands.

Since the islands were created by a series of lava flows, different parts of an island may have different ages. The oldest islands, San Cristobal and Espanola, emerged from the ocean about 3 or 4 mya. However, older islands of the group sunk beneath the surface of the water long ago. Sets of Islands have existed here for about 10 million years.

Exactly how and when the founding species of plants and animals reached the islands cannot be determined, but most were probably established sometime within the last few million years. Most of the founding animals rafted from South or Central America. Floating rafts of vegetation are occasionally seen floating off the coast of Ecuador.

As we study the plant and animal life of the Galapagos, we see examples of different stages of evolution. Some species that arrived recently are identical to those on the mainland of South and Central America, but earlier immigrants have evolved significantly from their relatives on the mainland.

The famous giant tortoises of the Galapagos (Spanish for tortoises) were once found on most of the major islands. Genetic studies show that the Chaco Tortoise of South American is their closest living relative, but the most recent ancestor may be extinct.

Also based on genetic studies, it appears that the founding tortoises first populated San Cristobal and Espanola about 3 mya and then spread to the other islands. Espanola and San Cristobal are the two islands closest to South America.

Lacking large herbivores like deer, tortoises filled that niche and became the grazers of the islands. As time passed the populations on the different

Figure 9.1 Galapagos Tortoises *The dome shaped shell of a tortoise from Santa Cruz Island (left), and the saddle shaped shell of a tortoise from the island of Espanola. [JCB]*

islands evolved into the varieties we see today.

Some scientists believe the tortoises are the same species, while others consider the varieties to be separate species. About 15 varieties have been recognized, but a third of them have become extinct due to the activities of humans. Five of the recognized varieties live on Isabela, the largest island of the Galapagos group. Five volcanic cones formed Isabela, and a different variety of tortoise inhabits each of the five regions around these cones. Here again we see a pattern of isolation leading to the evolution of different forms.

The species (or subspecies) of tortoises can be divided into two general groups, those with dome shaped shells and those with shells shaped like a Spanish saddle (Figure

9.1). The saddlebacked varieties live on the drier islands and because of the shape of their shell, they can reach higher up into the vegetation.

There are four species of iguanas on the islands that DNA studies indicate evolved from the same ancestor (Figure 9.2). One is the Marine Iguana that feeds on the algae growing on the shoreline rocks. It is the only truly marine lizard in the entire world! This iguana is so distinctive that it has been placed in a separate genus all by itself.

The other three iguanas are land species belonging to a genus that is also unique to the Galapagos Islands. One, the Santa Fe Land Iguana is found only on Santa Fe Island. Another, the Galapagos Land Iguana is found on the islands of Santa Cruz, Fernandina, and Isabela, all of which were connected

Figure 9.2 Iguanas *The Marine Iguana (left) is from the island of Espanola, and the Land Iguana (right) is from the island of Santa Fe. [JCB]*

Figure 9.3 Darwin Finches *The Large Ground Finch (left) specializes in eating larger harder seeds, while the Sharp-beaked Ground Finch eats smaller seeds. Both birds are found on Genovesa Island. [JCB]*

during the Last Glacial Maximum that ended about 20,000 years ago.

The third land iguana, the Galapagos Pink Land Iguana, is only found on the north end of Isabela around the Wolf Volcano. It was just discovered in 1986 and limited DNA tests suggest that it diverged from the other land iguanas about 5 million years ago.

Since the two groups of iguanas evolved from the same ancestor, the early population must have divided into two distinct populations, one feeding off the marine algae and another living off the land vegetation. These two groups are examples of adaptive radiation. Because they have diverged so much, the two groups must have separated very early in the history of the islands. Although their closest living relative has not been identified, the ctenosaurs of Central America and Mexico are a closely related group.

The most famous birds on the islands are the finches known as Darwin's finches (Figure 9.3). There are 13 species put into 3 to 5 genera, depending on which classification is used.

The radiation of a group of animals like Darwin's finches is explained very simply if one assumes that the original population became established on several islands. Since environmental pressures differ from island to island, these isolated populations eventually evolved into distinct species. If one of these new species migrated to another island, the two species would be unable to mate, but would live as two species on the same island. This process must have happened several times, since many of the finch species are presently found living together on many of the major islands.

Darwin's finches have specialized in several different ways. We find large-billed finches, small-billed finches, and even a bird called the Woodpecker Finch. This Woodpecker Finch has a short finch-like bill, but uses a cactus spine to pry insect larvae from dead tree branches in the same way woodpeckers use their long bills.

Two animals from the Antarctic region have been established on the islands and have evolved into distinct species. They are the Galapagos Penguin and the Galapagos Fur Seal (Figure 9.4), both with close relatives in South American. These animals are found only around the Galapagos group of islands, but since they can travel easily among the islands and interbreed, they have not differentiated into subspecies.

Besides the penguin, there are four other species of endemic sea birds, two

*Figure 9.4a **Galapagos Endemics** (clockwise from top left) Giant Prickly Pear Cactus (Santa Fe Island); Waved Albatross; Galapagos Fur Seal; Yellow Warbler, an endemic subspecies; Galapagos Mockingbird (San Cristobal Island); Swallow-tailed Gull; Lava Cactus; Large Painted Locust [JCB]*

Figure 9.4b Galapagos Endemics Galapagos Penguin (left), the only penguin occasionally found north of the Equator. [Mike Weston]; Flightless Cormorant [pitneymark]

gulls, a cormorant that has lost the ability to fly, and an albatross, (Figure 9.4). As we have already observed, flightless birds are especially common on islands, presumably because the lack of natural enemies makes flight unnecessary. In addition, flying requires a great deal of energy, and flightless birds do not need as much food.

Galapagos mockingbirds represent a group of endemic birds that have not evolved as much as the finches, presumably because they colonized the islands more recently. There are four different species living on four different islands or groups of islands. The Chatham Mockingbird lives on San Cristobal (also called Chatham), the Hood Mockingbird lives on Espanola (also called Hood), and the Charles Mockingbird did live on Floreana (also called Charles), but it is now extinct on Floreana and is only surviving on two very small islands

near Floreana. The fourth species (Galapagos Mockingbird) lives on most of the islands to the north and west of the three islands just mentioned (Figure 9.4). There are six subspecies of the Galapagos Mockingbird, which are restricted to specific islands. Their ranges do not overlap, as do the ranges of many of the finches. These four mockingbirds have been placed in the same genus with the mockingbirds of North and South America.

The Yellow Warbler is a bird in which we see only a slight degree of evolution (Figure 9.4). Considered to be the same species as those on the mainland, the Galapagos population has evolved into a different subspecies. In these birds, we observe a population in the early stages of evolution. Given enough time, they will evolve into a separate species. The Yellow Warbler represents a more recent immigration to the islands than do the mockingbirds or the finches.

Other endemic birds are a martin, flycatcher, rail, dove, and hawk. The majority of land birds are endemic, but there are a few land birds that breed on the islands that are not endemic. They represent recent immigrants that have not had enough time to diverge from their South and Central American relatives. From Darwin's finches to the birds that are just like their relatives on the mainland, we see birds in different stages of evolution.

Among the other vertebrates are seven or more endemic species of lava lizards, five endemic geckos, four endemic snakes, an endemic bat, and eight species of endemic rats although four have become extinct due to human activities.

DNA studies of the lava lizards have found that they arrived on two separate occasions. Carried from South America by the Humboldt Current, one arrival populated San Cristobal and Marchena while the other arrival populated the islands to the west. Their closest living relative is the Knobbed Pacific Iguana of Ecuador and Peru. It belongs to the same genus as the lava lizards of the Galapagos (Figure 9.5).

Generally, reptiles and mammals are uncommon on islands, but rats and snakes are sometimes trapped on rafts of vegetation and are more likely to reach an island than are larger mammals like deer and bear. The native rats of the Galapagos have not

Figure 9.5 Santiago Lava Lizard Unlike birds, the female lava lizards like this one are more brightly colored. [Alex Giltjes]

fared well with the coming of humans, and several species have become extinct because of competition with introduced Norway Rats and Black Rats.

It is estimated that about 2500 terrestrial and freshwater invertebrates are found on the islands and about 50 percent of these are endemic. About two-thirds of the invertebrates are insects, and there are about 70 endemic land snails belonging to 9 different genera.

There are almost 400 species of seed plants on the Islands and over 40 percent are found nowhere else in the world (Figure 9.4). Included in these endemic plants are 7 endemic genera. Even though genera are defined in a somewhat subjective fashion, the existence of endemic genera gives an idea of the degree of evolutionary divergence attained by these plants. Most of the endemic plants show close ties with South and Central American species, and as expected, the vast majority of non-endemic plants are of South and Central American origin.

The shallow waters around the Galapagos Islands harbor numerous interesting fish and other sea creatures. Most of these animals rarely venture into the deep ocean, so the shallow waters form an ecological *island* for many of them.

Over the years some shallow water creatures have found their way to the islands and populations have been established. As time passed, many evolved into new species. Of the more than 300 species of fish found along the shores of the islands, about one-sixth of them are endemic. There are also about 800 species of marine mollusks of which about one-sixth are endemic. In the shallow seas around these islands, we see the same patterns of evolution as is seen on the islands themselves.

We have spent some time discussing the flora and fauna of the Galapagos Islands to show how isolation and time have led to the creation of many new species and a few new genera. Along with the picture of evolution that emerges, we are reminded that the Galapagos Islands have been isolated for 3 to 5 million years. Since several endemic genera are present on the island group, this age may represent an approximate time for the evolution of a new genus.

There are many other islands where the principles described above are illustrated, but the Galapagos is somewhat unique in one respect. When humans colonize a new area, especially an island, much of the native habitat is destroyed and new species of plants and animals are introduced. These new creatures along with the destruction of the native habitat invariably upset the delicate balance of the island, and many of the unique island creatures became extinct.

As we have seen, several species of Galapagos animals have become extinct, but in comparison to what has happened on many other islands, the Galapagos creatures have fared a little better. One reason for this state of preservation is undoubtedly due to the small size of many of the islands and their limited water supply, which makes them uninviting to permanent human habitation. Today, four of the islands are inhabited, another supports a small air base, and the others have been designated as a national park. The protection of the unique wildlife of the Galapagos Islands for the enjoyment of future generations seems hopeful today, partly because they have become a popular tourist attraction. If the unique wildlife is not preserved, the tourist business will suffer. Unfortunately, the preservation of many of the unique creatures on other islands is not as hopeful.

The Hawaiian Islands

Situated near the middle of the North Pacific Ocean, the Hawaiian Islands are the most isolated group of islands in the world. North America, the nearest continental land mass, is located some 4,000 kilometers to the east, and the nearest island is 1600 kilometers away.

The Hawaiian Islands are reasonably large and contain many varied habitats. Consequently, several animals and plants that have managed to reach the islands provide excellent examples of adaptive radiation. Unfortunately, since the arrival of humans, many of the native animals have been exterminated.

When the first Polynesians arrived, in the fourth century AD, they cleared away much of the lowland forests and dramatically changed the environment. With the coming of later settlers and the introduction of nonnative species, the destruction continued, and today, few primitive areas remain.

This loss of habitat has driven many of the unique plants and animals to extinction. About 70 percent of the bird species have been exterminated because of human intervention. Much of what we know about the original life forms has come from earlier accounts and from the study of bones.

The eight major Hawaiian Islands lie on the eastern end of a 2500 kilometer long chain of islands, submerged reefs, and atolls (atolls are circular coral reefs built up around submerged volcanoes). Near the western end of the chain lies Midway Island, of World War II fame (Figure 9.6).

A process similar to the one that formed the Galapagos Islands also created the Hawaiian chain as the

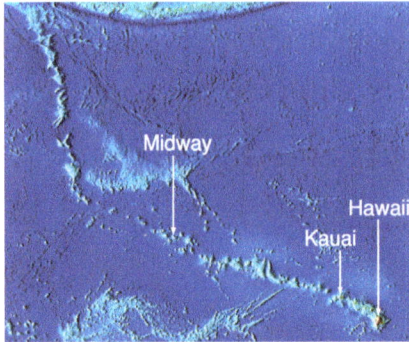

Figure 9.6 Hawaiian-Emperor Seamount Chain West of Midway the chain bends north and follows the Emperor chain of seamounts. The Aleutian trench is at the top of the picture. Hawaii and Midway are about 2,300 km apart. [NGDC/NOAA]

Pacific Plate moved over a hot spot. Lava from the hot spot builds up, creating an island. Presently, the hot spot is under the Big Island of Hawaii. Hawaii is the largest and most eastern island of the chain, and the only one with active volcanoes (Mauna Loa and Kilauea).

As the moving Pacific Plate carries the island off the hot spot, the volcanic activity on the island ceases, and over time, the great weight of the island slowly depresses the lithosphere. Since the islands sink and erode with age, the older islands (those located more to the west) tend to be smaller.

Obviously, each island of the Hawaiian chain has a different age. Since radiometric dating gives the age at which the rock was last solidified, recent lava flows on the Big Island will give ages close to zero. On the other hand, Midway, one of the older islands at the western end of the chain, is about 27 million years old.

The chain of islands is much longer than we have stated. West of Midway the chain makes a bend and continues northward forming the Emperor chain

of *seamounts*, which are submerged islands (Figure 9.6).

The oldest seamount is almost 70 million years old, however, it is very unlikely that many of the present day plants or animals had their start on islands of such an old age. Remember, throughout the history of the island chain, the older islands to the west have slowly sunk beneath the ocean while newer islands are created to the east. Therefore, if a plant or animal got established on one of the first islands, they would have had to move from island to island many times in order for the population to survive to the present day. As the number of required moves increases, we would expect it to be less likely that a population could survive to the present.

Radioactive dating of rocks from Kauai, one of the oldest of the eight major islands, indicates that it is about 5 million years old. However, at the time Kauai was forming, undoubtedly some of the more westerly islands were still large and above water. Therefore, a few of the plants and animals present on the eight major islands of today might have been established a little earlier on one of the older islands that have now disappeared. Establishing the maximum time some of the Hawaiian plants and animals could have been isolated will, therefore, have to be an educated estimate. Five to ten million years is probably a reasonable estimate for the age of the older colonizations.

Most of the Hawaiian organisms possess some attribute that increases the chances of their long distance dispersal. Some of these methods of dispersal include spores and seeds small enough to be carried great distances by the wind, seeds that are not easily destroyed by sea water and can drift great distances across the open ocean, seeds that are easily attached to the sea birds that frequent the islands, durable seeds that can pass through the digestive tract of a sea bird and survive to germinate, and snails, snail eggs, and other creatures that are small enough to be easily carried on the feet of sea birds.

Land birds are sometimes blown over long distances by storms, and presumably this is how the land birds reached the islands. Less than a dozen colonizations could explain the current land bird diversity.

The Hawaiian Islands have no native amphibians or reptiles, no native freshwater fish, and no native land mammals other than one species of bat. We would expect these animals to be absent since it is unlikely that non-flying animals could survive such a long ocean voyage.

As one would expect on such an isolated island group, the percentage of endemic species is extremely high. More than 95 percent of the native plants and animals of the Hawaiian Islands are found nowhere else. This area has the highest rate of endemism in the world!

Because of the length of its isolation, the number of endemic genera is impressive. Over 50 percent of the insect and land mollusk genera are endemic. More than a third of the bird genera are endemic, and about one-eighth of the flowering plant genera are endemic. One endemic group of birds, the honeycreepers, has diverged so significantly that they have been placed in a separate family. These remarkable numbers give an idea of the evolution that has taken place on the islands.

As with the Galapagos Islands, the shallow waters around the islands form ecological niches that are effectively isolated by the deep water of the surrounding ocean. Consequently, about 30 percent of the reef fish and 30

percent of the invertebrates are endemic to these unique habitats.

A species of seal known as the Hawaiian Monk Seal is an unusual endemic to the area. Its closest relatives are a species in the Mediterranean area and a species in the Caribbean. Unfortunately, the Caribbean species has been exterminated and the Mediterranean Monk Seal population has declined to about 500 individuals. The Hawaiian Monk Seal apparently reached the islands from the Caribbean before the Isthmus of Panama was completely closed, a few million years ago.

Monk seals are the most primitive of the living seals as evidenced by their ears and flippers. The tibia and fibula bones of their front flippers are not fused at their bases as they are in other seals. Their ears are more like those of fossil seals than other living seals. This unusual creature is a relict from the distant past, but it is precariously close to extinction. Although protected by law, it is very vulnerable to human intervention. The current population seems to be declining and could be less than 1500 individuals.

The land birds of the Hawaiian Islands are represented by eight families, five of which contain just one species each. From an evolutionary point of view, the most interesting are the endemic honeycreepers. The honeycreepers illustrate adaptive radiation more dramatically than Darwin's finches (Figure 9.7). Before humans arrived, there were about 60 species of honeycreepers (and many subspecies) found on the eight major islands, and one on each of the islands of Laysan and Nihoa.

Considering the extensive adaptive radiation of this group, the original colonization must have occurred very early, perhaps 5 to 10 mya. Some members have thick seed-eating bills,

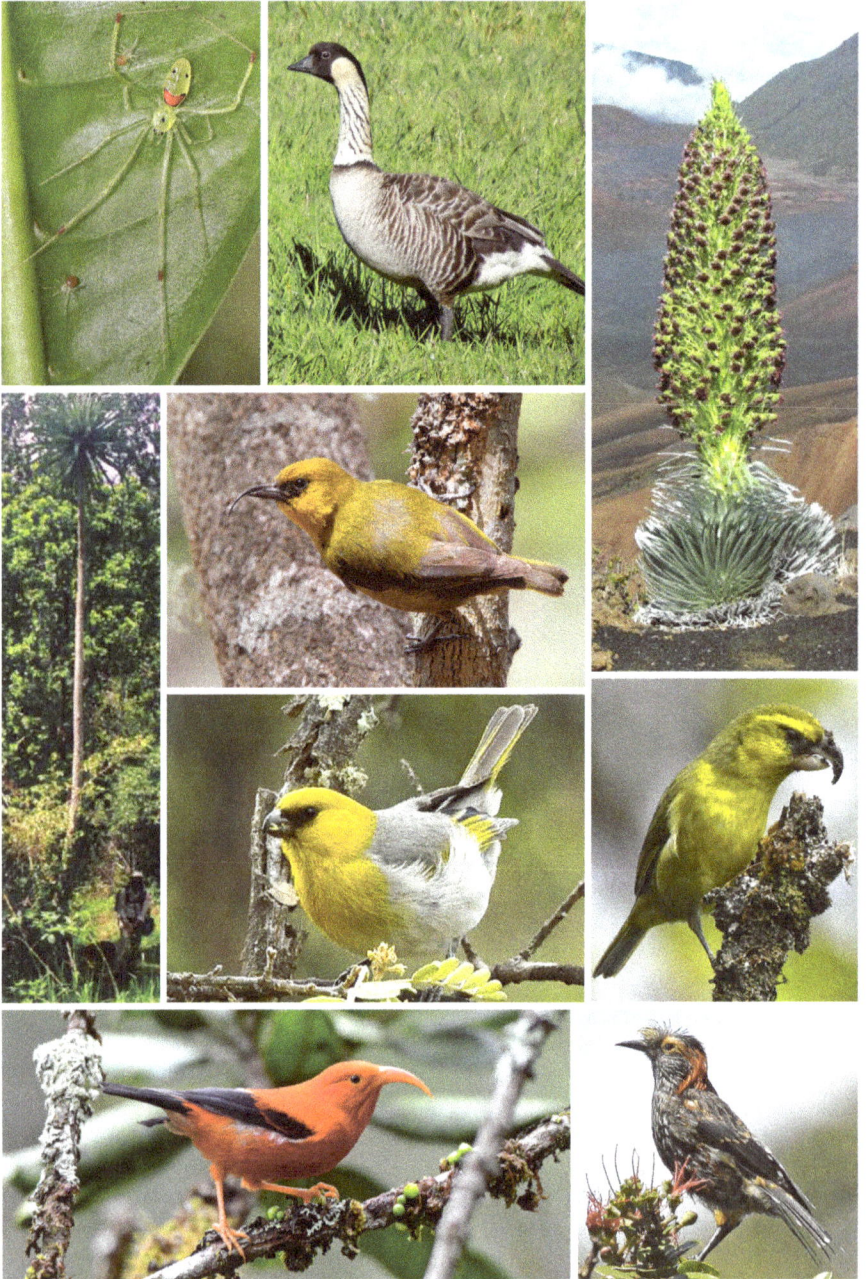

Figure 9.7a Hawaiian Endemics *(clockwise from top left) Happy Face Spider [Nathan Yuen]; Nene Goose [Noah Kahn/USFWS]; Haleakala Silversword [Paul Krushelnycky]; Maui Parrotbill [Eric Vanderwerf]; Crested Honeycreeper [Eric Vanderwerf]; Iiwi [Jim Denny]; A giant lobelia, more than 10 meters tall (note man at lower right). [Karl Magnacca]; (top center) Akiapola'au [Bob Gress]; (bottom center) Pilila [Eric Vanderwerf]; All the birds except the Nene are honeycreepers.*

Figure 9.7b Oahu Akialoa Subspecies of this unusual honeycreeper were found on Lanai, Oahu, and Kauai. They ranged from 12 to 22 cm in length, but about one third of this length was their bill. They are all extinct. [J.G.Keulemans]

shoulders of these otherwise black birds. They also killed thousands of honeycreepers, which they used to make feathered cloaks and helmets.

The destruction of the native forests and the introduction of foreign species also contributed to the extinction of the honeyeaters. Before they were exterminated, a different species had evolved on each of the islands of Molokai, Hawaii, Oahu, and Kauai (Figure 9.8). The last survivor of the group was the Kauai O'o, last seen in 1987.

The honeyeaters and honeycreepers represent only a few of the many interesting native birds that evolved on the Hawaiian Islands. However, if you have ever visited these islands, it is very likely that you did not see many of its rare native birds. Only about 17 species of honeycreepers are left. Of the 140 species of birds that were present when the Polynesians first arrived, half are now extinct and many

some have long thin bills and long tongues used to extract nectar from tubular flowers, some have hooked parrot-like bills used for scooping fruit, and others have thin bills for picking up insects. These widely different forms evolved by exploiting many of the niches that were vacant when the first immigrants arrived. The lack of competition allowed these early immigrants to evolve without experiencing pressure from other animals generally found in more populated areas.

Another bird family that displays a modest radiation is the honeyeaters, whose close relatives are found in Australia, New Zealand, and many islands of the southwest Pacific region. Unfortunately, all of these beautiful endemic birds have been exterminated. The Polynesians killed thousands for a few yellow feathers that graced the

Figure 9.8 Hawaiian O'o These beautiful birds was killed by the thousands for their feathers by the Hawaiians and later by the Europeans. They were last seen in 1934. [J.G.Keulemans]

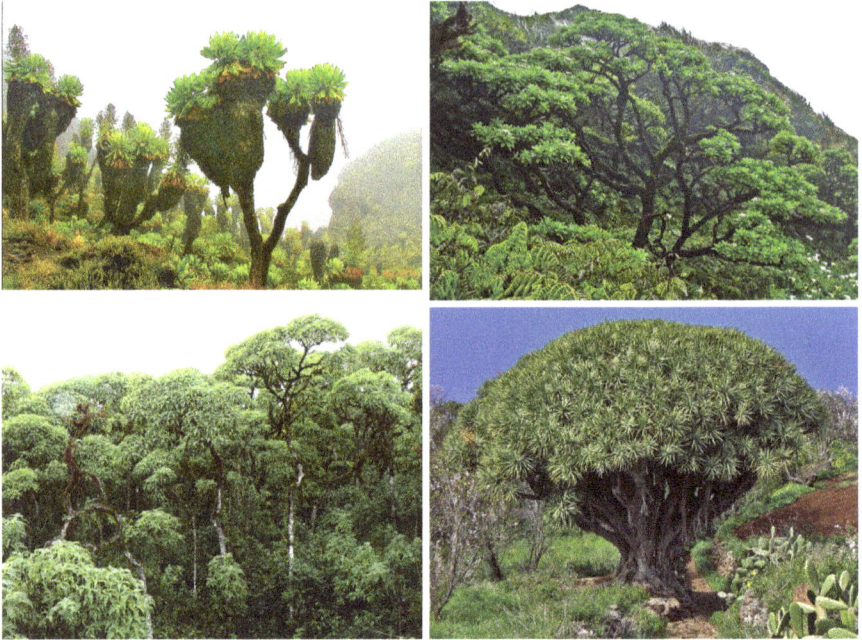

Figure 9.9 Arborescence *(clockwise from top left) Giant Groundsel, a member of the aster family found on Mount Kilimanjaro in Tanzania, Africa. [Brendon]; Black Cabbage Tree, a member of the aster family found on Saint Helena Island located near the middle of the South Atlantic Ocean. [Dan Fulwiler]; Canary Islands Dragon Tree a member of the asparagus family. [Zyance]; Tree Scalesia, a member of the aster family found on Santa Cruz, Floreana, Santiago, and San Cristobal of the Galapagos Islands. [Haplochromis]*

of those remaining are endangered. Humans have introduced most of the common birds on the islands.

An interesting evolutionary trend observed on many islands is the evolution of small herbs into trees, a phenomenon known as *arborescence*, which means *becoming a tree*. Most trees have large seeds that are not easily carried to islands, leaving most islands void of large canopy trees. On islands where large trees have not been established, small herbivorous plants often adapt to the role occupied by trees on the continents. This trend is illustrated by the lobelias of Hawaii, a group of plants that have diversified on the islands and provide another example of adaptive radiation. Elsewhere, lobelias are small flower

garden plants, but on the Hawaiian Islands, several have evolved into tall palm-like trees (Figure 9.7).

Another example of arborescence is found in the large treelike Hawaiian silverswords (Figure 9.7). They appear to have evolved from tarweeds, which are small herbs native to California. Another trend toward gigantism can be seen in the Hawaiian violets, some of which grow 2 to 3 meters tall.

On many islands like the Galapagos, St. Helena Island, the Canary Islands, and even Mount Kilimanjaro, plants belonging to the aster family (asters, sunflowers, and daisies), the bellflower family (the lobelias), and the asparagus family (asparaguses and agaves) have evolved into trees (Figure 9.9). These treelike plants do not have growth rings like true trees such as the oaks, maples, and conifers.

Lake Baikal

In the mountainous region of south central Siberia just north of the Mongolian border, lies Lake Baikal, the largest body of fresh water on the Earth. This long crescent shaped body of water is more than 600 km long and about 80 km across at its widest part. It was created by a rift, which is a structure formed when the crust is stretched and cracks open. The rift is still active and widens by about 2 cm a year. It is estimated to be about 25 million years old, making it the oldest lake in the world.

In surface area, Lake Baikal is only the seventh largest freshwater lake in the world, but it is by far the deepest continental depression in the world. At its deepest part, Lake Baikal is 1620 meters deep, 1285 meters of which are below sea level! Sediments in the deepest part of the lake may be as deep as 7 kilometers. United States and Russian studies of sediment cores have provided a record of the climate for the last 250,000 years. Future cores are expected to yield data back as far as 5 million years.

Figure 9.10 Gammarid A crustacean belonging to the same class as lobsters and crabs. There are about 300 species in Lake Baikal, a third of all the known species in the world. They are important scavengers and food sources for the fish. [Dave Gray]

Holding about one-fifth of the fresh water in the world, Lake Baikal is a giant island of water containing many endemic species of aquatic life. The lake drains into the Arctic Ocean by the Yenisei-Angara River system which shares some of the lake's aquatic life.

From the patterns we have seen elsewhere, it should not surprise us to learn that over 80 percent of the approximately 1500 species of animals living in the water are endemic to the area. About 25 percent of the almost 600 species of plants found in the lake are endemic. There are almost 90 endemic genera and even several endemic families of animals found in the lake. These endemic animals

Figure 9.11 Baikal Seal Known locally as Nerpa, it has unusually long claws. [Sergey Gabdurakhmanov]

Figure 9.12 Comephorid *This Big Baikal Oilfish is about 20 cm long. Since it has no swim bladder and is about 35 percent oil, it able to tolerate a wide range of pressures and can swim at all depths in the lake.*

include freshwater sponges, segmented worms, snails, clams, shrimp-like crustaceans called gammarids (Figure 9.10) flatworms (turbellarians), various insects, fish, and an endemic species of seal.

The endemic seal belongs to the same genus as some other arctic seals (Figure 9.11). It is likely that the Baikal Seal population reached the lake during a time when the Arctic Ocean extended further south than it does today. Such a time could have been when the glaciers melted near the end of the last ice age (approximately 10,000 years ago). The seals advanced up the Yenisei-Angara River system, which was shorter then. Although closely related, this seal is easily distinguished from its arctic relatives by the unusually long claws on its front flippers.

There are between 50 and 60 fish species in the lake and about 60 per cent are endemic. The lake is home for an endemic family (the Comephoridae), which contains just two species belonging to the same genus (Figure 9.12). The comephorids or golomyankas are unusual in that they do not have scales and give birth to live young.

A second family of fish, the Cottidae (called sculpins), is represented by two endemic subfamilies that contain a total of 24 species distributed among 8 genera.

The presence of an endemic family of fish, an endemic family of freshwater sponges, and an endemic family of snails attests to the ancient nature of Lake Baikal. Although we do not know when the ancestors of these groups colonized the lake, it could have been as long as 25 mya.

Lakes around the world are aquatic islands, many of which are home to endemic species. There are several

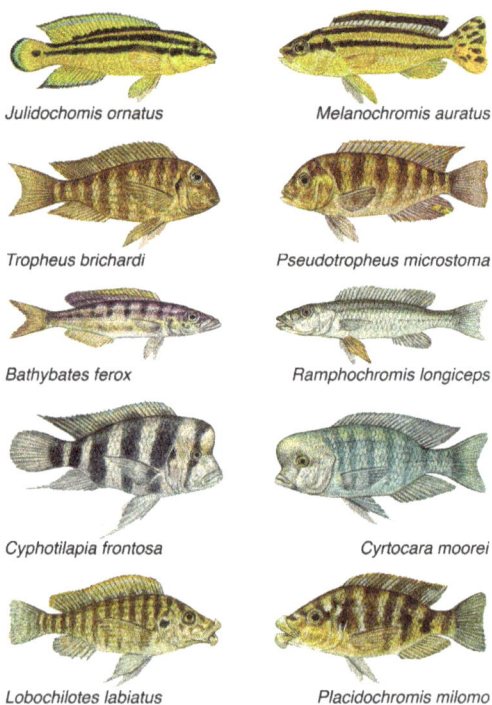

Julidochromis ornatus

Melanochromis auratus

Tropheus brichardi

Pseudotropheus microstoma

Bathybates ferox

Ramphochromis longiceps

Cyphotilapia frontosa

Cyrtocara moorei

Lobochilotes labiatus

Placidochromis milomo

Figure 9.13 Cichlids *The fish on the left are from Lake Tanganyika and the fish on the right are from Lake Malawi. Fish living in similar environments, but in different lakes have evolved similar forms. As expected, all the Lake Malawi cichlids are more closely related to each other than they are to any cichlid in a different lake. A similar statement is true for Lake Tanganyika. [Roberto Osti]*

large isolated lakes in Africa, which, as expected, are home to large numbers of endemic fish. Lake Tanganyika, a rift valley lake, is the second largest body of fresh water on Earth. It contains about eighteen percent of the world's fresh water and is home to some 400 species of fish, about 90 percent of which are endemic.

It also contains many species of jellyfishes, mollusks, crabs, shrimps, and various worms, a high percentage of which are endemic.

Lake Victoria contains about 300 native species of fish of which more than 90 percent are endemic. However, the introduction of nonnative species to the lake has driven many to extinction. Lake Malawi, another rift valley lake that is about two million years old, contains over 1000 species of fish, more than any other lake in the world. Over 90 percent are endemic.

Many of the fish in these African lakes belong to the cichlid family, a popular group of aquarium fishes. Well known members of the family include the popular aquarium species known as angelfish and the tilapia, the popular name for several species that are raised for food.

Cichlids are often used to illustrate adaptive radiation and *convergent evolution* (Figure 9.13). Animals that inhabit similar environments and search for the same food tend to evolve similar features. They evolve a very similar look even though they are not closely related.

Madagascar

Although we do not have space to do justice to all the interesting islands and lakes on Earth, several have creatures so unusual that they must be discussed. One is Madagascar, a large continental island lying almost 400 kilometers off the southeast coast of Africa.

About 180 mya near the early Jurassic, Gondwana began to fragment into two pieces as South America and Africa separated from Antarctica, Madagascar, India, and Australia. Madagascar and India split from Antarctica near the middle Cretaceous and Madagascar began to separate from India about 90 mya, carrying many dinosaurs with it. Madagascar has been isolated since that time.

Although Madagascar is technically a continental island, it was isolated so long ago that most of the original life forms, like the dinosaurs, have either become extinct or evolved into new species. Many of the present day plants and animals are the descendants of individuals that rafted from Africa. The Mozambique Channel, that separates Madagascar from Africa, probably formed during Cretaceous times or even as far back as the Triassic, more than 80 million years ago.

Having been an isolated world since the age of the dinosaurs, Madagascar harbors a large percentage of endemic plants and animals. On Madagascar 98 percent of the land mammals are endemic. Of the 12,000 or so species of flowering plants, 70 percent are endemic and 6 or 7 families are found nowhere else in the world. All but 2 of the 150 species of frogs are endemic, and 95 percent of the approximately 250 reptiles are unique to the island. These reptiles include one species of crocodile, 13 tortoises, 180 lizards, and about 60 snakes.

When humans first arrived about 2000 years ago, most of the island was forested or lightly wooded. Today, 90 percent of the island has been converted to fields, pasture lands, and man-made deserts. In many areas, after the forests have been burned, the soil erodes and simply disappears. Some

Figure 9.14 Ancient Giant One elephant bird species stood 3 meters tall and may have weighed 400 kilograms. It became extinct soon after humans reached the island of Madagascar.

say Madagascar is the most eroded land on Earth. Once the soil is destroyed, the forests cannot return. Sadly, humans have destroyed many of the unique animals as well.

The most famous birds of the island are perhaps the extinct elephant birds (Figure 9.14). There were at least four species, and the largest of these huge flightless birds stood nearly 3 meters tall and weighed about 400 kilograms. Elephant birds, like the ostriches of Africa, very likely represented relicts from the time when Gondwana was fragmenting.

The eggs of this giant bird were 30 cm long. One sold in 2013 for a little over one hundred thousand dollars. How much would a live bird be worth?

The elephant birds might have inspired the stories of the Roc, the giant mythological bird of the Sinbad legends. Elephant birds apparently survived on the island until about 200 years ago, but hunting and habitat destruction eventually caused their extinction.

Besides the extinct elephant birds, there are over 100 living endemic species of land birds, one-fourth of which belong to five families that are endemic to Madagascar.

Figure 9.15 Tenrecs (top) Tailless or Common Tenrec [Marco Poggioni]; Lesser Hedgehog Tenrec [Wilfried Berns]; Lowland Streaked Tenrec [Frank Vassen]

The unusual mix of mammal life on Madagascar suggests that the island became isolated before mammals were established on the island. The ancestors of the represented groups must have rafted from Africa.

The insectivores include two species of shrews that could have been transported to the island by early colonists and a very unusual family of mammals known as the tenrecs (Figure 9.15). Tenrecs are of special interest because they appear little changed from the earliest mammals and are one of the better glimpses we have of the early mammals that coexisted with the dinosaurs.

Fossils of tenrecs have been found in East Africa of Miocene age, and their closest living relatives are the otter shrews (family Potomagalidae) of West Africa. Evidence of their primitive nature is that they do not maintain a constant body temperature and have a single opening called a *cloaca* through which urine and fecal material is passed. Reptiles, birds, and the monotreme mammals also have cloacas.

Represented by 24 species, the tenrecs look superficially like long nosed rats, but DNA evidence shows they are more closely related to the hyraxes, elephant shrews, sea cows (which include the manatees and Dugong), elephants, and the Aardvark, all of which belong to the superorder Afrotheria. Although the members of

Figure 9.16 Giant Elephant Shrew One of the unusual members of the superorder Afrotheria. [John White]

Afrotheria appear to be quite varied, it is striking that several members have long snouts Figure 9.16).

Africa was essentially isolated between about 110 and 40 mya, and the Afrotheria evolved to fill various ecological niches during that period. The ancestors of the tenrecs likely reached Madagascar by rafting during this time.

Elephants and sea cows are the only members of the Afrotheria that left Africa. Elephants eventually reached Asia and North America, and some members of the sea cow group (order Sirenia) spread eastward into the seas around Southeast Asia, northern Australia, and northward to the Bering Sea, while others spread westward into the warm seas of the Caribbean.

Today there are four sea cow species, the Dugong which is found in the coastal waters around the Indian Ocean, and the three species of

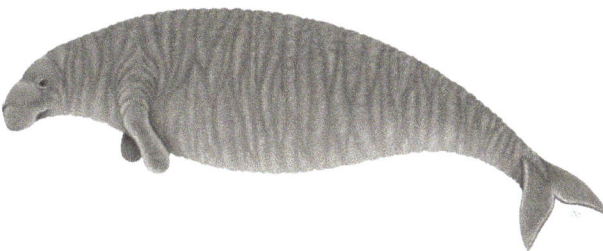

Figure 9.17 Steller's Sea Cow This large aquatic Afrotherian mammal lived in the Bering Sea. Steller said it was tame and gentle as are its living relatives. The last ones were killed in 1768. Its closest living relative is the Dugong. [Nioell/deviantArt]

manatees, one of which is found in the coastal waters and rivers of the Caribbean Sea and Gulf of Mexico, another lives in the fresh water of the Amazon Basin, and the third lives in the rivers of West Africa.

The largest sea cow on record was a 9 meter long, 4,000 kg animal known as the Steller's Sea Cow, which lived in the cold waters of the Bering Sea along the coasts of Alaska and Siberia (Figure 9.17). The Steller's Sea Cow was discovered in 1741 by Georg Steller, but just 30 years later it had been hunted to extinction.

Madagascar has 17 species of rodents that belong to 8 endemic genera. They all belong to the same subfamily which suggests that they all descended from one founder species that rafted to the island millions of years ago.

The only hoofed mammals currently on the island are the wild boars, that were introduced by early settlers. However, at least three species of dwarf hippopotamuses are known to have lived on the island, but they are only known from subfossils. *Subfossils* are relatively young bones from animals that died during the last few thousand years. The dwarf hippos became extinct sometime after humans colonized the island.

Figure 9.18 Ring-tailed Mongoose One of the ten endemic species of carnivorans on Madagascar. [Jeff Gibbs]

Water is a more severe obstacle for land mammals than it is for bats, and Madagascar is home to 6 families of bats, one of which is endemic.

The carnivorans are represented on Madagascar by 10 endemic species, and humans have introduced 3 more. All the endemic carnivorans are small catlike animals, and they all belong to the family Eupleridae, which is endemic to Madagascar (Figure 9.18). Molecular DNA studies indicate that they evolved from one mongoose ancestor that rafted from Africa about 20 million years ago.

The lemurs are a very unusual group of mammals that are endemic to the island of Madagascar. They are primates that are relicts of an early group of the primate order, whose ancestors rafted to the island from Africa about 60 mya. DNA studies have confirmed that all the lemurs are the descendants of a single colonization. Their earliest relatives are known from Eocene age fossils of Europe and North America (which were just separating at that time). Lemurs belong to the group of primates known as prosimians. They are more primitive than monkeys.

Lemurs are classified into 5 families, and there are about 100 living species placed in 15 genera (Figure 9.19). At least 15 species have been killed off since humans arrived on the island. As usually happens, the lemurs that were exterminated were generally larger and more unusual than the ones that remain today. Several of the larger species are called sloth lemurs, as they were slow climbers that fed on vegetation and hung from trees like South American sloths. The largest was about the size of a female gorilla. Unfortunately, bones are our only evidence of their existence (Figure 9.20).

Since mammals are not as mobile as birds, they sometimes make it easy to

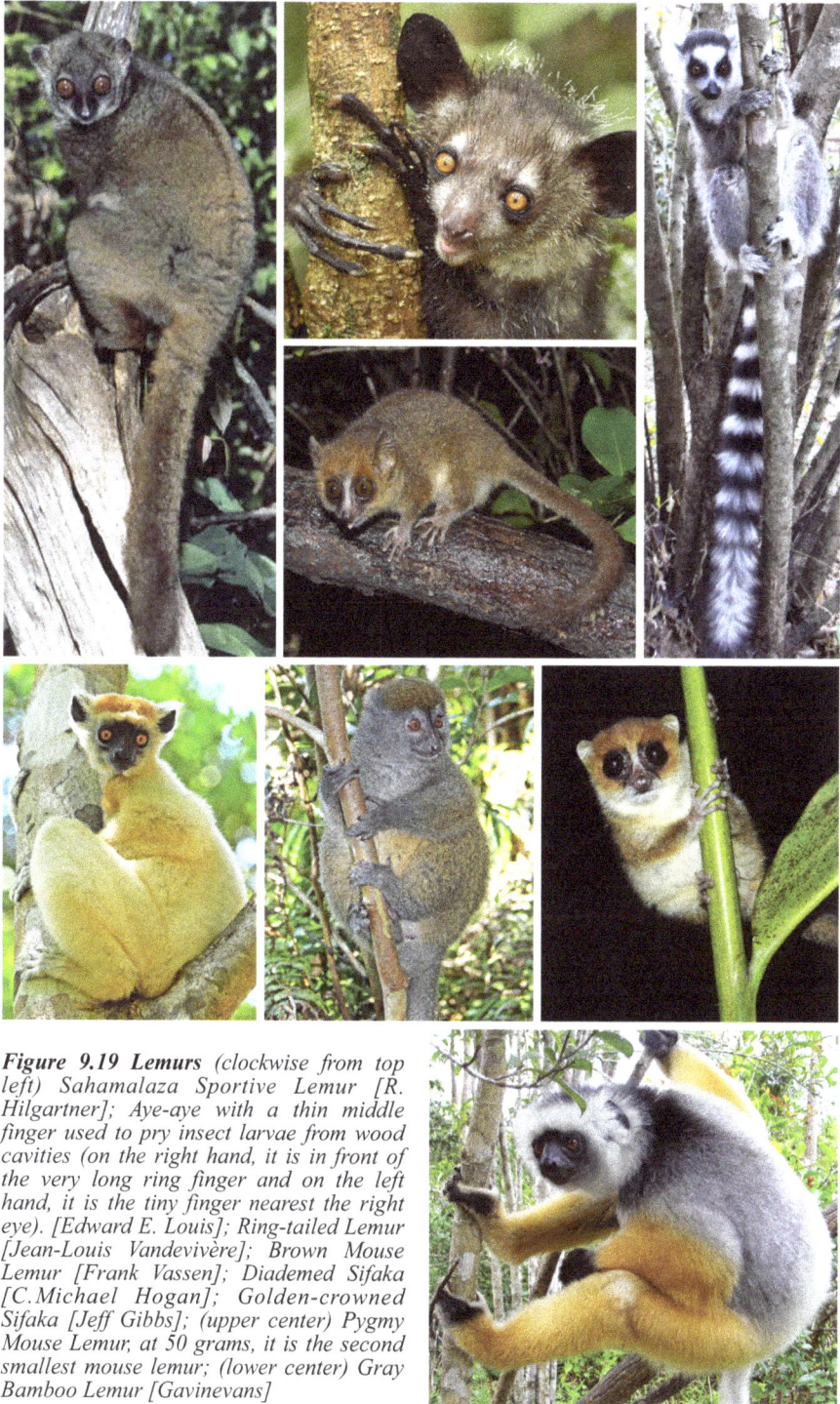

Figure 9.19 Lemurs *(clockwise from top left) Sahamalaza Sportive Lemur [R. Hilgartner]; Aye-aye with a thin middle finger used to pry insect larvae from wood cavities (on the right hand, it is in front of the very long ring finger and on the left hand, it is the tiny finger nearest the right eye). [Edward E. Louis]; Ring-tailed Lemur [Jean-Louis Vandevivère]; Brown Mouse Lemur [Frank Vassen]; Diademed Sifaka [C.Michael Hogan]; Golden-crowned Sifaka [Jeff Gibbs]; (upper center) Pygmy Mouse Lemur, at 50 grams, it is the second smallest mouse lemur; (lower center) Gray Bamboo Lemur [Gavinevans]*

Figure 9.20 Giant Lemur *The Giant Sloth Lemur weighed about 150 kilograms. It became extinct soon after humans reached the island. [Smokeybjb]*

observe the results of the evolutionary process. For example, by looking at a guide to the mammals of Madagascar, the pattern of isolation and evolution is often unmistakably displayed. Two closely related species of dwarf lemurs are found in Madagascar, but one species lives on the eastern part of the island while the other is found in the west. They are separated by a range of mountains. The obvious explanation is that the mountain range separated the original population and the two groups have evolved into distinct species.

In trying to unravel why so many of the unique creatures of Madagascar have become extinct, the changing climate is often mentioned. However, the major transformation of the island was the destruction of the forests by the early colonists. Most of the animals died because their homes were destroyed. The assault continues today, although people from all over the world are helping to protect these marvelous creatures. The rapidly expanding human population makes the prospects for long-term success very uncertain.

New Zealand

New Zealand was once a part of Gondwanaland and is, therefore, a continental island. It consists of two main islands and several smaller associated islands. Much of the land was laid down as sediment on the margin of Gondwana during the late Paleozoic and Mesozoic more than 250 mya. Throughout the later Cretaceous and Cenozoic, these sediments were reworked by stages of mountain building, glaciations, and erosion. Most of the South Island was covered with glaciers during the last ice age. The Tasman Sea, which separates New Zealand from Australia, probably opened during Cretaceous times.

Because New Zealand has been isolated for so long, we expect it to have many endemic organisms. Indeed, it has about 2,500 species of ferns and seed plants, 80 percent of which are endemic. There are 41 species of freshwater fish and all but 5 are endemic. Before humans arrived, there were about 250 species of birds and 70 percent were endemic. Today about half of the land birds are extinct.

There are several animals restricted to New Zealand that have their roots in Gondwana and are considered relicts. The Tuatara (genus *Sphenodon*) is the last surviving member of an ancient order of reptiles that arose in the Triassic Period (Figure 9.21).

There are two subspecies of the Tuatara, the Northern Tuatara and the North Brother Island Tuatara. They live on islands mostly to the northeast of the North Island. When the Polynesians arrived they brought dogs and Pacific Rats with them. The rats exterminated the Tuatara population on the main islands. Currently there is an effort to reintroduce them to the mainland in heavily fenced reserves.

Figure 9.21 Sphenodon This 80 cm long Northern Tuatara is a surviving members of an ancient lineage that appeared more than 200 million years ago. [Tourism New Zealand] It has a parietal eye as does the Carolina Anole on the right (the parietal eye is circled). [TheAlphaWolf]

Probably, the ancestors of the Tuatara were present on the islands when the split with Australia occurred. As we saw in the last chapter, the Tuatara is a true diapsid reptile, having two openings in its skull. Its teeth do not have sockets, but are fused directly to the jaw. An unusual and somewhat primitive feature of the Tuatara is the presence of a third eye (called a parietal eye or pineal eye) located in the middle of the forehead (Figure 9.21). It is believed that originally parietal eyes were not used for clear vision, but merely to sense light and dark. Although it is covered by scaly skin and is not functional, the degenerate eye does have a lens and a retina. Fossil evidence tells us that many early fish, amphibians, and reptiles possessed this feature. Several modern reptiles and amphibians also have a parietal eye.

All the other reptiles of New Zealand are either skinks or geckos, all of which are endemic. There are no snakes, turtles, or crocodiles on the islands. The reptiles include about 30 species of skinks and about 15 species of geckos, belonging to 3 endemic genera.

It is interesting that all the geckos and all but one of the skinks give birth to live young. Of the more than 650 geckos in the rest of the world only those in New Caledonia give birth to live young, and about 40 percent of the 1500 species of skinks in the world give birth to live young. The rest are of course egg-layers. The explanation for this unusually high rate of livebearers is most likely because New Zealand, especially the South Island, experienced extensive glaciations during the last ice age. In reptiles, giving birth to live young means that the mother carries the eggs inside her body until they hatch, a distinct advantage in a very cold area.

The first humans to reach New Zealand were Polynesians known as the Maori. They arrived around 1300 AD when the only mammals present were three species of endemic bats. One, the New Zealand Long-nosed Bat, is closely related to several species in

Australia. The other two are the only members of an unusual family called the New Zealand short-tailed bats. Unfortunately, one of the members became extinct in the late 1960s after a group of fishermen brought Black Rats to Big South Cape Island. This tragedy has been described as one of the greatest ecological disasters of the twentieth century for New Zealand. Not only were the bats exterminated, but many endemic birds became extinct. In 2006, a program was begun to rid the island of the rats.

The New Zealand Lesser Short-tailed Bat is the only surviving member of the group. They are unusual in that they spend much of their time scurrying along the ground looking for food like a mouse. They can fold their wings under a leathery membrane which allows them to move through the ground litter and dig tunnels while protecting their wings. Their thumbs are unusually long and have a large claw which is used for digging and climbing trees.

Because grazing mammals were absent, large flightless birds called moas (Figure 9.22) were able to diversify and fill the niches usually occupied by grazing mammals.

After the Maori arrived, they hunted the moas to extinction and were responsible for the extinction of several other species of large birds. One of these birds was the Haast's Eagle (Figure 9.22), which preyed on the moas. With a wingspan of almost three meters, the Haast's Eagle was the largest eagle in the world, but as the

Figure 9.22 Haast's Eagle and Moas The Haast's Eagle was the world's largest eagle. It is now extinct as are the moas. [J.Megahan]

moas were killed off, it also became extinct.

The ancestors of the moas and the kiwis reached New Zealand before the opening of the Tasman Sea about 60 to 80 mya. Molecular studies indicate that the closest living relatives of the kiwis are the Emu and cassowaries of nearby Australia and New Guinea.

However, the moas seem to be more closely related to a group of South American birds known as the tinamous. If this is true, the ancestors of the moas must have come from Gondwana when New Zealand broke off. Based on subfossil evidence, we know that when the Maori arrived there were as many as 7 species classified into three families. The largest grew about 3.5 meters tall and weighed over 200 kilograms (Figure 9.23).

The rats and dogs that accompanied the Maori caused the demise of many fragile creatures. Fires set by the Polynesians destroyed large areas of the primal forest, and when the Europeans arrived, they continued the destruction. Before the Polynesians reached the islands, forests covered approximately 80 percent of the land, but today only about 15 percent remains. As if the destruction of most of the forests was not enough, the Europeans introduced mammals that have completely upset the natural balance of life.

After the arrival of humans, almost 60 bird species became extinct. The Maori were responsible for about 40 and since the mid-1800 arrival of the Europeans, another 20 have been exterminated.

There are currently almost 50 endemic species of land birds in New Zealand, including five species of kiwis (Figure 9.24). Other interesting endemic birds include three parrots, the Kaka, the Kea, and a flightless parrot called the Kakapo (Figure 9.24). The

Figure 9.23 Upland Moa *At just under one meter tall, this species was one of the smallest moas. It lived in the South Island highlands and was one of the last species of moas to be exterminated by the Maori around 1500 AD. [George Edward Lodge]*

Kakapo is the world's largest parrot, but it is unfortunately very close to extinction. As we have learned, many island birds are flightless and New Zealand has several. Along with the kiwis and the Kakapo, there are flightless rails including the Weka and the Takahe (Figure 1.10). There are also several species of penguins that are endemic to the area.

Other than the moas, one of the New Zealand's greatest losses was the Huia, a unique wattlebird that was hunted to extinction by 1907 (Figure 9.25). They were killed for museum specimens and private collections as well as for their tail feathers which were used to decorate hats. Such senseless killing may seem totally ridiculous to us, but rhinoceroses are killed today because

Figure 9.24 New Zealand Endemics (clockwise from top left) Southern Brown Kiwi [Tourism New Zealand]; Mercury Island Tusked Weta [Chris Winks]; Kea, the only alpine parrot. [Mark Whatmough]; Saddleback, a member of the endemic wattled crow family. [Paddy Ryan]; New Zealand Rock Wren, a member of the endemic New Zealand wren family. [Rob Hutchinson/Birdtour Asia]; South Island Tomtit, an Australasian Robin. [Mark Jobling]; North Island Robin [Tony Wills]; Tui, a member of the honeyeater family. [Matt Binns]; Kakapo, the world's largest and only flightless parrot. [Josep del Hoyo]

some irrational people believe the horns have medical properties.

As one would expect on such an isolated island, most of the insects are endemic. About 90 percent of the more than 20,000 insects are found nowhere else in the world, and several endemic genera are present. Some insects are unusually large. The wetas, a type of cricket, are represented by over 100 species, one of which has a body length of about 10 centimeters, the heaviest insect in the world (Figure 9.24).

New Zealand has over 160 endemic earthworms, including a 1.4 m long giant named *Spenceriella gigantea.*

Today, New Zealand is home to four frog species belonging to the same endemic family, but two are critically endangered. Subfossils tells us that there were seven species before the Polynesians arrived, but three have become extinct in the past 1000 years.

They represent the most primitive members of the frog order as evidenced by several unusual features. For example, they do not have external eardrums and make very few sounds. They have one additional vertebrae, and they have tail-wagging muscles even though they have no tail. The tail muscles are remnants from the frog's ancestors, which like other amphibians possessed a tail.

They also lack what is called an early jump recovery. When they jump, they land on their belly and must reposition their legs for the next jump. More advanced frogs land with their legs in the ready position for the next jump, so they do not have to hesitate between jumps.

The ancestors of the frogs likely reached the area before Gondwana fragmented. Their closest relatives are two species of tailed frogs that are found in northwestern United States and southern British Columbia.

Figure 9.25 Huia The male (front) had a beak of average length, but the female's beak was unusually long, the most extreme difference of any bird in the world. They were killed for museum specimens and private collections and became extinct by 1907. [J.G.Keulemans]

An unusual animal that can also be linked to Gondwana is a species of velvet worm. We were introduced to this animal in Chapter 6 were we discussed it as a possible link between the annelids and the arthropods. Other species of velvet worms are found on the Gondwana landmasses of Africa, South America, and Australia.

As for the plant life of New Zealand, all the conifers are endemic as are more than 80 percent of the flowering plants. One of the largest trees in the world is an evergreen conifer called the Kauri, which is in the same family as the Norfolk Island Pine, the Monkey Puzzle Tree of Chile, and the Bunya-bunya of Australia (Figure 9.26). Kauri trees can grow 50 meters tall with a 5 meter diameter trunk. Unfortunately they produce valuable lumber, so most of the original forests have been cut down. The petrified wood in Arizona's Petrified Forest National Park is from

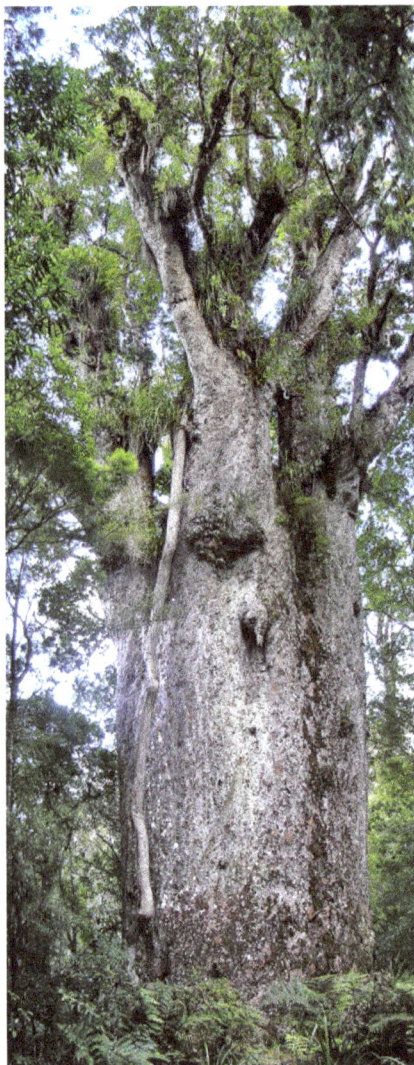

Figure 9.26 New Zealand Kauri *A conifer that is one of the largest trees in the world. [Miguel A. Monjas]*

members of this family. They were much more abundant during the Jurassic and Cretaceous.

The family that the Kauri belongs to is related to a family of conifers known as podocarps. The podocarps are a large family that first appeared in the late Permian and are now confined

mainly to the Southern hemisphere lands that were once part of Gondwana.

About 2 to 7 mya the mountains of New Zealand became high enough to create alpine habitats. Today, more than 90 percent of the alpine flora is endemic and there are even several endemic genera. These alpine species must have evolved during the past several million years. This time span for the evolution of a new genus agrees roughly with the time it took for new genera to evolve on the Hawaiian and Galapagos Islands. It seems that a few million years is enough time to produce organisms different enough to be placed into separate genera.

As we have seen, Madagascar and New Zealand along with several other large continental islands were already isolated before mammals could inhabit them. Today, the only mammals find on these islands are the descendants of those mammals that accidentally rafted or flew (in the case of bats) to the area. This limited dispersal resulted in two especially unique and interesting groups on Madagascar, the tenrecs and the lemurs.

Cuba, Hispaniola, and several of the larger islands of the Caribbean also fit into this category. There are no large native mammals on Hispaniola, but an interesting and primitive insectivore called the *Solenodon* is found there, and another species is found on Cuba (Figure 9.27). Unfortunately, the Cuban *Solenodon* is close to extinction.

The solenodons represent relics that somehow reached these islands at a very early time and have managed to survive while their relatives on the continent died out. They are one of the few venomous mammals in the world. Modified salivary glands in the lower jaw produce the venom, but it is not lethal to larger animals. Because they are primitive mammals, solenodons are one of those animal treasures that are

Figure 9.27 Hispaniolan Solenodon *This primitive mammal is endemic to the island of Hispaniola. It is about 30 cm long excluding the tail, and it is the only mammal with a venomous bite. [Jorge Brocca]*

of great value in helping us in our attempt to understand the evolution of the mammals.

World's Biogeographic Regions

After the dinosaurs disappeared, the mammals and birds experienced an explosive period of evolution. With the earlier breakup of Pangaea, many of the large landmasses were isolated to varying degrees. The great variety of plants and animals we see today is due to these long periods of isolation during much of the Cenozoic Era.

Even casual observers are aware that most animals and plants are found in specific regions of the Earth. The pouched kangaroos and Koalas are only found in Australia. Giant Pandas are restricted to the mountains of central China. Guanacos and the slow moving sloths live only in South America. Prairie dogs and Pronghorn Antelopes with their unique horns are found only in North America; while the Giraffe, hippos, and our close cousins the gorillas and chimpanzees only live in Africa.

Although a few boundaries are somewhat fuzzy, it is useful to divide the world into several biogeographic regions, each with its own unique life forms (Figure 9.28). The mammals and land birds of these regions are the most well known, but many other animals and plants are also representative of these regions.

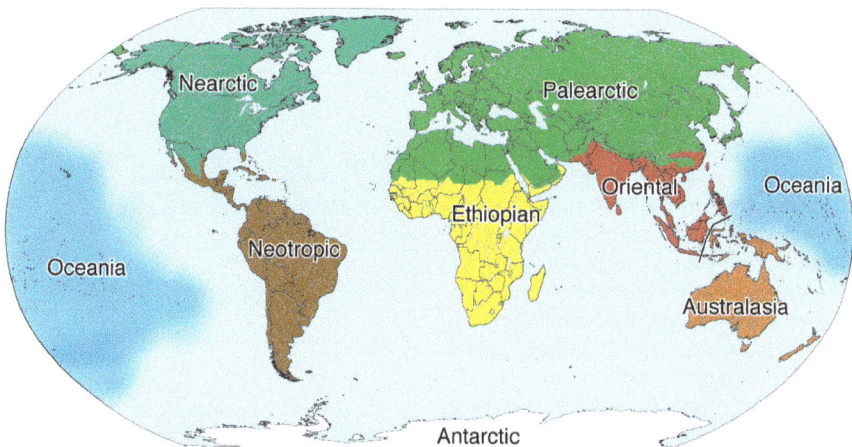

Figure 9.28 Biogeographical Regions of the World *The Earth has been divided into eight biogeographical regions based mainly on the animal life that inhabits each region. Because of the faunal differences he observed, Alfred Wallace drew a line separating the Oriental and Australasian regions. Today it is known as the Wallace Line. [Sandrine Tercerie]*

Although the early Cenozoic was an important time of rapid evolution, our knowledge of this very important time is limited since there are essentially no fossils of Paleocene age from Africa or Australia and very few from Europe.

As Gondwana broke apart during the Cretaceous, South America, Antarctica, and Australia remained attached or nearly so for some time. The exact details of the breakup are not known at this time, but the connections played an important role in the evolution of the mammals on these land masses.

Australasia

Australia is either the smallest continent in the world or the largest island. Except for Antarctica, it is the only continent that is completely isolated from all the other continents. Because of the duration of its isolation, it is home to an astonishing array of plants and animals that are found nowhere else in the world.

Australia along with New Guinea and several neighboring islands form a biogeographic region known as the Australasian region, or simply the Australian region. It also includes New Zealand and New Caledonia to the east.

Australia and New Guinea are the only places in the world where monotremes are still found. These primitive egg-laying mammals include the Platypus of eastern Australia and Tasmania, and four species of echidnas or spiny anteaters (Figure 9.29b). All four species of echidnas are found in New Guinea, and the Short-beaked Echidna, is also found in Australia.

After the egg-laying monotremes and marsupials were established in Australia, the continent was isolated before the placental mammals could get established there. For most of the Cenozoic Era, therefore, the marsupials

of Australia were able to evolve without competition from placental mammals. In all, there are about 200 species of marsupials in Australia

Figure 9.29a Australasian Mammals The Feather-tail Glider (top) is the only member in its family (Acrobatidae). This smallest gliding mammal has a head-body length of about 7 cm. It weighs just 10 to 15 grams and is very cute. [Stephen Mahony] The Musky Rat-Kangaroo (bottom) is also placed in its own family and has several primitive features like reptile-like scales on its tail and 5 toes on its hind feet. [PanBK]

Figure 9.29b Australasian Mammals (clockwise from top left) Eastern Gray Kangaroo and joey [Danielle Langlois]; Platypus, a monotreme. [Hanno Stamm]; Tasmanian Devil; Common Wombat; Eastern Barred Bandicoot [last three by JJ Harrion]; Banded Anteater [Martin Pot]; Koala [David Iliff]; Short-beaked Echidna, a monotreme. [JJ Harrison]

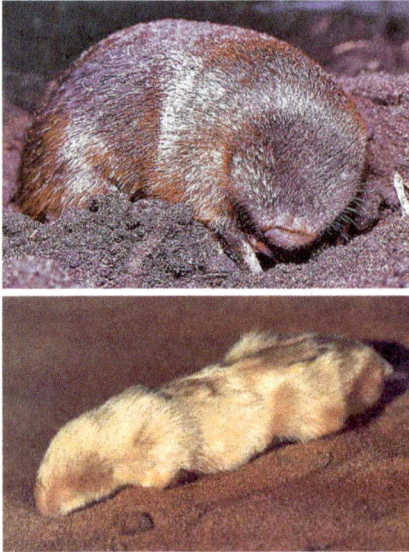

Figure 9.30 Two Moles *A species of golden mole (top) is a member of the suborder Afrotheria. [Killer18] The Southern Marsupial Mole fills a similar niche in Australia. [Bartus.malec]*

grouped into about 16 different families (Figure 9.29).

The earliest marsupial fossils are from North America and China and are of Cretaceous age, but at about the same time they are also found in South America. The place where marsupials originated is not known since the fossil evidence during this critical time is too fragmentary to make a definitive statement. One likely possibility is that marsupials evolved in either North America or China, but reached the other continents by rafting or island hopping.

The marsupials diversified in South America and during the early Cretaceous (approximately 140 mya) they reached Australia by way of Antarctica, which was not yet covered with ice and was still connected to both South America and Australia. As we would expect, fossil marsupials have been found in Antarctica. Australia and

Antarctica became separated about 40 mya.

The earliest marsupial fossils found in Australia are of teeth and bone fragments of early Eocene age (about 55 million years old), but the fossil record in very poor and there is some controversy about the date.

As the Australian marsupials adapted to fill various roles in the scheme of life, they evolved features that enabled them to function better. The kangaroos fill the role of grazers and are basically the antelopes, sheep, and goats of Australia

The diversity we observe in other marsupials illustrates the interesting evolutionary principle known as convergent evolution, the independent evolution of superficially similar organisms that are not closely related. There are marsupials that look and act like moles called marsupial moles (Figure 9.30), marsupials that act like mice, (called marsupial mice) and marsupials that live in trees like certain monkeys (the cuscuses). In addition, there is an anteater (the Numbat or Banded Anteater), the wombats that are the marsupial counterpart of the Groundhog or marmots.

The Sugar Glider is a marsupial that looks so much like a flying squirrel, which is a placental mammal, that many people have a hard time telling them apart (Figure 9.31). It is amazing that two mammals that are not very closely related could evolve into such similar looking animals. The basic building blocks of life provide the raw materials that allow evolution to solve a particular problem in many different ways.

Meat-eating marsupials evolved strong jaws and sharp, tearing teeth similar to those of the placental wolves and lions. For example, the pouched marsupial Thylacine or Tasmanian Wolf was a spectacular example of

Figure 9.31 Two Gliding Mammals
Spectacular examples of convergent
evolution are the Southern Flying Squirrel
(top), a placental mammal from North
America, and the Sugar Glider, an
Australian marsupial. The flying squirrel is
more closely related to all placental
mammals (like the elephants) than it is to
the Sugar Glider. [Joe McDonald]

Displaying an all too common lack of concern for other living creatures, the ranchers of Tasmania launched a campaign of extermination that was entirely successful, and a Tasmanian Wolf has not been seen in the wild since one was killed in 1930. The last captive animal died in the Hobart Zoo in 1936.

The quolls are a small group of six species of carnivorous marsupials that hunt other small animals. They are very similar in looks and habits to several members of the placental weasel family. Early on, the Tiger Quoll was called a Spotted Marten (Figure 9.33).

The most well-known Australian mammals are the monotremes and marsupials, but almost one-fourth of the native Australian land mammals are not marsupials, but rodents belonging to the largest family of mammals, the

convergent evolution that looked superficially like just another member of the dog family (Figure 9.32).

The Aboriginal Australians arrived in Australia about 50,000 years ago, and undoubtedly contributed to the extinction of many large marsupials including the Tasmanian Wolf which was extinct on the Australian mainland when the Europeans arrived.

On Tasmania, the Tasmanian Wolves were hunted to extinction in the early part of the 1900s so we will never again have the pleasure of seeing these wonderful and unusual creatures.

Figure 9.32 Two Carnivores Another
example of convergent evolution, the
Coyote (top) is a placental mammal that
hunts small animals, while the Tasmanian
Wolf or Thylacine was a masupial that also
preyed on small animals. Hundreds of
Thylocines were killed and they became
extinct by about 1940. [Coyote, Marilyn
Schiele; Thylacines, John Gould]

Figure 9.33 Two Cat-like Mammals *The American Marten, a member of the weasel family (above), looks and acts very much like the Eastern Quoll, a marsupial. The Eastern Quoll is common in Tasmania, but is extinct on mainland Australia. [marten, Tim Gage; quoll, Michael Barritt and Karen May]*

Muridae. Their ancestors must have rafted to this island continent about 5 mya, long after the marsupials had become established.

Many of the Australasian birds are unique. There are almost 40 families that are endemic to the region or largely so. The three species of cassowaries and the Emu form two of these families. They are large flightless birds that are relicts like the rheas of South America and ostriches of Africa.

The Magpie Goose, Kagu, and Plains-wanderer are the only members of their respective families. Other endemic families include lyrebirds, bowerbirds, the megapodes or mound-builders, whipbirds, logrunners, Australian wrens, butcherbirds, woodswallows, Australasian robins, honeyeaters, pardalotes, sittellas, Australian treecreepers, boatbills, whistlers, Australian mudnesters, and birds-of-paradise (Figure 9.34).

Although the marsupials are the representative mammals of the region, many other animals and plants are unique as well. Because of its long period of isolation, the region is home to perhaps the most bizarre flora and fauna in the world.

The eucalyptus are the signature trees, with over 700 species that are

Figure 9.34a Australian Endemic Birds *(left to right) Plains-wanderer [Patrick_K59]; Superb Lyrebird [Attis]*

Figure 9.34b Australasian Endemic Birds (clockwise from top left) Southern Cassowary [JCB]; Australian Brush-turkey, a megapode or mound-builder. [JCB]; Magpie Goose, which does not have webbed feet. [Scott Bowman]; Kagu, an endemic of the island of New Caledonia. [Tony Morris]; Emu [JCB]

Figure 9.34c Australasian Endemic Birds *(clockwise from top left)* *New Holland Honeyeater [JJ Harrison]; Pied Currawong, a butcherbird. [JCB]; White-browed Woodswallow [Aviceda]; Satin Bowerbird at his bower [JCB]; Scarlet Robin [JJ Harrison]; Superb Fairywren, one of the Australian wrens. [JCB]; Striated Pardalote [JCB]; Eastern Whipbird [JCB]*

endemic to Australia and surrounding islands. The tallest flowering plant in the world is the Australian Mountain Ash, which is a eucalyptus. It is just shy of 100 meters tall. Specimens that were likely the tallest trees in the world once grew in Tasmania, but they were cut down in the late 1800s. What a sad day, when someone destroyed the tallest tree on Earth.

Another notable group of trees and shrubs are called the southern beeches, the only members of the family Nothofagaceae. Members of this small group of about three dozen species are only found on the Gondwana land masses of Australia, New Zealand,

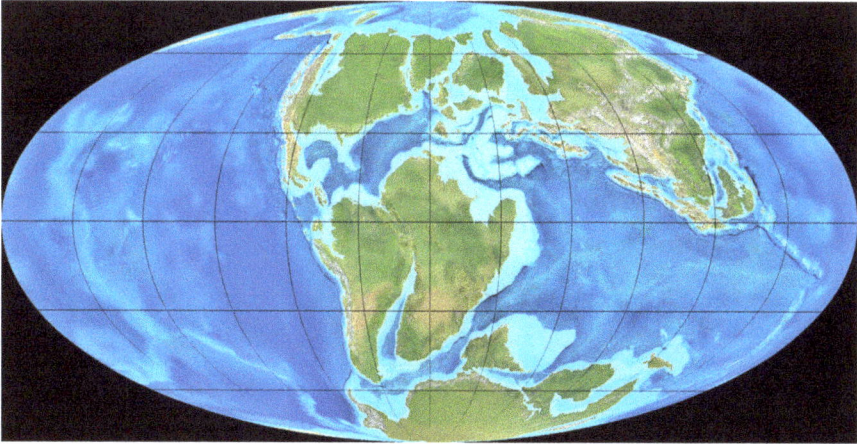

Figure 9.35 Earth During the Middle Cretaceous *120 million years ago, South America was separating from Africa. [Ron Blakey/Colorado Plateau Geosystems, Inc]*

New Guinea, and southern South America. Fossils have been found in Antarctica as expected.

Neotropic

The South Atlantic Ocean began to form about 140 mya as South America became isolated from Africa (Figure 9.35). After the dinosaurs became extinct, the only mammals present were marsupials, South American ungulates, and xenarthrans. South America remained isolated for most of the Cenozoic, but was connected to North America about 3 mya. This connection caused a great interchange of animals, and many of the more tropical species moved north into Central America. The deserts of southwestern United States and Mexico form a barrier that most of the tropical South American species have not crossed.

This partially isolated area of South and Central America is known as the Neotropical region. Surpassing even the Australian region in unique forms, the Neotropic or Neotropical region is one of the truly great storehouses of biological diversity on the planet.

The South American marsupials evolved spectacular examples during the continent's long period of isolation. There were several large flesh-eating marsupials that preyed on the South American ungulates. One saber-toothed marsupial called *Thylacosmilus* looked very much like the saber-toothed cats of North America that were placentals (Figure 9.36). It is another example of convergent evolution. *Thylacosmilus* first appeared in the Miocene, but became extinct soon after the Great American Interchange. It apparently could not compete with *Smilodon* which had invaded South America from the north (Figure 7.55).

Although these large marsupials are now extinct, even today, marsupials represent a large fraction of the mammal species of South America. Most of the 65 or so species are mouse-like or rat-like creatures that have been grouped into 3 families. They are, of course, unique to South America.

Of the three families, one is a little known group (the shrew opossums), which contains about 5 or 6 species.

Figure 9.36 Thylacosmilus *A marsupial saber-toothed carnivore that lived in South America during the Neogene Period. [Roman Uchytel] It had an unusual sheath like structure on its lower jaw to protect its large canine teeth. [photo, AMNH/Claire Houck]*

They live in the high Andes and feed mostly on insects as do placental shrews. Shrew opossums are about the size of a small rat, larger than their placental counterparts. They are placed in their own order.

One family of South American marsupials contains just one species, a mouse-like opossum called Monito del Monte, which unfortunately means *mountain monkey* (Figure 9.39). DNA studies show that Monito del Monte is the closest relative of the Australian marsupials. Molecular data indicates that the ancestor of all Australian marsupials came from South America about 50 million years ago.

Figure 9.37 Monkey Noses *New World monkeys, called platyrrhines, like the White-headed Capuchin (left) have nostrils pointing to the sides. [JCB] Old World monkeys, called catarrhines, like this Patas Monkey have nostrils that point down.*

The remaining family of South American marsupials (Didelphidae), contains the most species and is the most basil of all the marsupial families. One species, the Virginia Opossum, is the only marsupial that is established in North America.

Like marsupials, the xenarthrans have retained some primitive reptile-like features, suggesting that they represent an early branch from the main line of mammals. For example, they have small brains, a lower and poorly regulated body temperature, primitive teeth without an enamel covering, and vertebrae and ribs that are more reptile-like.

The ancestors of the New World monkeys and the South American rodents (classified as caviomorph rodents) reached South America sometime during the Eocene about 40 million years ago, probably by rafting from Africa.

New World monkeys (platyrrhines) can be told from their Old World cousins by their noses that have round widely separated nostrils pointing to the sides of their head (Figure 9.37). Old World monkeys (catarrhines) have nostrils that are close together and point downward.

In the tropical forests of South America, some of the monkeys evolved long tails that could be manipulated much like a third hand that can be used to help the animals swing through the trees. Tails that can be used as a fifth hand are called *prehensile tails*.

One classification scheme defines five families of New World monkeys (Figure 9.38). The smallest family is the one containing the night monkeys or owl monkeys. They have very large eyes, and all but one species is nocturnal. Most of the daylight hours are spent in hollow trees. They are relatively small, and most weigh between 0.5 and 1 kilogram.

Another small family contains about a dozen species of capuchins and squirrel monkeys. The capuchins are probably the best known of the South American monkeys. They have prehensile tails and are the most intelligent of the South American Monkeys. Squirrel monkeys do not have prehensile tails, but they use their long tail for balance.

The marmosets and tamarin monkeys belong to a family of about 25 to 40 species. This family contains the smallest South American monkeys.

The titis, uakaris and saki monkeys belong to a family of about 20 to 40 species, depending on the classification scheme used. They have long fur, and all but the uakaris have long tails, but none have prehensile tails.

The last family contains the howlers, spider monkeys and woolly monkeys. Members of this group are the largest South American monkeys, and the howlers are the loudest. All members of the family have prehensile tails.

One South American anteater, the Silky Anteater, also has a prehensile tail as does at least one porcupine, and several marsupials, including the Virginia Opossum.

Figure 9.38a Neotropical Monkeys (top to bottom) Pygmy Marmoset, world's smallest monkey with a body that is 12 to 15 cm long. [Malene Thyssen]; Woolly Spider Monkey with baby [Bart vanDorp]; White-nosed Bearded Saki [Rich Hoyer]

Figure 9.38b Neotropical Monkeys (clockwise from top left) White-headed Capuchin [JCB]; Spix's Night Monkey [napowildlifecenter]; Golden-mantled Tamarin [lowjumpingfrog]; Brown Howler Monkey [Peter Schoen]; Brown Titi Monkey [Cliff]; Bald Uakari [Eugenia Kononova]

Figure 9.39a Neotropical Mammals *(clockwise from top left) Brazilian Porcupine, a prehensile-tailed species. [Tim Wirtz]; Honduran White Bats [Leyo]; Patagonian Mara, of the cavy family. [Jason Hollinger]; West Indian Manatee [USFWS/Keith Ramos]; Monito del Monte is the only neotropical member of the Australidelphia superorder of marsupials. [José Luis Bartheld]; Bare-tailed Woolly Mouse Opossum [icelight]; Common Mountain Viscacha, of the chinchilla family. [Alexandre Buisse]; Cuban Hutia [Joe Burgess]*

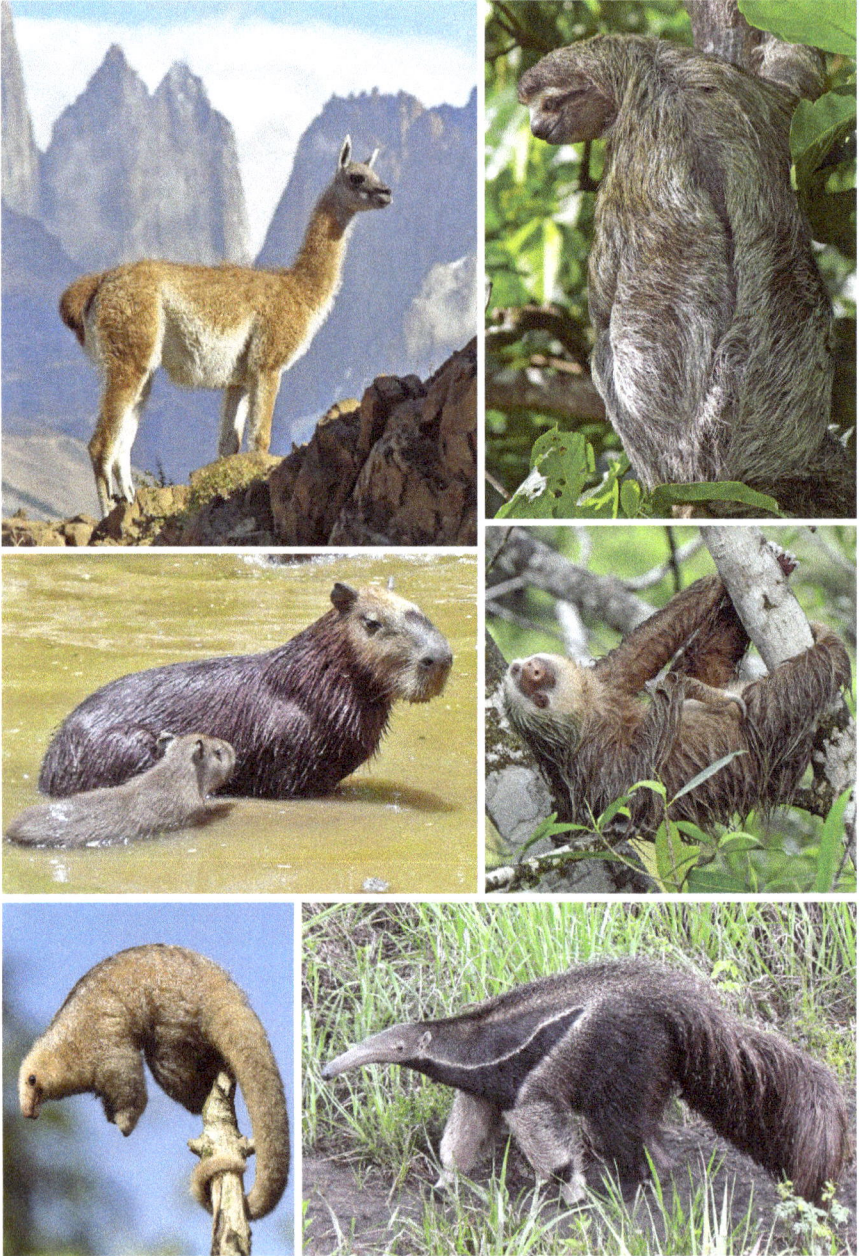

Figure 9.39b Neotropical Mammals *(clockwise from top left) Guanaco, a member of the camel family whose ancestors migrated from North America during the Great American Interchange. [Brian Pinkerton]; Brown-throated Three-toed Sloth, a xenarthran. [JCB]; Hoffmann's Two-toed Sloth, a xenarthran. [Geoff Gallice]; Giant Anteater, another xenarthran. [Anagoria]; Silky Anteater, a small anteater with about a 20 cm long body and 20 cm long tail and also a xenarthran. [Quinten Questel]; Capybara with baby, one of the unique caviomorph rodents and the world's largest rodent. [Mateus Hidalgo]*

Figure 9.39c Neotropical Mammals The Nine-banded Armadillo is the only xenarthran that has moved into the Nearctic region. [Hans Stieglitz]

In addition to the five families of New World monkeys, there are two families of South American anteaters, two families of sloths, one family of armadillos, six families of bats, the solenodons, which were discussed earlier (Figure 9.27), and about a dozen families of strange caviomorph rodents like the Nutria, agoutis, pacas, chinchillas, the New World porcupines, the capybaras, maras, viscachas, cavies or guinea pigs, and the hutias of Cuba and several other Caribbean Islands, (Figure 9.39).

As we saw, the South American ungulates trace their ancestry to an early branch of the odd-toed ungulates. Therefore, they belong to the Laurasiatheria group of mammals. The South American ungulates are another good illustration of convergent evolution as several evolved forms that resembled the horse, hippopotamus, rhinoceros, elephant, and camel. As we saw in Chapter 7, the entire group became extinct after North America and South America were connected near the end of the Pliocene. The first humans to enter the Americas caused some of the later extinctions.

As a result of the Great American interchange, members of only three endemic South American families moved north and ultimately prospered,

the North American Porcupine, the Virginia Opossum, and the Nine-banded Armadillo. Many more mammals moved south, a few of which belonged to the camel family, the pig family, and members of the cat, dog, bear, raccoon, and deer families. Also several new rodent families are now established in the south.

There are about 5,400 species of mammals in the world which are classified into about 150 families. About 32 of these families are unique to the Neotropical area and two more are mostly confined to the region.

The isolation of South America during the Cenozoic produced the most numerous and diverse collection of birds in the world! There are 200 to 230 families of birds in the world, and members of about 35 of these families are found entirely or mostly in the Neotropical. There are between 9,700 and 10,300 species of birds in the world, and about one-third are found in the Neotropical region.

South America's bird life includes unique birds like the Hoatzin, Oilbird, and Sunbittern, which are the only members of their respective families. Other endemic families are the big flightless rheas, the trumpeters, potoos, screamers, seriemas, motmots, toucans, tinamous, the curassows and guans, puffbirds, manakins, antbirds, most tanagers and hummingbirds, and the spectacular cotingas (Figure 9.40).

The two species of seriemas are believed to be the closest living relatives of the 1 to 3 meter tall terror birds that lived in South America during the Cenozoic and in North America after the Great American Interchange.

There are about 310 species of hummingbirds and all but a few that venture into North America are restricted to the Neotropical region. The smallest bird in the world is the

Figure 9.40a Neotropical Birds *(clockwise from top left) Greater Rhea [Lip Kee Yap]; Northern Screamer [Pete Morris]; Sunbittern [Jennifer Williams]; White-whiskered Puffbird [Len Blumin]; Hoatzin [JCB]; Red-legged Seriema, a close relative of the extinct terror birds that appeared in the Paleocene. [José Reynaldo de Fonseca]*

Figure 9.40b Neotropical Birds *(clockwise from top left) Pale-winged Trumpeter, one of the three species in the family. [Dubi Shapiro]; Great Curassow [Tristan Bantock]; Plate-billed Mountain Toucan [JCB]; Broad-billed Motmot [JCB]; Hairy-crested Antbird [Glenn Bartley]; Common Potoo is perfectly camouflaged and sticking up on the left of the photo. [Terry Simms]; Oilbird [Steve Garvie]*

Figure 9.40c Neotropical Birds *(clockwise from top left) Fiery-throated Hummingbird [JCB]; Blue-winged Mountain-Tanager [JCB]; Araripe Manakin [Rick elis.simpson]; Spotted Nothura, a member of the tinamou family. [Dario Sanches]; Long-wattled Umbrellabird, a spectacular member of the cotinga family. [Tomasz Doron]*

Bee Hummingbird of Cuba. It is only 5 cm long and weighs less than 2 grams (a penny weighs 2.5 grams).

Oilbirds are colonial birds that congregate in caves where they nest. As the young birds grow, they become very fat and can eventually weigh about one and a half times as much as the adults. They were once killed for

their oil which is where the name Oilbird comes from. Oilbirds are unusual as they are one of the few birds that use echolocation for navigation.

One of the most unusual birds in the world is the Hoatzin of South America. It is interesting because when the chicks hatch, they have two claws on each wing which they use for climbing. The claws are lost in the adult birds.

These claws may be an example of an atavism or throwback. Since birds evolved from dinosaurs that had claws on their front legs, the genes that orchestrate the growth of these claws are likely still present in birds, but have been disabled. If these genes are later reactivated, a throwback is produced.

Although tinamous are able to fly, they are most closely related to the group of flightless birds that includes the rheas, Emu, kiwis, cassowaries, ostriches, the extinct moas of New Zealand, and the extinct elephant birds of Madagascar.

There are about 6,300 species of frogs in the world classified into about 50 or so families. More than half the frog species in the entire world are found only in the Neotropical region! Amphibians are currently declining at an alarming rate and about one-third of the frogs are classified as threatened. Due to our inability to find them, it is believed that over 150 species have become extinct in the past 20 years.

As for the other amphibians, the Neotropical region contains almost half the salamanders in the world and half the caecilians (Figure 9.41). Caecilians are legless amphibians that spend most of their time burrowing in the soil. They look like earthworms or snakes, but have smooth skin. Their eyes are protected by a covering of skin.

Figure 9.41 Caecilian This neotropical amphibian, called a Mexican Burrowing Caecilian, is 30 to 50 cm long. The light dot at the far right is one of its vestigal eyes. [Franco Andreone]

Figure 9.42 Green-banded Urania Even though it looks like a butterfly, this South American endemic is really a day-flying moth. [JCB]

There are about 18,000 known species of butterflies in the world and 45% of them are found in the neotropical region. If we include moths the total number in the world goes up to about 170,000 (Figure 9.42).

As we cut down the South American rainforests and convert them into a cow pastures, we are destroying the homes of these unique animals, and many will be extinct in the near future. A great ecological disaster.

Ethiopian

As Pangaea fragmented during the early Cretaceous, Africa and South America separated while India moved north. Africa was isolated for about 100 million years.

Today, Africa is isolated from the rest of the world by water on the south, west, and east, but is connected to Asia in the northeast. However, the Arabian desert provides a barrier that many plants and animals cannot cross.

The region south of the Sahara Desert is called the Ethiopian or Afrotropical. Although there was some interchange between Africa and Asia during the Cenozoic, many plants and animals are still restricted to the region.

The diversity of the bird life of the Ethiopian is not as impressive as the

Neotropical, however, it is still the exclusive home of about 30 families of birds. The Secretarybird, the Egyptian Plover, a stork-like bird called the Shoebill and another named the Hamerkop, are the only members of their families. Additional endemic families are the ostriches, guineafowl, small rail-like birds called flufftails, mousebirds, wood hoopoes, ground hornbills, vangas, barbets, African sugarbirds, oxpeckers, whydas,

Figure 9.43a Ethiopian Birds (clockwise from top left) Southern Ground Hornbill [Drew Weber]; Vulturine Guineafowl [Usha Harish]; Bearded Barbet [Tristan Bantock]; Black Scimitarbill, a wood hoopoe. [Dubi Shapiro]; Sickle-billed Vanga [Dubi Shapiro]

Figure 9.43b Ethiopian Birds *(clockwise from top left) Common Ostrich [Stig Nygaard]; Hamerkop [Voidoffrogs]; White-necked Rockfowl [Michael Andersen]; Egyptian Plover [Steve Garvie]; Eastern Paradise-Whydah [Warwick Tarboton]; Yellow-billed Oxpecker [Steve Garvie]*

Figure 9.43c Ethiopian Birds (clockwise from top left) Cape Sugarbird [Alan Manson];
Blue-naped Mousebird [Doug Janson]; Red-collared Widowbird [Jacqui Herrington];
Secretarybird [Yoky]; Shoebill [Hans Hillewaert]

Figure 9.43d Ethiopian Birds Madagascar Flufftail [Dubi Shapiro]

widowbirds, and the rockfowl or bald crows (Figure 9.43). Much of the bird diversity is due to the great number of species found in the tropical forests of the Congo Basin.

About 22 families of mammals are found only in the Ethiopian region. In our discussion of Madagascar, we were introduced to the tenrecs, Madagascar carnivorans, and lemurs. In addition, endemic families on the continent include the elephant shrews (Figure 9.16), the Aardvark, the only member in its family, the giraffe family, the hippopotamuses, including the Pygmy Hippopotamus, and the galagos or bushbabies which along with the lemurs, belonging to the prosimian primate group. Hyraxes are squirrel-like mammals belonging to the Afrotheria superorder and there are seven families of rodents (Figure 9.44).

Africa is also home to many familiar species of mammals such as, two species of elephants (Bush or African Elephant and Forest Elephant), two species of gorillas (Eastern and Western Gorilla), Common Chimpanzee, Bonobo or Pygmy Chimpanzee, and great herds of zebras, wildebeests, antelopes, and the African or Cape Buffalo (Figure 9.44).

Figure 9.44a Ethiopian Mammals (top to bottom) Cape Buffalo [Nevit Dilmen]; Blue Wildebeest [Muhammad Mahdi Karim]; African Elephant [nickandme]

Figure 9.44b Ethiopian Mammals (clockwise from top left) Mountain (Eastern) Gorilla [Richard Ruggiero/USFWS]; Okapi, of the giraffe family. [Charles Miller]; Aardvark, a member of the Afrotheria. [Albert Herbigneaux]; Senegal Bushbaby [Bo Jonsson]; Yellow-spotted Hyrax, a member of the Afrotheria. [D.Gordon E.Robertson]; Spring hare is one of the two species in its endemic rodent family. (Pedetes); Pygmy Hippopotamus

Oriental

As the island continent of India moved northward during the late Mesozoic and early Cenozoic it eventually collided with Asia about 40 to 50 mya. With this collision, the Himalayan uplift began. The mountains provide a barrier that severely restricts the spread of many plant and animal species. Today, separated from the rest of the continent by this great range of mountains, India, southern China, and the southeastern portion of Asia form the Oriental biogeographical region. Included are Burma, Thailand, Vietnam, and the other countries and islands to the south.

Endemic bird families in the region include the Malaysian Rail-babbler and Bornean Bristlehead, the only members in their families. Other endemic families include the Asian barbets, ioras, fairy-bluebirds, leafbirds, most of the wren-babblers, most of the laughingthrushes, and 3 of the 4 species of treeswifts (Figure 9.45). One species of treeswift has spread south to a few islands in the Australasian region.

As its name implies, the Bornean Bristlehead has yellow-orange bristle-like feathers covering its head and red bristle-like feathers on its neck. It is an endemic to the island of Borneo and is about (25 cm long).

The Oriental region is home to many endemic mammals including the Bornean Orangutan, the Sumatran Orangutan, Tiger (several subspecies), Malayan Tapir, Indian Elephant, and three species of rhinoceroses (Figure 9.46a and 9.46b). The Sumatran Rhinoceros is very rare with less than 200 surviving. They are killed for their horns in the absurd belief that they are a potent medicine. This unique two-horned rhinoceros is the closest living relative of the extinct Woolly

Rhinoceros that roamed Eurasia during the Pleistocene Epoch.

Asian Elephants are smaller than the African Bush Elephant, but larger than the African Forest Elephant. Asian Elephants have smaller ears and their back arches up while their African cousins have a downward sagging back. The trunk of an Asian Elephant has just one finger-like projection, so they pick things up like you would if your fingers and thumb were glued together. However, the two species of African elephants have two projections on their trunk which can be used like a thumb and finger, making them a little more dexterous.

The smallest members of the even-toed ungulate order are commonly called mouse-deer, and they belong to the family Tragulidae. All are endemic to the Oriental region, except one member which lives in the rainforests of Central and West Africa. The smallest is just under half a meter tall and weighs only about two kilograms.

Five other very unusual families of mammals are endemic to the region. They include the tree shrews, which we introduced earlier as they probably resemble the earliest primates (Figure

Figure 9.45a Oriental Endemic Bird Families Chestnut-capped Laughingthrush [K S Kong]

Figure 9.45b Oriental Endemic Bird Families (clockwise from top left) Green Iora [Alan Ng]; Bornean Bristlehead [Ck Leong]; Whiskered Treeswift [Daniel Koh]; Asian Fairy-bluebird [K S Kong]; Golden-fronted Leafbird [Daniel Koh]; Blue-throated Barbet [J.M.Garg]; Malaysian Rail-babbler [Daniel Koh]

Figure 9.46a Endemic Mammals of the Oriental Region (clockwise from top right) Asian Elephant [Yathin S. Krishnappa]; Lesser Mouse-Deer [Bjørn Christian Tørrissen]; Phillippine Tarsier [Kok Leng Yeo]; Sunda Flying Lemur or Sunda Colugo [Lip Kee Yap]; Bornean Orangutan and baby [Johnny Wee]; Bengal Tiger with two cubs [Wikigringo]

Figure 9.46b Endemic Mammals of the Oriental Region *The Indian Rhinoceros is one of the two species belonging to the one-horned rhinoceros group. [Krish Dulal]*

8.35), the gibbons or lessor apes which were introduced earlier (Figure 8.36), a little prosimian primate called the tarsier, and a strange gliding mammal known as the colugo or flying lemur, although it is not closely related to the lemurs of Madagascar (Figure 9.45).

The completely aquatic Dugong, which is related to the elephants as mentioned earlier, is endemic to the region. They are the closest living relative of the extinct Steller's Sea Cow. Like the Steller's Sea Cow, the Dugongs have a notched tails (Figure 9.17), but manatees have rounded tails (Figure 9.39a).

The colugos or flying lemurs are one of several groups of gliding mammals (Figure 9.46a). They are not closely related to either the flying squirrels or the marsupial gliders, so they independently evolved the ability to glide. They are another excellent example of convergent evolution.

Tarsiers belong to the haplorhine group or dry nosed primates (Figure 9.46a). Other members of the haplorhine group are the monkeys and apes. Wet nosed primates or strepsirhines include lemurs, lorises, pottos, and galagos or bushbabies.

Tarsiers have huge eyes, allowing them to feed at night on insects and small lizards. Named for their elongated tarsus or heel bones, tarsiers are very good jumpers.

The region also has an endemic family of rodents called the Oriental dormice, although they are not true dormice like the character at the tea party in Lewis Carroll's novel *Alice's Adventures in Wonderland*.

A group of antelopes known as the four-horned antelopes is also endemic to the region. There are two members of the group (called a tribe). One member has only two horns and is called the Nilgai, which means *blue cow* in Indian. The other is the Four-horned Antelope. They may be the most primitive members of the Bovidae family.

Nearctic

North America, north of Mexico, is known as the Nearctic region, and it is currently isolated from Asia by the Bering Sea. As we have seen, Beringia connected these two great landmasses during much of the Cenozoic. Consequently, many similar species exist in the two regions. Some consider them to be the same biogeographical region called the Holarctic.

There are no endemic bird families and only three endemic mammal families in the Nearctic, but many species of mammals and birds are endemic to the region (Figure 9.47).

Figure 9.47a Nearctic Mammals *(clockwise from top left) American Bison [JCB]; White-tailed Prairie Dog [Devon Pike]; Musk Ox [Tim Bowman/USFWS]; Black-footed Ferret [Ryan Hagerty/USFWS]; Aplodontia or Mountain Beaver [Ronn Altig]; Pronghorn [James Leupold/USFWS]*

Figure 9.47b Nearctic Mammals (top) A Botta's Pocket Gopher with empty cheek pouches [Chuck Abbe] and with full cheek pouches. [Davefoc]

The three families of mammals that are endemic to the region are the Pronghorn and Mountain Beaver, which are the only members in their respective families, and the pocket gophers, although a few species of pocket gophers extend down into Central America.

The Mountain Beaver is a rodent that looks like a muskrat except the Mountain Beaver or *Aplodontia* has a short tail (Figure 9.47a). Its home is the mountains of far western United States and southern British Columbia. It is considered to be a primitive rodent because of the way its jaw muscles are attached to its cheek bone.

The Pronghorn is the last surviving member of a family that appeared about 20 mya during the early Miocene in North America.(Figure 9.47a) Although Pronghorns are sometimes called Pronghorn Antelopes, they are not true antelopes since they have unique branching horns which are shed annually.

True horns, which have a bony core covered by a sheath of compressed hair (make of keratin), are never shed, and do not have branches. Antlers differ from horns in that they are made entirely of bone, are shed yearly, and can have many branches.

The Pronghorn is perhaps the fastest land mammal in the world. It can sprint at speeds up to 60 miles per hour and can run for many miles at 30 to 40 miles per hour. Why do they need to be so fast?

When they were evolving in North America, they were the prey of a cheetah-like cat (*Miracinonyx trumani*) that was a fast sprinter. This cheetah-like cat, which is now extinct, was a very close relative of the Mountain Lion and a slightly more distant relative of the modern Cheetah.

Pocket gophers are endemic rodents distinguished by their large fur-lined cheek pouches that are used to carry food (Figure 9.47b). The pockets extend back almost to their shoulders.

The Nearctic and Palearctic regions are not as rich in diversity as the regions just discussed because they do not contain tropical forests, which are the homes of most of the Earth's diverse life forms.

Palearctic

The Palearctic region is by far the largest of the biogeographical regions. It includes Europe, Africa north of the Sahara, most of the Middle East, and Asia except for most of India and the region south and east of the Himalayas. The high mountainous regions of northern India are included in the Palearctic as is most of China except for the lowland areas in the far south and southeast.

Figure 9.48a Endemic Animals of the Palearctic Region *(clockwise from top left) Snow Leopard [C.P.R. Environmental Education Centre, India]; Male Tufted Deer with fang-like tusks. [Heush]; Ibisbill; Wallcreeper [both by Mike Nelson/Birding Ecotours]; Dunnock, a member of the accentor family. [Smalljim]; Bearded Reedling [Kev Goodling]*

Figure 9.48b Endemic Animals of the Palearctic Region (clockwise from left) Takin
[Gregory Moine]; Red Panda, a not too distant cousin of the Giant Panda and raccoons.
[Greg Hume]; Saiga Antelope [Xavier Bayod]

There are 4 endemic bird families in the Palearctic, three of which have only one member. The sole members of their respective families are the Ibisbill of the high Himalayan plateau and surrounding area, the Wallcreeper, and the Bearded Reedling. A sparrow-like family called the accentors is also endemic to the region (Figure 9.48a).

The Palearctic is home to just two endemic mammal families, a group of small rodents called the mouse-like hamsters and the Red Panda, a cousin of the bears and raccoons and the only member of its family.

However, there are many mammal species that are endemic to the region (Figure 9.48). A few examples include the well known Giant Panda (p. 376), the Snow Leopard of the high Himalayas, Saiga Antelope, Takin, Eurasian Lynx, Wisent or European

Bison, several deer species of deer including the Tufted, Red, Roe, and Fallow Deer (Figure 9.48), and the Amur or Siberian Tiger (p. 352).

The Saiga is a small antelope of the vast steppes of central Eurasia. It is only about 0.7 m tall at the shoulders. In 1950 they were two million strong, but today they are critically endangered as only about 50,000 remain. They are endangered because they are being killed for the horns of the males, which are used for traditional Chinese medicine. This is the same medical belief that thinks powdered rhinoceros horn will cure most ailments. There seems to be no shortage of ignorance in the world.

Some species such as the European Bison, and Eurasian Lynx belong to the same genus as their Nearctic cousins. This close relationship is due to the

Figure 9.49 Spectacled Cormorant *This largest known cormorant was a poor flyer and lived around Bering Island, which is located east of the Kamchatka Peninsula. It was common when it was discovered by Georg Steller in 1741, but people colonized the island in 1826, and it was hunted intensely for its meat and feathers. The last ones were killed sometime around the early 1850s. [Joseph Wolf]*

interchange that occurred between Asia and North America during the Pleistocene when the two land masses were connected at various times. As ice built up on the land, the sea level dropped and exposed the Beringia land bridge.

Holarctic

The combined Palearctic and Nearctic is called the Holarctic. A couple of unusual animals that once lived in the Holarctic were the Steller's Sea Cow that we discussed earlier, and the Spectacled Cormorant (Figure 9.49). This bird was the largest known member of the cormorant family, but

following an all to familiar story line, they were hunted by humans and became extinct about a hundred years after their discovery.

There are a few families of mammals that are found only in this arctic region, which include the Walrus (the only member of its family), pikas, beavers, and a rodent family that includes the birch mice, jumping mice, and jerboas (Figure 9.50).

The additional bird families endemic to the Holarctic region are the loons, auks, waxwings, bushtits, larkspurs, Snow Bunting, and most of the kinglets (Figure 9.50).

There are, however, many familiar animal species that are endemic to the Holarctic such as the Polar Bear, Moose or Eurasian Elk, Caribou, the eider ducks, and many members of the grouse subfamily (Figure 9.50).

The Polar Bear is the world's largest terrestrial carnivore. Most of its range lies north of the Arctic Circle where it spends much of its time hunting seals

Figure 9.50a Holarctic Endemic *The Bohemian Waxwing is one of three species of waxwings found only in the Holarctic region. [Randen Pederson]*

Figure 9.50b Holarctic Endemics (clockwise from top left) Polar Bear and cub [Alan D.Wilson]; Razorbill, an auk that is the arctic equivalent of a penguin, an example of convergent evolution. [Stephen Burch]; Greater Sage Grouse [USFWS]; Spectacled Eider [Laura L. Whitehouse/USFWS]; Common Loon [John Picken]; American Pika [Alan D.Wilson]; Walrus [Joel Garlich-Miller/USFWS]; Caribou [Dave Menke/USFWS]

on the sea ice. Although Polar bears can swim well over 100 kilometers, they need sea ice to hunt and rest. As we pump carbon dioxide into the atmosphere and the planet warms, the sea ice is shrinking and the Polar Bears may be yet another casualty of human activities. They still being killed because some sick people enjoy killing majestic animals.

The auks are considered to be a northern bird that is analogous to the penguins of the Southern Hemisphere. They fill a niche in the north that is filled by the penguins in the south. One big difference is that penguins have evolved flipper-like wings that are used only for swimming while most auks retained the ability to fly.

One species of auk, the Great Auk, was flightless and was the auk most like the penguins. They were about 0.8 meters tall and lived in the North Atlantic region. They apparently used their short wings for swimming much as their penguin counter parts do today, another example of convergent evolution. The Razorbill is the closest living relative of the Great Auk

Unfortunately, this most unusual bird was hunted to extinction to provide specimens for museums and private collectors. In 1844, two collectors killed the last nesting pair. In a final act of indifference, one of the collectors stepped on the last egg. So today we are able to observe poorly mounted specimens in a few museums, but observing a live bird in the wild is no longer possible (Figure 9.51). Clearly, we have lost one of the unique birds of the world because of human greed and selfishness. We might note that all those people who participated in the extinction of the Great Auk are dead, but their greed has deprived future generations from the excitement of viewing this wonderful bird.

Figure 9.51 Great Auk A remarkable Arctic flightless bird that was hunted to extinction by collectors. The last pair was killed in 1844. [photo, Kelvingrove Art Gallery and Museum/Mike Pennington]

Antarctic

The last remaining biogeographical region is Antarctica. About 40 million years ago, it drifted southward over the South Pole and separated from Australia. It may have remained connected to South America for a few more million years. As it became isolated by the Antarctic Circumpolar Current, ice eventually covered the continent.

Before it became covered with ice, Antarctica must have been home to many unusual forms of life. Because of its location between South America and Australia, most of the land mammals must have been marsupials, and indeed, fossil marsupials have been found in Antarctica. Today the animal life is limited by the harsh conditions.

Penguins and seals are two of the animals that make their homes in Antarctica. The earliest fossil evidence of penguin ancestors is from soon after the extinction of the dinosaurs at the end of the Cretaceous Period. Fossil evidence suggests that penguins evolved near southern New Zealand and Antarctica, which were relatively close at the time.

The tallest penguin known is from the late Eocene of New Zealand, and it was about 1.7 meters tall. Today, the Emperor Penguin is the largest, standing about 1.2 meters tall. It is the only penguin that breeds and raises its young during the Antarctic winter (Figure 9.52).

Since penguins are well adapted to life in a cold climate, they would

Figure 9.52 Antarctic Region Two adult Emperor Penguins and a juvenile. The Emperor Penguin is about 1.2 meters tall, the world's largest. [Ian Duffy]

undoubtedly do well in the Arctic, but they have not managed to travel that far. However, as we have seen, they have managed to reach the Equator and are established on the Galapagos Islands, where they have evolved into a distinct species.

A Principle of Evolution

We have looked at a small sample of the interesting areas of the world, but the principle learned is obvious. Plants and animals that become established in isolated areas eventually evolve into distinct species. We would predict that many of the plants and animals living on an island or in an ancient lake are unique, having evolved to the point of becoming separate species.

Even if we knew nothing about Australian reptiles, our knowledge of the evolutionary process would enable us to predict that the vast majority of the reptiles would be unique to Australia. Reptiles were present before the breakup of Pangaea. The reptiles of Australia have been evolving in isolation for millions of years, and we would expect that many unusual and unique species inhabit the continent.

As we might expect by now, New Guinea, a large continental island, also has many endemic plants and animals. The number of native mammal species is approximately 180, and they belong to the same four orders as the Australian mammals. There are rodents (about 50 species), bats (about 70 species), marsupials (almost 60 species), and four species of spiny anteaters (egg-laying monotremes). About two-thirds of the mammals of New Guinea are endemic, and more than eighty percent of the marsupials and rodents are endemic. New Guinea and Australia were connected several times during the ice ages of the

Figure 9.53 Birds-of-Paradise (clockwise from top left) Raggiana Bird-of-paradise [markaharper1]; Greater Bird-of-paradise, who is upside down and whose white beak and eye can be seen at the lower right. [Andrea Lawardi]; Wilson's Bird-of-paradise [Dubi Shapiro]

Pleistocene when the ocean level was significantly lower, which explains why their mammals are so similar.

More than 50 percent of the lizards are endemic, and almost half of the breeding land birds are endemic. One of New Guinea's most famous bird families is the bird-of-paradise family. A few members of this bird group are found on adjacent islands, and three species live in nearby Australia.

These birds are so spectacularly beautiful that the early Spaniards felt they must be visitors from paradise (Figure 9.53). During the last half of the nineteenth century and on into the twentieth century, approximately two million of these birds were killed so their feathers could be used to decorate woman's hats and other garments. Apparently many humans had little respect for these *visitors from paradise*. Fortunately, the demand for feathers is no longer great enough to make feather hunting profitable, but deforestation might prove to be just as deadly to these beautiful birds.

Every island in the world that is isolated by more than a hundred

Figure 9.54a Island Endemics *(clockwise from top left) Juan Fernandez Firecrown, a hummingbird now restricted to Robinson Crusoe Island of the Juan Fernandez group. [Peter Hodum]; Broad-billed Tody, Hispaniola [Ron Knight]; Mayotte Sunbird, Mayotte and surrounding islets [Dubi Shapiro]; Marquesan Imperial Pigeon, Marquesas Islands [Samuel Etienne]; Black-billed Streamertail, a Jamaican hummingbird. [Dominique Sherony]; Yellow-shouldered Blackbird, Puerto Rico [Mike Morel/USFWS/Southeast]; Palawan Peacock-Pheasant, Palawan Island [Llimchiu]; Comoros Olive Pigeon, Mayotte and Comoros Islands [Dubi Shapiro]; (center) Yellow-billed Amazon, Jamaica [Wayne Sutherland]*

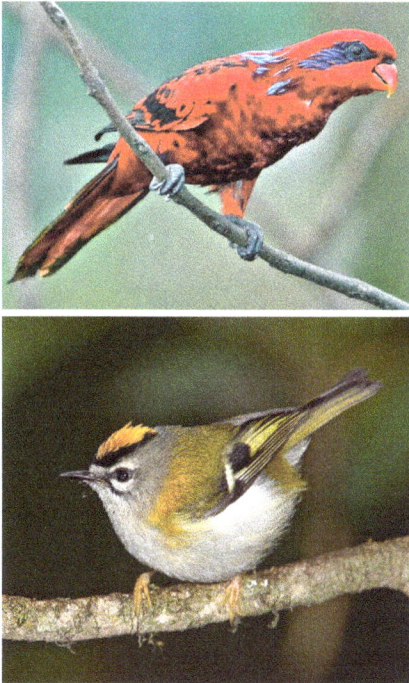

Figure 9.54b Island Endemics (top to bottom) Blue-streaked Lory, Babar and Tanimbar Islands, located west of New Guinea. [Doug Janson]; Madeira Firecrest, of Madeira Island located off the African coast and north of the Canary Islands. [Andrew Moon]

kilometers or so of water and is large enough to support life will likely have endemic animals living on it. Along with those we have already seen, Figure 9.54 shows just a small sampling of the many endemic birds of some of these islands.

The island of Tasmania is separated from continental Australia by about 160 km of ocean and approximately one-fifth of its vascular plants are endemic as are many of its animals.

Sri Lanka, a large island off the southeastern tip of India, boasts of about 550 land vertebrates, of which 25 percent are endemic. In addition, about 25 percent of its flowering plants are endemic. We see the same story

repeated, given enough time, isolated populations of organisms eventually evolve into new species.

Jamaica, the third largest Caribbean Island (Cuba and Hispaniola are larger) has 28 species of endemic birds. Only two native land mammals are known from the island, the Jamaican Hutia and the extinct Jamaican Rice Rat.

Almost every island in the Pacific has a few species of endemic birds. Unfortunately, as humans colonized the Pacific islands many birds were driven to extinction. It is estimated that about 1300 species of birds living on various Pacific islands became extinct after humans reached the islands. We have already seen several examples of this in New Zealand and the Hawaiian Islands. In all, about 10 per cent of the bird species in the entire world were wiped out after humans reached these islands during the past 3,500 years. Unfortunately, the same pattern is now being played out on the larger land masses. Especially in the Neotropical, Ethiopian, and Oriental regions, where the greatest biodiversity is found.

A recent tragedy in the neotropical region is the extinction of the world's largest woodpecker. The Imperial Woodpecker, last seen in 1956, was found in the mountainous coniferous forests of Mexico where they were killed for use in folk medicine and because loggers said they harmed trees. The local residents were given poison to kill the birds. It is hard to imagine two more idiotic reasons for wiping out such a spectacularly stunning bird. If they were still alive, Imperial Woodpeckers would be a valuable resource because birdwatchers from all over the world would gladly pay to see these amazing birds (Figure 9.55).

Figure 9.55 Imperial Woodpecker *It was once the world's largest woodpeker, but it is now extinct. [John Livzey Ridgway]*

The Earth Is an Island

Although the study of islands has enabled us to examine evolution in action, there is an extremely depressing side to our study. The destruction caused by the activities of humans is particularly evident on islands. Humans, however, are animals with the capacity to learn from experience and to adjust their actions accordingly. There are many important lessons to be learned by studying the consequences of upsetting the balance that was established on these islands.

The entire Earth can accurately be viewed as an island. Although we receive light energy from the Sun that supports life on this planet, the Earth is effectively isolated from the rest of the planets in the Solar System and the other stars in the galaxy. A brief study of the other planets in the Solar System is enough to reveal that the Earth is unique in its ability to support life. No other planet is capable of supporting the rich array of life that is so familiar to us on Earth. The life forms on Earth are found nowhere else in the Universe. They are all endemic to this planet.

Venus, our nearest planet, is much too hot to harbor life. Its carbon dioxide atmosphere produces an extreme greenhouse effect, making the planet's surface temperature hot enough to melt lead. Mars, our next nearest neighbor, is about one-fourth the size of Earth and has such a low surface gravity that most of its carbon dioxide atmosphere has been lost. The surface pressure is so low that liquid water cannot form. Water exists only as ice or water vapor. The other planets are no more hospitable than are our two nearest neighbors.

This planet that spawned life is the only refuge for the plants and animals that evolved here. How we care for the environment of this planet will determine not only the quality of life we have, but whether life as we know it will be able to survive. Indeed, humans are now able to destroy the very environmental conditions that make life possible. The destruction of the environment can come in many forms and can be upon us almost before we are aware of the danger.

Nuclear war is an obvious threat, but subtle dangers are even more serious. Some of the many serious problems facing the world today are water and air pollution, depletion of the ozone layer, toxic chemicals, deforestation, the killing of wildlife, the greenhouse effect (also called global warming), famine, and disease. A solution to any one of these problems, however, is not possible unless the human population is stabilized. A continually growing human population will eventually lead to unimaginable disasters.

The Polynesians populated many of the remote islands of the Pacific. Their story could hold a message for the people living on Island Earth today. What happens to a population that must live in a limited area? Although the Earth is larger than a Pacific island, it is still limited in area.

The islands were colonized by a small band of settlers who had sailed for many days. The only diseases present were those brought by the settlers, so life was relatively free of sickness. With a mild climate and an abundant supply of food and water, the population increased rapidly. The trees were cut down to provide areas for growing food and as the population grew, food became harder to find. The increasing population put more strain on the environment.

One such island that was colonized by the Polynesians is Rapa Nui (also called Easter Island). It is an island of about 55 square miles and is about 2,000 kilometers from the nearest island. Supported by fishing and farming, it is estimated that the population grew to 10 to 15 thousand at its peak. This successful population built the famous Easter Island statues.

When the Polynesians first arrived, the island was forested and supported a large colony of nesting sea birds. However, the increasing population eventually cut down the last trees and destroyed much of the vegetation. As wood for strong canoes was no longer available, fishing trips became more difficult and the local waters were over harvested. The sea birds were killed for food and the colony eventually disappeared. Many unique plants and animals were also exterminated.

The population could not migrate to alleviate their overpopulation problem, and periods of drought were especially cruel to the people whose numbers had pushed the environment beyond its sustainable limit. Wars and famine became more common, and some of the warring clans may have even practiced cannibalism. The civilization eventually crumbled, leaving many of the giant statues unfinished, a mute testimony to a once great civilization. When the European explorers arrived in the 1700s, the population had been reduced to about 4 thousand.

A similar story can be told of rabbits on Laysan, an atoll of the Hawaiian chain. Humans introduced the rabbits, and as their numbers increased dramatically, they reduced the island to a desert. Several endemic animal species were exterminated because of the devastation, including the Laysan rail and the Laysan honeycreeper.

An expedition from Hawaii's Bishop Museum in 1923 managed to kill many of the rabbits and the remaining ones starved to death. Today, some of the island vegetation has recovered from the rabbit plague, however, several unique island species of plants and animals have vanished forever.

What is the lesson we should learn? No area, not even the entire surface of planet Earth can support an increasing population. If we are not able to control the number of humans, nature will eventually limit the population to a number that the planet can support.

In nature, the population size of a group of organisms is controlled by the food supply or predators. Animals with no natural enemies are generally kept in check by disease or competition for food that results in starvation. Humans are animals with no natural enemies, and with modern medicine we have greatly reduced deaths that would normally result from diseases.

However, humans also can make decisions to control their numbers by less violent means than starvation and disease. Whether we take charge of our future or simply choose to let nature run its course remains to be seen.

The question is not if the human population will stop growing, for there is no doubt that the human population will eventually be stabilized. The question which we should be giving serious thought to is the means by which this stabilization will be accomplished. There are already way too many people in the world if most want to live like the average American or European. People who claim that the world does not have a population problem are ignoring the evidence.

As we saw in Chapter 1, having a stable population requires that, every couple have only two children to replace themselves. However, in countries where the population is growing rapidly, even if couples immediately began limiting their family size to two, the population would continue to increase for many years. The increase will occur because a large fraction of the population is still well below the childbearing age.

For example, Tanzania is a typical country with a very rapid growth rate of about 2.8 percent. As a result, more than half the population is under 20 years of age (Figure 9.56). If its growth rate remains at 2.8, the population of Tanzania, which was 45 million in 2012, will double in the next 26 years! Tanzania is not particularly unusual in this respect. There are many countries with similar population profiles. Even a well-developed country with many resources would find it difficult to provide for such a rapid increase.

The population profile of the United States looks very much like that of a country with a stable population (Figure 9.56). However, the US population as of 2014 was 319 million, and it was still growing at a rate of 0.77 percent. This growth rate adds about

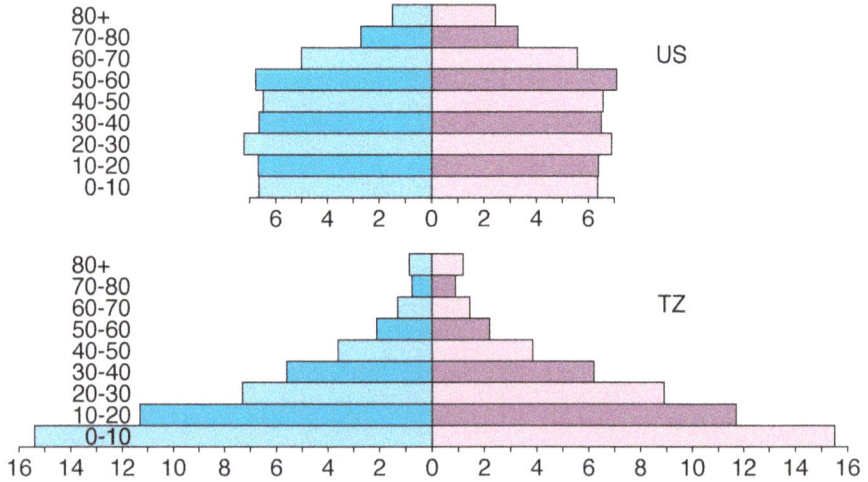

Figure 9.56 Population Pyramid *The two bar graphs show the percentage of people in each 10-year age group for the United States (top) and Tanzania (bottom). Males are in blue and females are in pink. For example, about 31 percent of Tanzanians are between 0 and 10 years of age, but the corresponding number is about 13 percent for the United States. [2014 US data from the CIA World Factbook; 2012 census by the Tanzanian National Bureau of Statistics]*

2.5 million people to its population each year. If that rate of growth continues, the US population will double to 640 million in 95 years.

The increase in the population of the United States is higher than all but about seven other countries in the world. Why is the US population growing so fast? There are three main reasons. First, the United States has the third largest population in the world. Only China and India have more people. Since the population is so large, even a modest *rate* of increase adds many people to the population.

A second reason is that the US has a higher birth rate than most developed countries like China, Japan, Russia, South Korea, and all the European countries except Ireland.

The third reason is immigration, which accounts for about one-third of the population increase.

Some people look at all the empty space in places like Nevada and say we have plenty of room, and our growing population is not a problem. The trouble with this view is that it fails to take into account what is required to supply the needs of a typical American.

United States has about 4.5 percent of the world's population, however, it uses more than 20 percent of the coal, oil, natural gas, wood, paper products, aluminum, copper, and many other materials. The United States uses more than 20 percent of the resources produced by the entire world! That means we are responsible for 20% of the rainforest destruction.

As can be seen from the data, a growing United States population has a much greater environmental impact on the world than the growing populations of many underdeveloped countries. However, every country needs to stabilize its population if the natural world is to survive.

Since the human population must ultimately stop increasing. A question of critical importance to future generations is how badly we will damage the environment before the population is stabilized. As the world's most ingenious animals, humans can destroy the very environment that supports our needs. The human population is already too large to save many of the plant and animal species that are threatened with extinction.

As forests are destroyed, the animals that make them their home cannot survive. It has taken millions of years for the plants and animals of the rainforests to evolve. If the trees are cut down and the soil erodes away, the area will not recover. Will our descendants learn about rhinoceroses and elephants from picture books or will these wonderful creatures survive beyond the present population crisis?

Many environmental problems are like a train racing toward an obstacle. Unless the brakes are applied early enough, it will not be possible to avoid a collision. It is very possible that problems like greenhouse warming and overpopulation will be difficult to deal with even if we were to immediately stop contributing to the problem.

When we decide not to respond to a potential problem, the consequences could be serious. Let's explore this dilemma using a hypothetical problem.

Suppose you notice a problem with your body and get an exam. One of your examiners says you have a serious disease and will die unless you are treated soon. You are told that the treatment will be involved and painful. You get a second opinion. This examiner says you are fine, and you do not need anything. With conflicting opinions, what should you do?

You only have two choices, you can either get the treatment or you could decline. However, these two choices

352

lead to four possible outcomes, and the outcomes are not all equal.

First, if you have the treatment, but you did not have the disease, you would have gone through the treatment for nothing, costing you pain, and inconvenience. As a second outcome, you could decide to have the treatment and you could have the disease, so your life would be saved.

The third possibility is that you decide not to get the treatment, and you do not have the disease so everything is fine. The fourth choice is that you decide not to get the treatment, but you did have the disease so you die.

Now its time to make a decision. Getting the treatment or doing nothing are not equivalent decisions! Treating your disease would be unpleasant and possibly even unnecessary, but it could save your life. On the other hand, doing nothing could be fatal.

Overpopulation and some of the problems associated with it like global warming are problems of this nature.

Since doing nothing might welcome disaster, the intelligent response would be to assume there is a problem and act accordingly, even if you are not sure. If we are wrong, solving the problem will be an inconvenience, but not a disaster.

If you are in the gas or oil business, you might claim global warming is a hoax since replacing coal and oil with solar might hurt you financially.

If the consequences of doing nothing about things like global warming and overpopulation are relatively long term, a selfish response would be to do nothing, and let our children and grandchildren deal with these growing problems. However, if a problem is serious, a delay could have fatal consequences. Let us hope we have more vision and compassion for ourselves, for our children's future, and for the Earth and all its plants and animals.

In the near future, will there be room for Siberian Tigers to roam free?

Chapter 10: Genetic Ancestry

Natural Selection

We have investigated much of the evidence that supports evolution, and have concluded that all present life forms are modified descendants of previously existing life forms. The story of change is clearly written in the fossil record, and the unique life forms found on isolated islands attest to the power of the evolutionary process.

All life forms share certain chemical similarities with the more closely related organisms being the most similar. Even the way we classify organisms based on their anatomical similarities speaks of the closeness of their relationship. Evidence from widely different sources tells the same story of change. There can be little doubt that the vast numbers of life forms on this planet are the result of evolution over hundreds of millions of years.

Although the public generally perceives Darwin to be the father of evolution, as we have seen, the idea that evolution had taken place was formulated before Darwin's principle of natural selection. Geologists recognized the distinct ordering of the fossils and were able to infer their relative ages well before radiometric dating established absolute ages. Long before Darwin, many scientists realized that life had changed dramatically with time. For example, the life forms of the Cambrian were very different from those of later periods. There were no fish, no reptiles, and no mammals in the Cambrian just to name a few. Sharks, which shed their teeth and produced abundant fossils in later periods, were not found in the earlier periods. Other examples are way too numerous to reiterate here. The clear conclusion from even a cursory view of the fossil record is that life forms have changed dramatically through the ages.

Darwin observed this evidence, but his contribution to evolution was the formulation of the principle called natural selection, the mechanism that he believed was responsible for the observed transformations. Natural selection works on the idea that those traits that help an organism live to the reproductive age will be passed on to the next generation. On the other hand, it is less likely that those traits that create a disadvantage will be passed on to the next generation.

Darwin was unaware of the genetic nature of these traits, but we now know that changes in these traits are caused by mutations in the strands of DNA, the information storage system in each cell that contains the recipe for making the organism. Changes in the DNA strands result in the production of different proteins and consequently, changes in the structure and function of various parts of the individual.

Every population of organisms contains a great deal of genetic variability because there can be several different recipes for any given type of protein. This diversity can be measured directly by comparing proteins from many members of a given population. One way of studying this variability is to measure what fraction of the population has two different genes or protein recipes for the same type of protein. Remember that chromosomes come in pairs, so each of us can have two different recipes for a given protein.

An individual who has two different genes is said to be *heterozygous* at that gene position. For example, if you have the gene for blue eyes and the gene for brown eyes, you are said to be heterozygous for the eye color gene. Studies of vertebrates have indicated that at any given gene position, more than 6 percent of the population is heterozygous. For invertebrates and plants, the percentages are about 13 and 17 respectively. Since there are tens of thousands of different types of genes, there is a large variation in the genetic makeup of a given population.

Although Darwin was unaware of the genetic source of variability, he deduced that populations have a great potential for change by observing domesticated plants and animals. The great variation we see today in the different breeds of dogs for example, demonstrates that wild populations contain enormous potential for change.

While some new mutations have undoubtedly been incorporated in many of the current breeds, all dogs are descendants of wolves that were captured and selectively bred by early man. The wolf was domesticated perhaps 20,000 years ago, but the modern varieties of dogs we see today are likely the result of only about 500 years of selective breeding. In a natural setting, a changing environment could have caused similar changes through natural selection.

Punctuated Evolution

There has been a great deal of discussion concerning the rate at which evolution proceeds. Some scientists, often called gradualists, feel that the evolutionary process can best be described as a gradual transformation of one form into another, taking place gradually over thousands or even millions of years. This model of evolution is known as gradualism (Figure 10.1).

Others feel the fossil record demonstrates that evolution is much more sporadic. Proponents of this sporadic process are often called punctuationists. They believe that most species remain relatively stable with little change over relatively long periods of time. When changes do occur, they happen quickly during geologically short time periods. Therefore, the fossil record is punctuated with sudden changes that occur during these short periods of rapid evolution. This idea of relatively

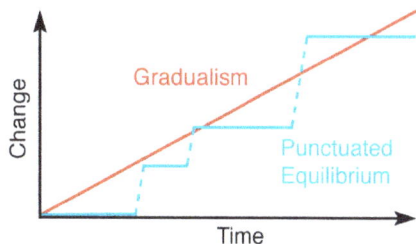

Figure 10.1 Punctuated Equilibrium Versus Gradualism Gradualists believe most changes occur gradually over time (red line). Punctuationists believe most species remain fairly stable and then change over a relatively short time (blue line). [JCB]

short intervals of evolution is usually described as punctuated equilibrium (Figure 10.1). Although both the gradualistic and punctuated points of view must have some validity in different situations, much discussion has centered about which one is more important.

Since the fossil record itself is not continuous in time and space, differentiating between the gradualist and the punctuated points of view is not as easy as it may sound. For example, 1000 to 10,000 generations of mollusks may represent an instant in the rock record. While many volumes have been written on this debate, it should be pointed out that the argument is over the rate at which evolutionary changes occur. Neither side doubts that the changes did occur.

Neutral Evolution

If we incorporate Darwin's idea of natural selection with our current knowledge of genetics, we would say that individuals that carry an advantageous mutation are more likely to survive and reproduce, thereby increasing the frequency of that advantageous mutation within the population.

A point mutation in a gene (a base change in an organism's DNA) can result in an amino acid change in the corresponding protein molecule. Although each amino acid has its own unique chemistry, many have similar properties. Replacing one water-loving amino acid with another may have little effect on how the protein molecule functions. A mutation that does not confer an advantage or disadvantage to the individual carrying it is called a neutral mutation. In our discussion, we are only referring to mutations in the sperm and egg cells that can then be passed on to the offspring. That is, mutations that occur in the gametes.

Many scientists believe that a significant number of all mutations are selectively neutral or nearly so. As we will see, there is some observational evidence to support this view. There have been some vigorous discussions among scientists about whether neutral mutations or advantageous mutations (natural selection) are more important in evolution.

Even though it carries no advantage, a neutral mutation has a certain probability of eventually spreading throughout the population. If a gene spreads throughout the population so that every individual has two copies of the mutated gene we say that the gene has taken over the population. It can be shown that it is possible for a mutation to take over a population even if it is slightly detrimental. Although we will not go into the mathematical details, we will attempt to understand qualitatively how a mutated gene can become established in a population.

Suppose a new mutation appears in one member of a certain population. On average, we expect that one half of his or her offspring will inherit that mutation, but since chance is involved, more than half could end up with the mutation. If the mutation is neutral, such that all the individuals of the population have an equal chance of surviving, then the number of mutated offspring that survive to bear young will also be a matter of chance. As can be seen, chance is involved at each stage of the process, and flipping an appropriate number of coins could simulate the events, or they could be simulated with a computer program that randomly picks every mating and each surviving offspring.

After several generations, a significant fraction of the population could carry the mutation. For a stable population, the probability that a mutated protein recipe will spread throughout a population (take over the population) is inversely proportional to the number of individuals in the population.

If we let N be the number of individuals in the population (more accurately, the size of the breeding population), the probability that two copies of a given mutation will eventually be carried by every individual in the population is just $1/(2N)$. The 2 comes from the fact that each individual carries a pair of genes for a given protein recipe, so the number of genes that must carry the mutation is twice the size of the population. This result seems correct if we reflect on it for a moment. The larger the population, the smaller will be the probability that a given mutation gets established in the population.

The average time required for a mutation to eventually take over the population can also be calculated with a computer program. These computer calculations show that if a mutation eventually takes over the population, the average length of time for it to happen is $(3.5N)$ generations. To be more precise, the calculations show

that 50 percent of the mutations that eventually take over the population will do so within $(3.5N)$ generations. This result also seems reasonable since we would expect that the time required for a mutation to take over the population should depend on the size of the population.

How quickly do new mutations take over a population? Let g represent the average number of new mutations in each gamete (egg or sperm). Then, the total number of mutations introduced into the population at each new generation will be $(2Ng)$. The 2 in the expression is because each offspring is formed by the union of one egg and one sperm, both of which could contain new mutations.

To calculate the total number of these introduced mutations that eventually take over the population, we merely multiply the number of mutations that are introduced into the population $(2Ng)$ by the probability that one of these mutations will be established $(1/2N)$. This product is simply g. Therefore, the average number of mutations that eventually take over the population each generation is just g.

At first sight this result may seem incorrect since it does not depend on the size of the population. However, although the probability of a new mutation becoming established goes down with the size of the population, the number of new mutations introduced each generation goes up with the size of the population. Therefore, the size of the population cancels out when we take the product of these two probabilities.

What would happen if new mutations were not continually introduced into a population or if no mutations became fixed in the population? For example, if all mutations were lethal, then new mutations could not spread throughout the population. (In reality, if all mutations were lethal, few offspring would survive since most individuals possess many mutated genes.) If mutations were not being continually established in a given population, the population would eventually become homogeneous. In other words, except for the differences between males and females, every individual would be a clone. There would be no individual variations in the population.

This homogenization would happen since a few chromosomes are lost each generation. In a stable population, each mating pair will have an average of two surviving offspring. It is highly unlikely that the two offspring will not share some of the same chromosomes. Boys get their Y chromosome from their father so two brothers will have copies of the same Y chromosome, and since girls get one X chromosome from their father (who only has one X chromosome), two sisters will have a copy of the same X chromosome.

The probability of a couple having a boy and a girl that are not genetically related to each other (do not share any copies of the same chromosome) is about one in 70 trillion, so most siblings do share at least one chromosome. If the two offsprings share one or more chromosomes from one of the parents, its homologous pair will not be passed on to the next generation. Eventually, all the homologous pairs will become identical and everyone in the population will have identical sets of chromosomes, except for the X and Y chromosomes. Except for differences in sex, all the individuals in the population would have identical traits.

The computer program that calculated how long it would take for a gene to spread throughout a population will also calculate how long it would take for the population to become homogeneous without mutations. A homogeneous population is one in which one gene of each type (or one chromosome of each type) has taken over the population. In other words, the homogeneous problem is identical to the gene take over problem.

Therefore, there is a 50 percent chance that a population would become homogeneous within about (3.5N) generations if no new mutations were introduced into the population. The observation that populations are not homogeneous, is evidence that mutations are being continually introduced into the various populations of organisms that inhabit the Earth. Studies have determined that about 30 mutations are present in every sperm cell.

Human Evolution

Since we know how mutations cause evolutionary changes, we can ask if mutations are frequent enough to account for the differences in species that we observe. Some studies have been done in this area, and we have an approximate idea of the rate at which DNA mutations take place. The measured mutation rate in the protein coding DNA in humans is at least 30. In other words, most people in the world grew from a fertilized egg containing at least 30 newly mutated genes. The total number of mutations per person is much greater than this since many mutations occur in the noncoding region of the DNA. These mutated genes will be passed on to their children.

Recall that several different codons specify the same amino acid. Because of this redundancy in the genetic code, it takes about two point mutations to change an amino acid. So, each person has about 15 amino acid changes due to their 30 mutations.

Is this mutation rate high enough to produce the observed variety of organisms we see in the world? To gain some insight into this question, we can look at some protein differences. As we have seen, the proteins of different species are slightly different. The alpha hemoglobin molecules of man and horse differ by 18 out of 146 amino acids (about a 12 percent difference) while man and chimpanzee have identical hemoglobin molecules. Apparently a few differences in the 25,000 or so genes of these animals is enough to produce the observed variations.

We will now do a rough calculation to see if the mutation rate is high enough to have produced the observed differences. The genomes of humans and chimpanzees have been sequenced and the differences in their protein coding DNA are found to be 1.2 percent. To get an idea of just what this number means, the DNA of two average humans differs by about 0.1 percent. The differences we see in people are due to this 0.1 percent difference in their DNA.

Biologists use the term *genotype* to describe the genetic makeup of an organism, and they use the term *phenotype* to describe the physical appearance of the organism. The phenotype of an organism is strongly influenced by its genetic makeup, but environment can also affect the phenotype of the individual. For example, you can change your physical appearance by exercise, or your skin tone by Sun exposure.

As we just saw, the proteins of chimpanzees and humans differ by about 1.2 percent. Assuming an average protein contains about 500 amino acids, 1.2 percent gives an average of about 6 amino acid differences for each protein. Since there are about 25,000 protein producing genes in humans and chimpanzees, there must be about 150,000 amino acid differences in all the proteins of these two organisms. Assuming their mutation rate is about the same, humans and chimpanzees have each accumulated about 75,000 amino acid changes since they diverged from their common ancestor some 6 or 7 million years ago. This rate calculates to be about one amino acid change every 90 years. If we assume an average generation is about 20 years long, the accumulated mutation rate must have been about one amino acid change every 4 or 5 generations.

As we saw above, the mutation rate in humans produces about 30 mutations (15 amino acid changes) per person. For neutral mutations, the number of mutations that become fixed in a population per generation is equal to the mutation rate per gamete. Therefore, if all were neutral we would expect 30 mutations (or about 20 amino acid changes) to be fixed every generation.

This number is much higher than the one amino acid change every 4 or 5 generations that was just estimated from the protein differences between humans and chimpanzees. Of course, all mutations are not neutral. The important ones that differentiate us from chimpanzees are definitely not neutral. However, these numbers show that the mutation rate is more than sufficient to explain the observed differences.

We know that the human population has evolved substantially during the past several million years. What will future humans look like? Science fiction stories often depict the humans of the future as having larger heads (presumably smarter) than present day humans. Will this be the case? Is the human population still evolving? Of course, we cannot predict the future, but there are good reasons to suspect that human evolution will be rather sluggish in the future.

As we mentioned above, if a neutral mutation eventually spreads throughout a population, it will take about 3N generations. Since a human generation is about 20 years long, and there are over 7 billion people in the world, it will take about 400 billion years for a neutral mutation to spread throughout the human population. Of course, an advantageous mutation could spread throughout the population more quickly, but with a population of over 7 billion, the time for a mutation to be fixed is still quite a long time.

Now we are able to understand why many scientists believe that the evolutionary process molds small isolated populations more quickly. Since the time for any mutation to be fixed in a population is proportional to the size of the population, new mutations will spread more quickly in small populations. Therefore, evolution will occur at a faster rate in a small population than in a large population. A small group of individuals isolated from the main population may evolve relatively quickly into a new species.

The Molecular Clock

Although the rate at which mutations become established in a population is not a completely random process, the fact that many mutations are neutral has led some scientists to speculate that the rate of change of the amino acids in a protein can be used as a crude molecular clock. Once calibrated with the aid of the fossil record, this molecular clock could be used to obtain the approximate times when two species diverged.

To understand how this method works we will consider the alpha chain of hemoglobin. The first step in the process is to determine the number of amino acid differences on the chain for several vertebrates. Next, we plot the number of amino acid differences against the time when these animals diverged (as determined from the fossil record). We see that the data lies very close to a straight line (Figure 10.2).

The plot tells us that for the alpha chain of hemoglobin, there have been 90 amino acid substitutions in about 350 million years. This number is about one amino acid difference for every 4 million year interval since the two species diverged.

We should emphasize that because the process is random, we do not expect exactly one amino acid difference for each 4 million year interval since the species diverged. For example, the alpha hemoglobin molecules of humans and cows differ by 17 amino acids, which implies 68 million years since the two species diverged. However, because of the random nature of the process, we

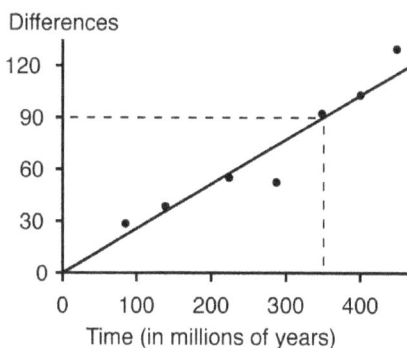

Figure 10.2 The Molecular Clock The graph shows the amino acid differences in alpha hemoglobin for several vertebrates. The times are from the fossil record. We see from the graph, 90 substitutions occurred in 350 million years, an average of about one substitution every 4 million years. [data: Motoo Kimura, see credits page 415]

would not be surprised if the time were as short as 40 million years or as long as 100 million years.

Since the problem is purely random, it could be simulated by throwing several coins and assuming an amino acid changes every time all the coins come up heads. You cannot predict exactly when the group of coins will come up heads, even though the average time may be 4 million years.

Another process that could affect the time is rapid change due to natural selection. When a significant evolutionary change occurs, such as the development of flight by birds, natural selection might cause other changes relatively quickly. There is evidence that the molecular clock does speed up at times and slow down at other times. However, even though it does not run at an entirely constant rate, the molecular clock is valuable as an approximate timekeeping device.

Chromosomal Aberrations

As we expect, the chromosomes of the great apes (and humans) are quite similar. However, studies of the banding patterns on their chromosomes indicate that several rearrangements have taken place during their evolution. Some scientists have speculated that chromosomal rearrangements (also known as chromosomal aberrations) play an important role in the evolution of species. Although quite different from point mutations, chromosomal aberrations are usually classified as mutations. About one in 500 humans is observed to have a chromosomal rearrangement.

Several types of chromosomal aberrations have been identified that produce changes in the chromosome structure. An extremely important chromosomal aberration is called *duplication*. Duplications occur when a segment of the chromosome is repeated. For example, abCDEfgh might become abCDECDEfgh. Duplications produce duplicates of gene recipes. The copied gene and the original gene can then evolve independently, eventually producing two different but related genes. We will look at an example of this in a moment when we discuss the globin family of genes.

Sometimes a small portion of a chromosome is moved to another position in the same chromosome or to a different chromosome. Such a move is known as a *translocation*. Although translocations may not greatly affect the protein recipes, the rearrangement may affect the way the genes are expressed.

An *inversion* is a segment of DNA that has been inverted within the chromosome. Inversions are severe changes since the protein recipes on the inverted segment will be read in the reverse order and will generally bear no resemblance to the original recipe. For example, the original chromosome might have been abCDEFgh, but after the inversion the chromosome might become abFEDCgh.

There are several chromosomal aberrations that have been detected when comparing human and chimp chromosomes. For example, human chromosome #4 and its analogous chimp chromosome, show an inversion around the centromere (Figure 10.3).

It is possible for two chromosomes to fuse and form one chromosome. This happened in the human ancestral line after chimpanzees and humans diverged some 5 to 8 million years ago. Humans only have 23 pairs of chromosomes, but the other great apes have 24. Chromosome #2 in humans was formed by the *fusion* of two chromosomes that are still single in the other great apes (Figure 10.3). These two chimpanzee chromosomes have nearly identical DNA sequences to those in human chromosome #2. A chromosomal fusion is present in about one in a thousand babies.

In some cases a person ends up with an extra chromosome and has some sort of abnormality like Down syndrome (also called trisomy 21). Most people with 45 chromosomes, however, are normal.

The fusion of two chromosomes has been observed in many other mammals as well. For example, the House Mouse

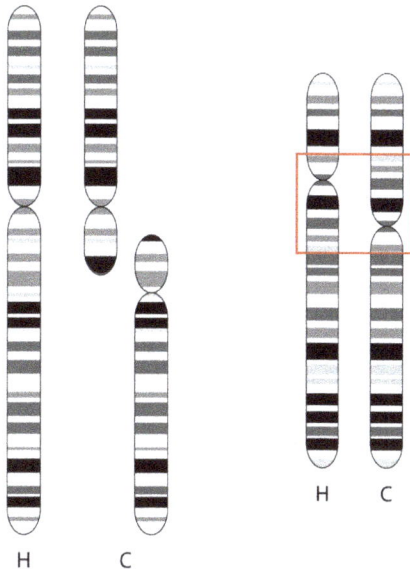

Figure 10.3 Chromosome Fusion Human chromosome #2 (far left) is compared with two similar chromosomes in chimpanzee. The human chromosome was produced by the fusion of these two chromosomes after the human chimpanzee split. The banding is produced by staining the chromosome (called G-banding). Regions of densely packed DNA stain darker than loosely packed regions. The centromeres are located at the constrictions. The #4 chromosomes of human and chimp (far right) show an obvious inversion around the centromere (red box). [drawn using data from: Ensembl release 75: Feb 2014]

normally has 40 pairs of chromosomes. However, on the island of Madeira, about 650 km off the coast of Morocco, a study found that six isolated populations have evolved significantly. They are distinct because they have different numbers of chromosomes due to fusion. One group had only 24 pairs, meaning there must have had 8 fusion events in its history. The time for this evolution to take place must have been less than 600 years since the island was discovered in 1419, and the mice were carried to the island by ships.

Chromosomal aberrations have played an important role in the evolution of humans. Although the DNA of chimpanzees and humans are only 1.2 percent different, there are several differences in the arrangements and duplications of the genes on the chromosomes. When the location of genes on the chromosomes of humans and chimpanzees are compared they are found to be about 2.5 percent different. Although we do not currently know how these rearrangements affect the development of the organisms, they are undoubtedly more important than the amino acid difference in the structural proteins. Chromosomal rearrangements can affect the development of an organism.

Normal meiosis produces gamete cells (like eggs or sperms) with half the number of chromosomes found in the normal body cells, but sometimes meiosis does not function properly, and the number of chromosomes is altered. During meiosis the members of a homologous pair are supposed to separate, but occasionally a pair does not separate, and gametes are formed with either one extra chromosome or one less chromosome (conditions known as *aneuploidy*).

Normal organisms generally have two haploid sets of chromosomes, one set from the father and one set from the mother. A breakdown of normal meiosis, however, can occasionally produce organisms with three or more haploid sets of chromosomes (conditions known as *polyploidy*). Chromosome duplication appears to play an important role in evolution. A new species can be formed in one generation if the number of chromosomes are doubled. Especially in plants, this condition can lead to new organisms with twice the usual number of chromosomes.

Since polyploids are generally unable to fertilize with the normal population, they represent a new species. This process of species formation appears to be more common in plants, but it is also found in a few groups of invertebrates. Approximately one-third of the flowering plant species appear to have originated from some form of polyploidy. Large numbers of domestic plants have been produced by polyploidy.

The proteins and RNA molecules that regulate the growth and development of an organism are much more difficult to study than structural proteins. Consequently, our knowledge of these important molecules is not as extensive.

As we saw earlier, some DNA codes for RNA molecules that play important roles in regulating the development of an organism. It is very probable that the most significant differences between organisms are differences in the regulatory genes rather than differences in structural genes.

The human head more closely resembles the head of a baby chimpanzee than it does the head of an adult chimpanzee. In other words, the human head seems to be similar to a chimpanzee head that did not continue to develop, but retained its fetal pattern into adulthood. When the rates of development of different parts of the body are altered by evolution we refer to it as *heterochrony*, which means *different timing*. It is possible that the human head is the result of mutations in the regulatory genes or some type of chromosomal rearrangement that affected the relative rate at which the head develops.

There are many other evolutionary examples that appear to be explained by heterochrony. For example, the simplest members of the chordate phylum share many features with the larval forms of the echinoderms. Scientists have speculated that chordates evolved from an echinoderm larvae that retained its larval form even after maturing sexually. Heterochrony is also observed in several species of salamanders. These salamanders mature sexually while retaining their larval forms. The mudpuppies are examples. This type of heterochrony that involves differences in the growth rates of somatic cells and germ cells is called *neoteny*.

Plant growth occurs near the middle of the stem in the early stages of seed germination. As the seedling grows larger, however, growth is restricted to the tip of the shoot and roots. In what appears to be another example of neoteny, grasses evolved by retaining the ability to grow near the base of the stem instead of the tip as other plants do. This type of growth has a distinct advantage when grazing mammals like deer and antelope are continually nibbling off the tips of the plants.

Regulatory genes may also evolve to inhibit the expression of genes that are still present in certain organisms. For example, whales may still have genes that are able to produce hind legs, but they may have regulatory genes that suppress the expression of these hind-leg genes. Humans may have the genes for body hair like the other apes, but certain regulatory genes may be responsible for suppressing these body-hair genes.

Chimpanzees and gorillas have 13 sets of ribs. Most humans have 12 sets of ribs, but 8 percent of the population has 13 sets. We all probably have the structural genes to make 13 sets of ribs, but regulatory genes may suppress the production of set number 13 in most individuals.

Evolution of Genes

A gene evolves when one or more of the nucleotides in its DNA sequence change. Some of these DNA changes will cause a different amino acid to be substituted into the protein molecule. By analyzing these changes, we can independently check the evolutionary trees determined by other means such as the study of fossils and the comparison of anatomical structures. In addition, protein analysis can sometimes lead to the discovery of new information about the relationships between organisms.

One protein molecule that has been studied extensively is hemoglobin. As we have seen, human hemoglobin is a complex structure composed of four protein molecules (two alpha chains and two beta chains). The sequences of amino acids for the alpha and beta chains are similar enough to suggest that they evolved from a common ancestor protein that was duplicated. After the duplication, the two protein recipes evolved independently. Slightly more than 40 percent of the amino acids in the alpha and beta chains of human hemoglobin are the same, much too high to be a mere coincidence.

To understand how a certain type of protein is analyzed and compared, we will consider a hypothetical example of a protein molecule composed of 10 amino acids. Suppose five of these proteins are analyzed from 5 different species of organisms. For example, we might analyze the alpha chain of hemoglobin from a human, a horse, a dog, a cat, and a rabbit. Further, suppose that the five proteins are found to have the sequences of amino acids as shown in Table 10.1.

The letters A, B, C, etc., represent one of the 20 different amino acids. For example, A might stand for alanine, B might represent lysine, etc. In a real protein, the chain of amino acids would be much longer than our hypothetical ten chain molecule. For example, the alpha chain (molecule) of hemoglobin is 141 amino acids long.

To analyze the proteins, we compare pairs of amino acids and construct a table listing the number of amino acids that are different in each given pair. For example, if we examine proteins #1 and #2 we see that they differ by just 2 amino acids. Protein #1 has P and G in the locations where #2 has D and N. Proteins #3 and #2 differ by 6 amino acids, etc. Table 10.2 summarizes the comparisons.

As can be seen from the table, proteins #1 and #2 are very closely related (red), and proteins #4 and #5 are very closely related (blue). Next, we see that protein #3 is more closely related to #4 and #5 than it is to #1 and #2 (violet). The simplest family tree

#1:	A	B	C	P	M	F	G	B	I	J
#2:	A	B	C	D	M	F	N	B	I	J
#3:	A	O	C	D	E	F	G	H	O	P
#4:	A	B	N	D	E	K	G	H	O	J
#5:	F	B	C	D	E	K	G	H	O	J

Table 10.1 Amino Acid Sequences of Five Hypothetical Proteins Note that proteins #1 and #2 differ in two locations (PG in #1 versus DN in #2), and #4 and #5 differ in two locations (AN in #4 versus FC in #5).

	#1	#2	#3	#4	#5
#1	0				
#2	2	0			
#3	6	6	0		
#4	6	6	4	0	
#5	6	6	4	2	0

Table 10.2 Amino Acid Differences *The table shows the number of amino acid differences between each pair of proteins for the five hypothetical proteins listed in Table 10.1.*

that can be constructed from the data is shown in Figure 10.4.

This family tree is the simplest tree that can be constructed using the data given. Although others are possible, it is the most probable. The study of amino acid sequences does not tell us the number of mutations that have occurred since some mutations do not result in an amino acid change. Also, some amino acid changes require two base mutations, and a few require three mutations. Since the amino acid changes do not accurately reflect the number of mutations, a more accurate picture of the evolution of a protein can

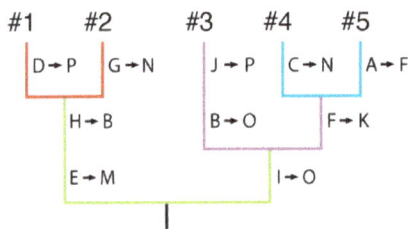

Figure 10.4 Family Tree *This tree was constructed using the data from the five hypothetical proteins in Table 10.2. #1 and #2 are closely related (red), and #4 and #5 are about as closely related to each other (blue). #3 is more distantly related to #4 and #5 (violet). #1 and #2 are even more distantly related to #3, #4, and #5 (green). The mutations are listed to the right at the approximate positions where they must have occurred. For example, at the lower right, O replaces I in #3, #4, and #5. [JCB]*

be obtained by sequencing the DNA that codes for the protein. Several studies of this type have been carried out, and we can expect to learn a great deal in the future as the volume of sequencing data grows.

One widely studied protein is cytochrome c, a molecule that plays an important role in the production of energy in the cell. Although there is a slight variation in its length, a typical molecule is approximately 100 amino acids long. In the organisms that have been studied, approximately three-fourths of the amino acids have been found to vary. Apparently only one-fourth of the amino acids are critical if the molecule is to function correctly. Family trees have been constructed from the cytochrome c data, and as expected, they follow the general pattern of family trees that are constructed using data collected from the fossil record and anatomical studies.

Similar family trees have been constructed using other proteins. In Table 10.3 we have compared the beta hemoglobin chains of six different animals. The data can be used to construct a family tree, that is shown in Figure 10.5. The length of a given branch in the tree is proportional to the average of the numbers in the appropriate row. For example, the average of the four numbers in the chicken row is about 47, and the average of the shark row is about 92. Therefore, the length of the shark branch is almost twice as long as the chicken branch.

A family tree constructed from the sequencing of a single protein might not agree exactly with a family tree based another protein or on DNA data. Slight disagreements may be because proteins often differ by only a few amino acids, and random fluctuations can be significant. For example, the beta hemoglobin chains of humans and brown lemurs differ by 25 amino acids, but gorillas and brown lemurs differ by only 22 amino acids. Do these numbers mean the gorilla is more closely related to the brown lemur than we are? No, because the difference is well within the random fluctuations expected. If we were to compare another protein, we might find that the brown lemur protein is a little more similar to the human protein than it is to the gorilla protein. To avoid incorrect conclusions, small numbers must be interpreted with statistical methods. A more accurate family tree would result if it were based on an average of several different proteins instead of just one.

Incidentally, the family tree shown was based on how similar the beta hemoglobin molecules are in these six animals, but it exactly agrees with the tree that we would construct based on the fossil record. There are 945 possible trees that can be constructed using six organisms. Figure 10.6 shows a few other possible trees for six organisms. The chance of getting the one shown is one out of 945. A remarkable result if the organisms were not related as indicated by the tree.

If we increased the number of organisms to just 10, there would be 34,459,425 possible trees. Yet we can say with confidence that the tree would agree with the one derived from the fossil record.

If we used a molecule like cytochrome c, which is found in plants, animals, and many one-celled organisms, we could easily generate a tree of 20 organisms. There are over 200,000,000,000,000,000,000 (200 quadrillion) possible 20 organism trees, and yet that tree based on cytochrome c would agree with the tree we have constructed from the fossil record and taxonomy. The obvious conclusion must be that these organisms are all

	human	monkey	cow	platypus	chicken	shark
human	0					
monkey	8	0				
cow	25	27	0			
platypus	34	35	41	0		
chicken	45	44	55	42	0	
shark	94	91	94	93	89	0

Table 10.3 Amino Acid Differences for Beta Hemoglobin Beta hemoglobin is 146 amino acids long. Statistically, all the numbers in a given row are the same. For example, the fifth row data does not say a chicken is more closely related to a human than a cow. [data: Richard E.Dickerson and Irving Giles (1983), credits p 415]

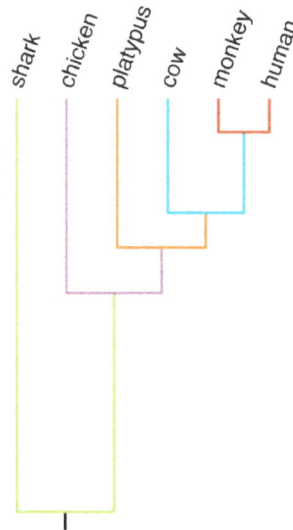

Figure 10.5 Beta Hemoglobin Family Tree This tree was constructed using the data in Table 10.3. The colors correspond to those in the table. As expected, humans and monkeys are more closely related to each other than either is to a cow. Humans, monkeys and cows are placental mammals and are more closely related than any one is to a monotreme (Platypus). Mammals and birds are more closely related to each other than they are to a cartilaginous fish (shark). [JCB]

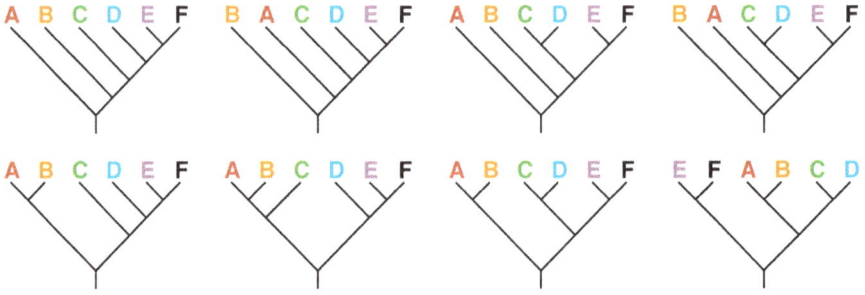

Figure 10.6 Possible Trees *A few of the 945 possible 6 organism family trees. The tree at the upper left is equivalent to the beta hemoglobin tree in the last figure. [JCB]*

related, and what we see is the result of millions of years of evolution.

In vertebrates, hemoglobin is the structure in red blood cells that carries oxygen from the lungs to the cells of the body and returns carbon dioxide to the lungs. Hemoglobin is formed from two alpha protein chains and two beta protein chains. As we have observed previously, the similarities between the beta and alpha chains are too great to be explained by pure chance. Undoubtedly, the two chains evolved from a common ancestor protein. Because of their similarity to myoglobin, we conclude that all three evolved from an ancestor protein.

Myoglobin is a single protein molecule that is used to store oxygen in cells. This globular protein molecule acts as a holder for a molecule known as a heme. The heme contains an iron atom that is able to hold an oxygen molecule. Like myoglobin, each alpha and beta chain also holds a heme, so four hemes are present in each hemoglobin structure.

There are several different forms of hemoglobin that are made of different chains. These different hemoglobins are present at various times in the development of the organism. Myoglobin and the protein chains that form the various hemoglobins belong

to a family of related proteins known as the globins.

To test the idea that these globin chains are related, the amino acid sequences for myoglobin and the various types of hemoglobin chains can be compared. For humans, six hemoglobin chains and myoglobin are compared in Table 10.4. Because of the small numbers, we expect some fluctuations. For example, 36 probably should be grouped with 39, 40, and 41 instead of 30. The evolutionary tree constructed from the globin data is shown in Figure 10.7.

This evolutionary tree suggests that the gene for myoglobin was duplicated (point 1 on Figure 10.7) in the very distant past. Subsequent duplications (at each branch point of the tree) eventually led to the six other protein chains. Supportive evidence comes from the study of other vertebrate myoglobin and hemoglobin. The lamprey, a jawless fish, does not have hemoglobin, but has several forms of myoglobin that carry oxygen in the blood. When these myoglobin molecules are not carrying oxygen, two molecules can combine to form a double unit, but this combination dissociates when oxygen is bound to the molecules.

Jawed fish and all higher vertebrates have hemoglobin that is composed of four chains (two alpha and two beta). This tells us that the first duplication that eventually led to the hemoglobin molecule must have occurred very near the time when the jawed fish lineage split off the jawless fish (point 2 on Figure 10.7). From the fossil record, we know this split occurred about 450 mya.

According to our tree, the next duplication on the beta branch (point 3 on Figure 10.7) occurred fairly early also, perhaps 200 million years ago. This duplication eventually led to the production of the gamma chain (after another split at point 4). Two gamma chains combine with two alpha chains to form the main type of hemoglobin found in the fetus. Fetal hemoglobin has a higher affinity for oxygen and is better suited for the transfer of oxygen from the mother. Fetal hemoglobin is replaced by adult hemoglobin soon after birth.

The epsilon chain is found in the embryonic yolk sac, but in humans it is replaced by the gamma chain in the first month of pregnancy. It is also found in a variety of other mammals.

Small amounts of the delta chain have so far been found only in the hemoglobin of primates (including the new world primates), so it must have been formed by a duplication at least 40 million years ago (point 5 on Figure 10.7). Approximately 2 to 3 percent of adult hemoglobin is made of two alpha and two delta chains. Since it resembles the beta chain so closely, there is little doubt that its origin was a duplication of the beta chain.

The only duplication in the alpha branch is the duplication that produced the zeta chain (point 6 on Figure 10.7). Judging from the diagram, this duplication occurred about 300 mya. The zeta chain is found in embryos during the first few weeks of development, but is replaced by the

	myoglobin (M)	alpha (α)	beta (β)	gamma (γ)	delta (δ)	epsilon (ε)	zeta (ζ)
myoglobin	0						
alpha	109	0					
beta	114	80	0				
gamma	114	85	39	0			
delta	113	81	10	41	0		
epsilon	116	88	36	30	40	0	
zeta	109	61	92	87	90	88	0

Table 10.4 Amino Acid Differences for Human Globin Molecules *Note that delta and beta have 10 amino acid differences (red), fewer than any other pair, so they must be very closely related (they evolved from a common ancestor globin molecule most recently). Next, we see that epsilon and gamma have only 30 differences (blue), so they must be more closely related than they are to the other globins. The next closest grouping is beta-gamma, beta-epsilon, delta-gamma, and delta epsilon (violet). The next closest are alpha and zeta (green). A family tree is shown in Figure 10.7. [data: A.Efstratiadis et al. (1980), credits p. 415]*

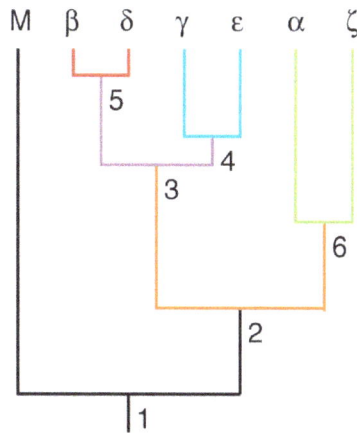

Figure 10.7 Evolution of the Globin Molecules *This tree was constructed using data in table 10.4. From left to right, the molecules are myoglobin (M), beta (β), delta (δ), gamma (γ), epsilon (ε), alpha (α), and zeta (ζ). Lengths of vertical lines are proportional to the time. [JCB]*

alpha chain near the end of the first month.

With the collection of additional data, much will be learned about the evolution of the various hemoglobin molecules. In the process, it is expected that we will be able to construct better family trees of the various animal groups.

The early evidence that supported evolution came from studies of paleontology, biogeography, and anatomy. Continued work in these areas will add a great deal to our expanding knowledge. However, the information that could be gained from the sequencing of DNA molecules is almost beyond comprehension. The vast amount of information stored in DNA has the potential of answering many questions about the relatedness of various organisms and the time of their divergence from a common ancestor. A wealth of new evolutionary information is waiting to be discovered by DNA sequencing studies.

Adaptation

Mutations in the DNA of an organism can produce changes in the protein molecules. These changes can be visible as differences in the structure of the organism. For example, the Okapi (Figure 9.44b) looks like a thick-bodied antelope. Its closest relative, the Giraffe, not only has different markings but also has a much longer neck and legs. These observed differences are due to variations in the DNA recipes of the two animals.

A structure that enables an organism to better exploit its environment is often called an adaptation. For example, a duck's webbed feet are better suited for swimming than the feet of a chicken, so people usually say that a duck's webbed feet are adapted for swimming.

Unfortunately, the word adaptation can be a misleading term. To many, it seems to imply that the organism needs certain things, which are subsequently provided because of the organism's desire for the change. An adaptation is a modification that has evolved, not because of an organism's desire, but because of the natural selection of advantageous mutations. This process produces some organisms that are better suited to live in their particular habitat.

Because of the evolutionary process, organisms often change in ways that make them better suited to the environment. The word adapt is often used to describe the evolution of a feature that enables an organism to better exploit its environment. Instead of saying ducks evolved webbed feet, we often say that ducks have adapted to their environment, and webbed feet are one of their adaptations. Adapted is used as a synonym for evolved. If plants or animals are unable to adapt to their surrounding conditions, the species may die out. Over the billions of years since life first appeared on this planet, many groups of organisms have been unable to adapt to the ever-changing environmental conditions, and they have become extinct.

The study of adaptations is one of the most fascinating topics in the world of living things. Adaptations are often the very things that make organisms so interesting. There are numerous examples. The feet, wings, and bills of various species of birds have been adapted to many different modes of life (Figure 10.8). Birds of prey have evolved sharp claws for catching and killing their food. Their hooked beak is adapted for tearing flesh. Many of these birds soar on the rising air currents as they scan the ground for food. They have evolved wide wings for this purpose. The majestic flight, powerful looking hooked beak, and sharp talons of the eagle are the adaptations that have inspired many people to think of eagles as a symbols of strength.

Other birds of prey, like the falcons, have long narrow wings that allow these birds to chase their prey at great speeds. They are among the fastest flyers in the bird world. Owls have evolved wings that cut through the air very quietly, enabling these hunters to silently search for prey. All birds have keen eyesight, a necessity since many must catch their food while flying at great speeds, while others must be able to maneuver through the branches of trees.

Many birds have adapted to a way of life in or near the water. Ducks, geese, and swans have flattened bills that they use to strain plants and small water insects from the water. Many water birds have evolved webbed feet to aid in swimming. Other water birds like the herons and cranes do not have webbed feet, but their long legs allow these birds to wade in shallow water as they search for food. They have long pointed beaks for spearing the fish, frogs, and the other water creatures on which they feed.

Carnivorous mammals like dogs and cats are adapted for hunting other animals, but in quite different ways than the birds of prey. The bodies of these mammals are adapted for speed, allowing them to run down their prey. They have long canine teeth that are used for stabbing, and they have special shearing teeth (carnassials), that are used to slice through the meat of their prey.

Whether they are the hunters or the hunted, an animal that is hard to see often has a distinct advantage. *Camouflage* has evolved in many different forms. One example is coloration that helps the animal blend in with its natural surroundings. The brown coat of a rabbit makes it difficult to see in dry grass. The Snowshoe Hare is brown in the summer, but its coat changes to white in the winter and blends with the snow (Figure 10.9), an adaptation that greatly increases the hare's chances of surviving the winter. Like all adaptations, these did not

Figure 10.8 Adaptations *The Green Hermit (left) is a hummingbird whose long bill and tongue allow it to drink nectar from inside a deep flower. Its white tongue can be seen sticking out beyond its bill (insert). A Red-footed Booby (right) is a diving bird that feeds on ocean fish. Its pointed bill and webbed feet are important adaptations. [JCB]*

evolve because the hares needed them, but those hares with the most advantageous coloration were more likely to survive to the reproductive age and pass their protective coloration genes to the next generation.

Sometimes the shape and coloration of an organism helps it hide. Some insects look like thorns, others look like moss-covered sticks (Figure 10.9), and still others resemble leaves. Some butterfly larvae are easily mistaken for bird droppings because of their shape and coloration, and some fish like the Leafy Sea Dragon are easily mistaken for algae (Figure 10.9). The lichen covered branch of a tree allows the Mossy Leaf-tailed Gecko to virtually disappear from view (Figure 10.9).

All insects are not camouflaged, however, and many are even very brightly colored. Many of these are bad tasting or even poisonous to the animals that might otherwise feed on them. Their bright coloration seems to be an adaptation to warn potential predators.

Insects are not the only animals that seem to be brightly colored as a warning to potential predators. The Gila Monster, one of the two poisonous lizards in the world, is orange and black. The poison dart frogs of Central and South America are the most brightly colored members of the frog world (Figure 10.10), apparently a warning of their poisonous nature. The Indians of the region would tip their blowgun darts with this frog poison, hence the name poison dart frogs.

As another form of protection, some insects seem to be impersonating more dangerous members of their group. This resemblance is called *mimicry* (Figure 10.10). There are harmless flies that are easily mistaken for honeybees, other flies resemble wasps, and some moths look very much like bumble bees. Undoubtedly many of these insects are avoided because of their adaptive deception.

Monarch Butterflies do not taste good because the larvae feed on milkweeds. The Viceroy is also bad tasting and looks very much like the Monarch. If a predator tastes one of the butterflies, it will avoid both species in the future since they look so much alike.

Some animals apparently defend themselves by trying to frighten away their potential predators. An unusually large number of insects and fish display a pair of colorful circles on their wings or bodies (Figure 10.10). In some cases the gaudy circles are on the underwings of insects and only show when the insect flies or spreads its wings. These circles look very much like a large pair of eyes, an adaptation that must have frightened away many potential predators.

In the desert regions of the Americas, many organisms have adapted to the harsh environment. Cactus plants have adapted to these dry areas by acquiring a waxy covering that cuts down on the loss of water. When water is available, the plant is able to expand as it takes in large quantities of water for use during the dryer parts of the year. The leaves have been modified into needles that do not allow water to escape, and the needles also discourage animals from eating the plant.

In the harsh desert environment of the southwestern United States, the Gila Monster has a tail that is adapted for food storage. When food is available, the Gila Monster eats more than it needs, and its tail grows very fat. During times when food is scarce, the Gila Monster uses the food stored in its tail to survive until food is available again.

For billions of years, organisms have been evolving in ways that better

Figure 10.9 Camouflage *(clockwise from top left) The Snowshoe Hare's white coat in winter blends with the snow. [D.Gordon E.Robertson]; In summer, its coat becomes brown, making it hard to see when the hare is sitting in dead grass. [Walter Siegmund]; The Mossy Leaf-tailed Gecko is a native of Madagascar. It is upside down in the photo, and you can see one of its black eyes and three of its toe pads near the bottom of the picture. [JialiangGao]; A Leafy Sea Dragon is a fish that lives in the ocean off the southern coast of Australia. It is well camouflaged in seaweed. [JCB]; A moss-mimic stick insect from Costa Rica. Its antennae can be seen at the upper right, along with one of its legs resting on a sprig of moss. [Twan Leenders]*

Figure 10.10 Warning Coloration and Mimicry (clockwise from top left) The Wasp Mantidfly (Climaciella brunnea) looks very much like a stinging wasp, but it is a harmless lacewing relative (order Neuroptera). Praying mantises are in the order Mantodea. [Kim Fleming]; With its large eyespots, the Spicebush Swallowtail larvae could be mistaken for a snake. [Ryan Hagerty/USFWS]; Most poison dart frogs are brightly colored. The bright red color of this Strawberry Poison Dart Frog may be a warning to predators. [JCB]; The Four-eyed Butterflyfish with a large eye spot could be mistaken for a very large fish, and it may further confuse a predator since the spot is on the tail, making its rear end look like its head. [Dwayne Meadows/NOAA]; This butterfly with large eyespots could be mistaken for an Owl. [Yoav David]; What looks like a bee on a flower is really a harmless fly.

enable them to survive in their environment. The life forms we see today are the result of this evolutionary process. Fossils tell us that many interesting groups of plants and animals are gone forever. Other groups are represented by a few lone survivors such as the Platypus, Tuatara, and the *Latimeria*.

The Groaning of Creation

Early humans appeared only a few million years ago, but in recent times they have greatly altered the Earth. Because of the activities of humans, many of the Earth's unique creatures are becoming extinct at an alarming rate. This extinction will soon rival the great extinctions that brought the Permian and the Cretaceous Periods to a close. Evolution has created an interesting and diverse array of living organisms that we alone seem to be capable of enjoying. Ironically, we are in the process of destroying a large part of it.

Traveling through space on an insignificantly small planet called Earth, we now realize that life has been evolving for billions of years on this planet. Every organism is a member of one large living community, all of whom share a common ancestor somewhere in the distant past. We are surrounded by life forms that are the result of billions of years of the creative process called evolution. As humans destroy this creation, the loss diminishes us all.

Humans have evolved into the great exploiters of the world. Throughout recorded history, there are numerous examples of one group of humans trying to benefit by dominating or making slaves of another group. Humans have a history of not only exploiting their fellow humans, but

other members of the animal kingdom as well.

We are only beginning to discover a few of the unusual abilities of some of the other members of the animal kingdom. Our closest relatives, the chimpanzees, use sticks and stones as tools for performing simple tasks. Many of the traits we normally tend to associate with humans such as intelligence, tool making, language, and long lasting social bonds are widespread in other groups of animals. Chimpanzees, however, are still hunted for food, and mothers are killed so their babies can be captured as pets.

Some people have the impression that humans are the only animals that are able to reason. They believe that other animals operate only on instinct. However, instincts are responses that have been programmed into the brain before birth. Baby mammals, including human babies, instinctively suck on objects placed in their mouth because their brain has been preprogrammed for that reaction. We do not understand how the brain is programmed, but learning is simply the process of programming the brain for a new task, something that all higher animals are able to do. If a brain can be programmed before birth (instinct), it very likely can be programmed after birth (learning).

Since so little is known about how the brain operates, it would seem rather presumptuous to make too many claims about the fundamental differences between the human brain and the brains of other animals. To be sure, humans have more developed brains than other animals, but there also seems to be a great deal of variation in the abilities of different humans. We know very little about the abilities of other animals. We do not even know how to test most of these things at present.

Today, as humans press in on the last natural areas of the world, they clear the trees and exploit the land in a futile attempt to satisfy the desires of an ever-increasing human population. The animals that once made these lands their home are killed or forced to live on ever shrinking reserves. Some justify this practice because they feel that human needs are more important, and that other animals have no claim to the land. Many place a low priority on the protection of animals that are not of obvious benefit to humans.

The beauty and symbolic value that other life forms provide is something that we all enjoy. What would the world be like without elephants, whales, birds or butterflies? Fantasy books have been written about mystical creatures that inhabit imaginary worlds, but we live in a world full of real amazing creatures, many of which are more mystical and intriguing than any that have been invented by the human mind. Actually, many imaginary creatures in books are modifications of the real animals that have inhabited this planet.

Many people feel that jobs are more important than saving a species of animal or a part of the environment. However, jobs are only temporary as are individuals. Many types of jobs become obsolete and new jobs are continually being created. If someone looses their job, they can be educated and trained in a new area. However, if an organism is exterminated or the environment is destroyed, it is lost forever. In the not to distant future everyone reading these words will be dead, but what will the environment be like when it is passed on to future generations?

The Mona Lisa is a painting that many feel is worth saving. To be sure, it is a great creation, but if it were lost, there are hundreds of artists in the world who could make reproductions so closely resembling the original that only a few experts could identify them as copies.

The last Tasmanian Wolf on Earth has been killed, but all the artists, scientists, and technicians in the world could not even begin to create a copy of this wonderful creation. They are gone forever and will never again walk the Earth, exterminated because some ranchers felt they interfered with their job of raising sheep and chickens. Today those ranchers are also dead, and a breeding pair of Tasmanian Wolves would be worth the price of millions of sheep. Was their extermination worth that price? Many people do not miss the Tasmanian Wolf, and some may even say good riddance. Some, of course, would say the same thing about the people belonging to certain ethnic or religious groups.

Darwin noticed that organisms produce more offspring than can survive to reproduce. Nature's general plan seems to be that many die before they are able to reproduce. In this way, the size of a given population remains relatively stable, and the population is constrained from expanding beyond the carrying capacity of the environment. If a population gets too large for its food supply, many offspring either die of starvation or are eaten by predators.

Humans, however, have upset the balance of nature by decreasing their death rate. Without a corresponding reduction in the birth rate, the human population will expand beyond the capacity of the Earth to support their numbers. The human population has already grown well beyond the Earth's ability to sustain it at a level to which most of us aspire.

There were more than seven billion humans inhabiting the Earth in 2013, and there will be eight billion by 2020. Part of the price that will be paid for this ever increasing population will be the extermination of many of the plants and animals that add variety and beauty to our lives.

Another price that most people will pay for this increase in the human population will be a lowering of our standard of living. The Earth may be able to provide enough food for more people than are alive today, but some may feel there is more to living than just having enough food to survive.

What is the advantage of pushing the human population to the Earth's carrying capacity? Even if the carrying capacity of the Earth is larger than many believe, wouldn't it be better to stabilize the population at a much lower number? The human population cannot continue to grow indefinitely, and the consequences of exceeding the Earth's carrying capacity are quite severe. What could possibly make this a risk worth taking?

Ultimately, the price humans will pay for not controlling their numbers is the destruction of the very environment that enables the Earth to support them. If we wait, the population will be reduced by disease and starvation.

The overpopulation of the Earth has generated many environmental problems such as air, water, and chemical pollution, destruction of our forests, and global warming. None of these problems can be solved unless the human population is stabilized.

Today, forests are being destroyed at an alarming rate as humans attempt to raise more cattle and grow more crops. With humans already pressing in on the last unoccupied land areas of the world, many rare creatures of this Earth will become extinct. Large numbers will be exterminated as the human population expands to fill and exploit every available space on the planet. Another great extinction is in progress. This one is being caused by humans.

Is there any hope for these life forms? There can be no doubt that the explosive human population must eventually stop growing and even drop in the future, but the longer it takes, the more plants and animals will be lost forever.

To save the creation, society must work swiftly to stabilize and then reduce the human population. Balancing the birth rate with the death rate would stabilize the population. The human population is expanding because modern technology has altered the balance of nature by decreasing the death rate. The population will eventually be stabilized, either by increasing the death rate or by lowering the birth rate. Those who do not elect to reduce the birth rate are choosing to let nature increase the death rate. A stable population would eventually be established if every couple limited their families to two children. The quicker stability is accomplished, the less damage will be done to the environment, and the less suffering future generations will have to endure.

Some people undoubtedly feel that no one has the right to impose limitations on the reproductive rights of humans. Reproduction is viewed by many as a basic freedom. However, knowledge of simple mathematics should be sufficient to convince any rational person that population growth cannot continue forever. Furthermore, those who choose to continue contributing to this growth are, by their actions, imposing severe limitations on the freedom of future generations to choose their own lifestyle. The quality of life will diminish as the population grows. Which freedom is more basic? It really does not matter what we think, since all freedoms will eventually be lost if the population continues to grow.

Another thing that must be done to save the creation is to help preserve those plants and animals that are in danger of becoming extinct. If their extinction can be postponed until the human population is stabilized, perhaps some will be saved.

The movie Jurassic Park is about a theme park containing real dinosaurs. The makers of the movie spent tens of millions of dollars to create the illusion that dinosaurs were again living on the Earth. Today, we are slowly destroying the real Jurassic Parks of planet Earth, places like Hawaii, Madagascar, Borneo, New Guinea, New Zealand, Australia, the Amazon, the great wildlife herds of Africa, the old growth forests of North America, and the jungles of India, Southeast Asia, Africa, and South America.

In recent times, the Tasmanian Wolf, the elephant birds, the moas, the Great Auk, Imperial Woodpecker, and many other wonderful animals have been exterminated by humans and are gone forever. Even with the help of concerned people throughout the world, many more plant and animal species will be lost in the near future because of humans. These unique creatures are the products of millions of years of evolution. As they are lost, who will shed a tear? Undoubtedly, the Giant Panda does not comprehend the fragile existence of its species, and if one day the last of its kind is found dying in a zoo, the animals shedding tears will not be the Pandas.

In the future will they only be safe in zoos?
[Zoological Society of San Diego]

Glossary

acanthodian: [*acanthodes*, thorny] An early group of jawed fish (sometimes called the spiny sharks), that lived from later Silurian to early Permian times. They had many fins that were supported by spines and were the first jawed fish to appear in the fossil record. Their relationship to other groups of fish is not fully understood.

Acanthostega: [*acanthodes*, thorny; *stega*, plate] The genus name of an aquatic animal from the late Devonian (about 370 mya) It is perhaps the most primitive amphibian known. It had 8 fingers, a powerful fishlike tail, internal gills like a fish, and hips that were not firmly attached to its spine. Its weak legs were probably used for moving through rotting vegetation on the bottoms of lakes and wetlands. Along with *Ichthyostega*, this is one of the earliest known amphibians. See *Ichthyostega*.

accrete: [*accret*, grown] A word used to describe how a planet forms and grows by collisions with other objects.

Acheulean tool: [after St. Acheul, France] A name used to describe the 150,000 to 1.5 million year old tools associated with *Homo erectus* and early forms of *Homo sapiens*. They were characterized by carefully shaped teardrop shaped hand axes and replaced the earlier Oldowan tools.

actinopterygian: [*actis*, ray; *pterygion*, fin] A member of the Actinopterygii class and the superclass Osteichthyes (the bony fishes). They are commonly known as the ray-finned fish because their fins are supported by a large number of parallel bones. Their fins contain few muscles, so the movement of the fin is controlled by muscles in the body of the fish. This group is divided into the chondrosteans and the neopterygians.

adaptation: [*adaptere*, to fit] A structure or behavior that helps an organism function more successfully in its environment. For example, the webbed feet of ducks. An adaptation is the result of natural selection.

adaptive radiation: The divergence of a group of organisms to fill a variety of different niches.

advanced: See primitive.

Aegyptopithecus: [*Aegypt*, Egypt; *pithekos*, ape] Genus of primate not generally considered an ape, but this cat sized primate may represent a form similar to the direct ancestors of the apes. Fossils have been found in the Fayum Depression in Egypt and are about 30 million years old.

aerobe or aerobic: [*aer*, air; *bios*, life] An aerobe is an organism that requires molecular oxygen to live. It is said to be aerobic. See anaerobe.

Afrotheria: [Africa; *theria*, beast or wild animal] A superorder of mammals that evolved from a common ancestor in Africa between 110 and 40 million years ago. Elephants, tenrecs, elephant shrews, hyraxes, sea cows, and the Aardvark belong to this group.

age: See eon.

agnathan: [*a*, without; *gnathos*, jaw] A fish belonging to the class Agnatha, the jawless fish. They are the oldest and most primitive fish. The first clear fossil evidence comes from the early Silurian, but fragmentary evidence is of Ordovician age. Agnathans never developed paired fins. Living members include the lampreys and hagfish.

allele: Alternate forms of a gene that control a particular characteristic. Alleles occupy corresponding locations or loci on homologous chromosomes. The gene for brown eyes and the gene for blue eyes are alleles that determine eye color.

alpha particle: Fast moving helium nuclei, also called alpha rays or alpha radiation, which can harm tissue or cause mutations. To become more stable, some radioactive elements emit alpha particles. Since this emission reduces the size of the nucleus by two protons and two neutrons, the atom is changed into a different element.

amber: Tree resin that has hardened over time. Insects, spiders, and even frogs and lizards are sometimes trapped in the resin and are preserved as it hardens into amber.

Amia: Genus of Bowfin. See Bowfin.

amino acid: The building blocks of protein molecules, they are organic acids that, in living organisms, come in 20 different forms.

ammonite: [from Amon, one of the chief gods of the ancient Egyptians who is usually depicted wearing ram's horns] A group of extinct mollusks, most of which had shells that looked like the coiled horn of a ram. The shells were composed of many sutured sections, and the sutures evolved into very ornate lobed patterns. Ammonites first appeared in early Devonian times, but became extinct at the end of the Cretaceous. The nautilus closely resembles the ammonite.

amniote: [*amnion*, fetal membrane] An animal (either a reptile, bird, or mammal), so named because of the presence of a fluid-filled sac that encloses the embryo. This sac is called the amnion, and the fluid is called the amniotic fluid. Neither structure can be readily seen in a chicken's egg. The evolution of the amniotic sac allowed reptiles and their ancestors to be completely free from the water, even for reproduction.

amphibian: [*amphi*, both; *bios*, life] A vertebrate belonging to the class Amphibia. Most living members spend part of their life in the water living like a fish, and part of their life on land living more like a reptile. Amphibians are not amniotes, which means their eggs do not have a fluid filled sack enclosing the embryo. Without water, their jellylike eggs dry out quickly so they must return to the water to lay their eggs. The earliest known amphibians are *Acanthostega* and *Ichthyostega*. Both lived in late Devonian times. Living members of the group include salamanders, frogs, toads, and the caecilians or apodans.

amphioxus: [*amphi*, both (sides); *oxys*, sharp] A name used to describe several species of small fish-shaped chordates (also called the lancelets) that live in coastal waters. Since they have a notochord but are not vertebrates, they are animals that bridges the gap between the invertebrates and the vertebrates. They likely resemble the fishlike creatures that were the ancestors of the first fish.

anaerobe or anaerobic: [*an*, without; *aer*, air; *bios*, life] An organism that either cannot tolerate molecular oxygen or does not use molecular oxygen. Oxygen is toxic to obligate anaerobes, but facultative anaerobes are able to grow with or without the presence of molecular oxygen. See aerobe.

anagenesis: [*ana*, up; *genesis*, origin] A term applied to the evolution that occurs in a lineage between branching points. That is, the evolution of a population of organisms over time that does not split into two or more groups. See cladogenesis.

analogous: [*analogia*, relationship] Structures that are used for the same function, but do not share a common evolutionary history are said to be analogous (or homoplastic). Since they evolved independently, we would not expect them to be similar anatomically. For example, the wings of a bird and the wings of a butterfly do not share a common evolutionary history, so they are analogous structures. The wings of a bird and the wings of a bat are also analogous structures since they had different evolutionary histories as wings. Structures that share a common evolutionary history are said to be homologous. See homologous and homoplastic.

anapsid: [*an*, without; *apsid*, arch] A member of the group of reptiles that have no openings, or fenestrae, in their skull. The first reptiles belonged to this group.

angiosperm: [*angeion*, receptacle; *sperma*, seed] A member of the division (phylum) commonly known as the flowering plants. The name refers to the fact that their seeds are enclosed in an ovary that develops into a structure we call the fruit.

annelid: [*annulus*, ring; *eidos*, form] A member of the phylum Annelida whose members are the segmented worms. Earthworms and leaches are examples. Marine annelids or polychaete worms superficially resemble centipedes. See Arthropoda.

anthracosaur: [*anthrax*, charcoal; *sauros*, lizard] An amphibian of the order Anthracosauria. They lived from the early Carboniferous to the end of the Permian. Since they evolved many reptile-like features, it is likely that they were the ancestors of the reptiles.

anthropoid: [*anthropos*, man; *eidos*, form] Monkeys and apes (called simians) belong to the infraorder Anthropoidea (or Simiiformes).

antibody: See antigen.

antigen: [from the beginnings of the two words **anti**body and **gen**erator.] A piece of a bacterium, virus, pollen, or other microorganism (usually a protein or polysaccharide) that causes the body's immune system to produce or generate antibodies. These antibodies are proteins that bind to the antigens and neutralize them.

anuran: [*a*, without; *oura*, tail] A member of the amphibian order Anura. They are the frogs and toads.

apodan: [*a*, without; *pod*, feet] See caecilian.

arborescence: [*arborius*, tree] Herbaceous island plants sometimes evolve into trees-like forms. This condition is called arborescence.

Archaea: [*archaios*, ancient] A collection of prokaryotes so different chemically from Bacteria that they are placed in the separate kingdom called the Archaea. Members include the methanogens (bacteria that produce methane), halophiles (that live in very salty environments like the Dead Sea), and thermoacidophiles (found in hot acid environments).

archaeocete: [*archaios*, primitive or ancient; *ketos*, sea monster (whales)] The group of early whales that are the ancestors of modern whales. They had legs, differentiated teeth, and lacked echolocation. The first ones appeared in the early Eocene and most died out by the early Oligocene.

Archaeopteris: [*archaios*, primitive or ancient; *pteris*, fern] A genus of tree from Middle Devonian times (about 400 mya) that is a *missing link* in the plant world. It represents a mosaic of fern and seed plant features. This tree displayed spore bearing fernlike leaves, but it had wood like that of a seed bearing gymnosperm (the wood shows annual growth rings).

Archaeopteryx: [*archaios*, ancient; *pteryx*, wing] The genus name of the earliest known fossil bird. It lived in the late Jurassic (about 150 mya) in what is now southern Germany. It had feathers, but a lizard-like head with socketed teeth, a slender tail with movable vertebrae, and unfused wrist and hand bones with clawed fingers. It lacked a keel suggesting that it may not have been a strong flyer. The 6 known specimens range in size from that of a pigeon to a chicken.

Archean: [*archaios*, ancient] The second geologic eon from 2,500 to 4,000 million years ago that followed the Hadean Eon and was followed by the Proterozoic Eon. The Archean began approximately when the Earth's crust started to form. Before the Archean, circulation of the Earth's surface driven by high internal temperatures and meteoritic bombardment likely prevented the formation of solid areas of significant size.

archosaur: [*archon*, ruler; *sauros*, lizard] A reptile belonging to the superorder Archosauria, which includes: dinosaurs, crocodiles, pterosaurs, and thecodonts. Thecodonts were the group that included the ancestors of the dinosaurs.

arthropod: [*arthron*, joint; *pod*, foot] Members of the phylum Arthropoda that include, insects, centipedes, millipedes, spiders, crustaceans (crabs, lobsters, shrimp, and crayfish), and extinct trilobites. The arthropods probably represent a polyphyletic group. For example, insects and centipedes may be more closely related to annelid polychaete worms than they are to crustaceans.

articulation: [*articulus*, connecting part] A joint where two moveable bones meet, such as the jaw. In mammals, the jaw articulation is between the dentary and squamosal, but in reptiles it is between the quadrate and articular bones.

artiodactyl: [*artios*, even; *dactylos*, finger] A member of the mammal order Artiodactyla, commonly known as the even-toed ungulates. Members include pigs, hippopotamuses, camels, giraffe, sheep, goats, cattle, deer, and antelope.

asthenosphere: [*asthenes*, weak; *sphaera*, ball] The layer of the Earth below the lithosphere that is partially molten. Motion in the asthenosphere causes the overlying lithospheric plates to move. See lithosphere.

Australopithecus: [*australis*, southern; *pithekos*, ape] The genus name of several species of early hominids that lived in Africa between about 3.6 and 1 mya. One of the early species was the direct ancestor of humans.

autotroph: [*autos*, self; *trophe*, food] An organism that can make its own food from inorganic material. Some bacteria, algae, and green plants are autotrophs.

Aysheaia: See Onychophora.

Bacteria: [*bakteria*, rod (referring to the shape of the first observed types)] The kingdom name for the prokaryotes that are not Archaea.

Baltica: The name given to Northern Europe east of the Ural Mountains. This area existed as an isolated land mass before the formation of Pangaea.

basalt: [*basanos*, touchstone] A fine gained igneous rock that forms much of the ocean bottom. It has less silicon and higher abundances of iron, magnesium, and calcium than granite. This mix makes basalt slightly more dense than granite. Basalt has a density of about 3.0 times that of water while granite has a density of about 2.7.

belemnite: [*belemnon*, javelin; *eidos*, form] An extinct mollusk (class Cephalopoda), similar in shape to the squids, but not closely related. They possessed a chambered internal shell used to adjust their buoyancy, and a pencil shaped shell at the rear of the animal that was often preserved. They appeared in the early Jurassic and died out during the great extinction at end of the Cretaceous.

Beringia: [after Bering Sea after Danish explorer Vitus Jonassen Bering] The name given to the land mass that provided an intermittent connection between Asia and North America during the later part of the Cretaceous Period and much of the Cenozoic Era. The connection allowed an interchange of plants and animals between the two regions, explaining why these areas share so many similar life forms today.

beta particle or radiation: Fast moving electrons or positrons that are emitted by atomic nuclei as a neuton is converted into a protron or visa versa. Many radioactive elements emit beta rays, a process that transforms the atom into a different element. Since beta particles have a great deal of kinetic energy (energy of motion), they can harm body tissue or cause mutations.

bichir: A chondrostean (primitive ray-finned fish), which belongs to the family Polypteridae as does the Reedfish. The ten or so species have thick, diamond-shaped, bonelike scales (ganoid scales), and they live in several rivers of Central Africa. They have gills, but are also air breathers with lungs and paired nostrils.

blastoid: [*blastos*, bud; *eidos*, form] An echinoderm that looks like a nut or a flower bud attached to a stalk that anchors them to the sea bottom. They appeared in the Silurian and died out at the end of the Permian.

Bowfin: The Bowfin or *Amia* and the gars are the most primitive of the neopterygians (modern ray-finned fishes). The Bowfin and gars have a more primitive asymmetrical tail and a swim bladder that acts as a lung, allowing them to breathe air.

brachiopod: [*brachion*, arm; *pod*, foot] An animal that looks superficially like a bivalve clam. However, the two valves are not symmetrical, but differ in both size and shape. The internal structure of the animal is quite different also. Brachiopods first appeared in the Cambrian and were very common throughout the Paleozoic, but declined near the end of the Era. More than 30,000 fossil types are known, but only about 250 living species remain.

bryophyte: [*bryon*, moss; *phyton*, plant] A division (phylum) of nonvascular plants whose members include the mosses, liverworts, and hornworts. They must live near water because their sperm cells must swim to the egg cells. For that reason, they are sometimes called the amphibians of the plant world. A young moss plant looks very much like a filament of green algae, more evidence that suggests they evolved from a type of green algae.

Burgess Shale: [after Burgess Pass in British Columbia] A fossil bed in western Canada that was deposited in middle Cambrian times. The deposit is unusual in that it was formed by mud slides that trapped and preserved many soft bodied animals.

caecilian: [*caecus*, blind] A member of the order Apoda (meaning *without feet*), one of the three orders of living amphibians. Caecilians are burrowing worm-like animals that live in tropical regions. They are of particular interest because they have retained a few patches of scales. Early amphibians must have had scales like their fish ancestors, but frogs and salamanders have lost their scales.

Cambrian: [from *Cambria*, the Roman name for Wales] The first of the six time periods of the Paleozoic Era, the Cambrian extended from about 541 to 485 mya. During the Cambrian, trilobites were the dominant life forms in the sea. Life was not yet established on the land. It was followed by the Ordovician Period.

Carboniferous: Named after the great deposits of coal (composed mainly of carbon) that were laid down during this time, it is the fifth period of the Paleozoic Era. It extended from about 359 to 299 mya and is divided into two subperiods, the Pennsylvanian (323 to 299 mya) and the Mississippian (359 to 323 mya). The first reptiles appeared near the end of this period. It was followed by the Permian and was preceded by the Devonian Period.

carbonization: A fossilization process where the organism is compressed and the volatile material is lost, leaving only a thin film of carbon. Plant fossils are often of this type, and great detail can be preserved. They are sometimes called carbonized compressions.

carnassial: [*carno*, flesh] Mammals belonging to the order Carnivora have specially modified knifelike teeth called carnassials that are used for shearing flesh. The fourth upper premolars and the first lower molars are the modified teeth.

carnivoran: See carnivore.

carnivore: [*carno*, flesh; *vorare*, to devour] A general term describing an animal or plant that eats animals. A mammal belonging to the order Carnivora is more properly called a carnivoran. There were two groups of terrestrial flesh eating mammals in early Cenozoic times, the carnivorans and the creodonts. See creodont.

catarrhine: [*cata*, downward; *rhinos*, nose] A member of the group of primates that includes the Old World monkeys and apes. Members of this group have a narrow nose with closely spaced nostrils, which, as their name implies, are pointed downward.

Cenozoic: [*cainos*, new; *zoon*, life] Following the Mesozoic, it is the third and last era of the Phanerozoic Eon. It began about 66 mya with the extinction of the dinosaurs and continues to this day. It is sometimes known as the age of mammals since they became one of the dominant life forms during this era. The Cenozoic is divided into three periods, the Paleogene, Neogene, and Quaternary.

centromere: [*centrum*, center; *meros*, part] the region on a chromosome where the spindle fibers are attached during mitosis and meiosis.

cephalopod: [*cephal*, head; *pod*, foot] A member of the class Cephlopoda. They are mollusks that include the squids, cuttlefish, nautiloids, and octopuses. Many have complex eyes and a well-developed nervous system. Their muscular foot has been modified into tentacles. Cephalopods most likely evolved from a monoplacophoran type ancestor in the late Cambrian. See *Neopilina*.

Chelonia: [*khelone*, tortoise] A reptile order whose members are the turtles, numbering about 325 living species.

chert: A fine grained rock formed in soft sediments by the accumulation of microcrystals of silicon dioxide. The microcrystals form into irregular shaped nodules and as the nodules grow within a sediment, they can incorporate some of the surrounding material and organisms. Therefore, chert can contain small fossils. Darker colors of chert are often called flint.

chlorofluorocarbons or CFC's: Chemicals that were once used in refrigeration equipment, for certain types of cleaning, and in the production of certain plastics. These gaseous chemicals do not react easily, and many eventually end up in the ozone layer where they are broken down by sunlight. The chlorine released acts as a catalyst, converting ozone back into molecular oxygen, leaving the ozone layer depleted. See ozone.

chlorophyte: [*chloros*, grass green; *phyton*, plant] A member of the division (phylum) Chlorophyra of the plant kingdom whose members are commonly called green algae. They are believed to be the ancestors of green land plants since both groups contain chlorophyll a and b, true starch, and cellulose in their cell walls. Fossils believed to be of lime-secreting chlorophytes have been found in rocks of Cambrian age, but they may have evolved even earlier.

chloroplast: See plastid.

Chondrichthyes: [*chondros*, cartilage; *ichthys*, fish] One of the four classes of fishes, commonly called cartilaginous fish that includes the sharks, skates, and rays. Since the group diverged from other fish in the late Silurian, they do not have a swim bladder. The swim bladder is an air filled sack that adjusts to keep a fish from rising or sinking. It is a modified lung that evolved in the bony fish lineage after the sharks had split off. Sharks became an important animal in the sea during the Devonian.

chondrostean: [*chondros*, cartilage; *o s t e o n* , b o n e] A g r o u p o f actinopterygians (ray-finned fishes), usually known as the primitive ray fins. They first appeared in the late Silurian, and diversified greatly near the end of the Devonian. They were very successful during the rest of the Paleozoic and the first part of the Mesozoic, but were then replaced by more advanced ray finned fish (neopterygians). Surviving members include the sturgeons, paddlefish, bichirs and the Reedfish. Their non symmetrical tail is evidence of their primitive nature.

chordate: [*chorde*, cord or string] A member of the phylum Chordata. Animals that, at some stage of their development, possess a flexible rod or notochord running down the back of the animal below the nerve chord. Vertebrates are chordates since they have a notochord in their embryo stage, but vertebrae soon develop around the notochord.

chromist: [*chroma*, color] A member of the brown algae group Chromista (which some call a kingdom). They include the brown algae, diatoms, and water molds.

chromosome: [*chroma*, color; *soma*, body] In eukaryotes, their long double helix strands of DNA are wound around protein molecules (histones) to form compact rod shaped structures called chromosomes.

clade: [*clados*, branch] A group of species that includes an ancestor and all its descendants, a monophyletic group. The group could be a kingdom, a family, or any other taxonomic rank. See monophyletic.

cladogenesis: [*clados*, branch; *genesis*, creation] A term used to describe the evolution of two new species by the splitting of a lineage. See anagenesis.

cladogram: [*clados*, branch; *gramma*, thing drawn] A branching diagram or family tree that shows how organisms evolved from a common ancestor.

class: Phyla (or divisions) are divided into classes. Species belonging to a given class are more closely related to other members of their class than to members of other classes. For example, any two members of the bird class are more closely related to each other than either is to any member of the mammal class. See species.

Cnidaria: [*cnide*, nettle] The phylum, also known as Coelenterata, whose members include the jellyfishes, sea anemones, and corals. They have stinging cells called nematocysts on their tentacles.

coccolithophore: [*kokkos*, grain; *lithos*, stone] A single-celled flagellated protist and an important component of marine plankton. They are protected by a covering of very small calcium carbonate plates. When the organism dies, these plates become part of the ocean bottom ooze and are called coccoliths. Coccoliths first appeared in Jurassic times and continue to flourish today. The White Cliffs of Dover were formed from the remains of these organisms.

codon: A sequence of three nucleotides in a DNA molecule. Successive codons specify the amino acid sequence of a protein molecule. When a messenger RNA molecule (mRNA) is transcribed from the DNA, the codons are referred to as mRNA codons.

coelacanth: [*coilos*, hollow; *akantha*, spine] One of the two members of the order Coelacanthiformes, an order of the class Sarcopterygii (the lobe-fined fish). See latimeria.

Coelenterata: See Cnidaria.

coelurosaur: [*coilos*, hollow; *oura*, tail; *sauros*, lizard] Theropod dinosaurs of the infraorder Coelurosauria. (Theropoda is a suborder of the order Saurischia.) Theropods retained a bipedal mode of locomotion. Fossil evidence suggests that birds evolved from a coelurosaurian dinosaur.

colugo: Either of two species of gliding mammals belonging to the family Cynocephalidae and order Dermoptera. They are also known as flying lemurs, but they are not closely related to lemurs. The Philippine Flying Lemur is found on Mindanao and a few islands north of there. The Sunda Flying Lemur or Malayan Colugo is found in several countries in Southeast Asia. They are one of several groups of mammals that independently evolved the ability to glide.

condylarth: [*condylos*, knuckle; *arthron*, joint] A mammal belonging to the extinct order Condylarthra, which included a diverse collection of herbivores. First known from the later Cretaceous, various condylarths are probably the ancestors of the artiodactyls (even-toed ungulates), perissodactyls (odd-toed ungulates), and the extinct South American ungulates (notoungulates) which evolved from the odd-toed ungulates. The group may be polyphyletic.

conjecture: Although this word is not an official term, we have suggested that a simple statement for which there is not an overwhelming amount of supportive evidence should be called a conjecture (instead of a theory).

convergent evolution: The evolution of superficially similar organisms that are not closely related. For example, the marsupial Sugar Glider of Australia looks very much like the placental flying squirrels of North America, but they are not closely related. Some euphorbias of Africa are plants that look very much like the cactus plants of the Americas, however, they are not closely related.

Cooksonia: See Psilotopsida.

cosmological red shift: The stretching of the wavelengths of light which is produced by the expansion of space. Stretching the wavelength of a visible light wave will make the wave redder in color.

cotylosaur: [*cotyle*, cup; *sauros*, lizard] A member of the order Cotylosauria, the most primitive order of reptiles. The name refers to the cup shape of their vertebrae. Cotylosaurs are often referred to as *stem reptiles*, since it appears that all the reptile orders evolved from this group. They lived during late Carboniferous and Permian.

covalent bond: A chemical bond formed when two atoms share one or more electrons. In a hydrogen molecule, the electron spend most of their time between the two nuclei, creating a negative region which holds the two positive nuclei together.

creodont: [*creas*, flesh; *odont*, tooth] Any of the extinct, flesh-eating mammals of the order Creodonta. Creodonts had special shearing teeth called carnassials as do mammals of the order Carnivora, but different teeth were modified in the two groups, indicating that they had evolved independently. Both creodonts and the ancestors of modern carnivores were present in the early Cenozoic, but the last creodonts became extinct in the Pliocene.

Cretaceous: [*creta*, chalk] Named after the many chalk deposits that were laid down during the period, it lasted from about 145 to 66 mya. It is the third and final period of the Mesozoic Era. Dinosaurs continued their domination, and the flowering plants appeared. A great extinction at the end of the period eliminated the dinosaurs and many other life forms. It was followed by the Paleogene and was preceded by the Jurassic Period.

cusp: [*cuspis*, point] A cusp or tubercle is a bump on the chewing surface of a tooth. One group of early insect eating mammals (the triconodonts) had molar teeth with three tubercles, and another group (the multituberculates) had molars with many cusps.

cyanobacteria or cyanophyte: [*cyanos*, blue; *phyton*, plant] A member of the division (phylum) Cyanophyta, sometimes called the blue-green algae, and one of the divisions of the kingdom Bacteria. Fossils as old as 3,500 million years are believed to represent members of this group. They started as single cells but evolved filamentous forms (many cells attached in a long strand) about 2,000 mya. The oxygen they release during photosynthesis was the original source of oxygen in the Earth's atmosphere. DNA evidence tells us that a cyanobacteria invaded the cells of early organisms and evolved into the chloroplasts of the higher plants.

cycad: A gymnosperm belonging to the division (phylum) Cycadophyta.

Cycads look like palm trees, but palm trees are angiosperms (flowering plants). Cycads first appeared in the Permian and were common during the Mesozoic, but they declined as the flowering plants became more abundant. Today, about 400 species are found in various tropical regions of the world.

cynodont: [*cyon*, dog; *odont*, tooth] A member of the suborder Cynodontia, the largest and most important group of therapsid reptiles. Cynodonts lived during the Triassic, and most were carnivorous. Mammals evolved from this group of reptiles. *Cynognathus* [meaning *dog jaw*] was a cynodont.

cytoskeleton: A scaffolding of microfilaments and microtubules that act like muscles and bones within a cell's cytoplasm. These filaments and tubules are composed of proteins and their function is to control the structure and movement of a cell.

deductive reasoning: Logical reasoning that uses one or more general statements (called premises) to formulate a specific conclusion or prediction. The reasoning used in mathematical manipulations where the rules of logic are used to reach conclusions based on certain general assumptions. Using the premises of a scientific theory, deductive reasoning is used to make predictions. Incorrect predictions imply that the theory is wrong.

Denisovian: [after the Denisova Cave in the Altai Mountains of Siberia] An archaic human that represents a subspecies of *Homo sapiens*. It is represented by two teeth and a finger bone. However, DNA analysis shows it was a closer relative of Neanderthal than to modern humans. Some people of Southeast Asia still have Denisovian DNA.

density: The density of an object is the mass of the object divided by its volume. We usually compare densities to water. A liter of water has a mass of one kilogram, but a liter of iron is about eight times more massive. Therefore, we say iron has a density of eight, compared to water that has a density of one.

deuterostome: [*deuteros*, second; *stoma*, mouth] As a fertilized egg divides and grows, it forms a fluid filled sphere of cells called a blastula. One side of the blastula then pushes in and a three layered cup shaped structure called a gastrula is formed. The outer layer becomes the ectoderm (skin and nervous system), the inner layer becomes the endoderm (digestive tract), and the middle layer becomes the mesoderm (connective tissue). A second opening soon develops so the embryo looks like a fat hollow tube. In protostomes, the mouth develops from the first opening, and the second opening becomes the anus. In deuterostomes, the mouth develops from the second opening and the first opening becomes the anus.

Devonian: [after Devonshire, a town in England] The fourth period of the Paleozoic Era, the Devonian extended from about 419 to 359 mya. The Devonian is sometimes known as the age of fishes since these animals greatly diversified and became the rulers of the seas during this period. The plants also diversified greatly and spread over the land. Amphibians appeared in the later part of the period. It was followed by the Carboniferous and was preceded by the Silurian.

diapsid: [*di*, two; *apsid*, arch] An animal (either a reptile or a bird) that possesses two openings on each side of the skull (called fenestrae), that facilitate the attachment of the jaw muscles. Members of this group include the crocodiles, dinosaurs, and birds. The skulls of lizards and snakes have been further modified.

diatom: [*dia*, crosswise; *tomos*, divide] Single-celled protists that belong to the Chromista kingdom and carry on photosynthesis. They have a pillbox-shaped covering that contains silica. When they die, their coverings or shells can accumulate in great numbers, forming deposits called diatomaceous earth. They first appeared in the Cretaceous, and today they are the most abundant component of plankton.

dinosaur: [*deinos*, terrible; *sauros*, lizard] A general term used to describe any of the extinct terrestrial reptiles that possessed certain skeletal characters. Many had an erect posture and adaptations to bipedal locomotion. They had a diapsid skull with an additional opening in front of the eye cavity. Based on their hip structure, the dinosaurs are divided into two orders, the ornithischians and the saurischians. Although the name means *terrible lizard*, dinosaurs were not lizards, and many were very small.

dip angle: The angle the Earth's magnetic field lines make with the ground. Different locations on the Earth have different dip angles. Certain iron compounds line up with the Earth's magnetic field when magma cools, and the dip angle can be used to determine the latitude of the rock when it solidified.

diploid: [*diploos*, double] A cell having two sets of chromosomes, one set from the mother and one set from the father. Our normal body cells (called somatic cells) are diploid cells, but eggs and sperms are haploid cells. See haploid.

Diplura: [*diploos*, double; *oura*, tail] An order of hexapods with chewing mouthparts that are enclosed in a cavity in the head of the animal. They do not have eyes or wings and as their name implies, they have two tails (called cerci). Some have pincher-like tails, and most are less than 5 mm long. They live in the ground under leaves and rocks.

dipnoan: [*dis*, twice; *pnein*, to breathe] A lungfish belonging to the order Dipnoi. Dipnoi contains 6 living species. One species lives in Australia, one in South America, and 4 species live in Africa. See sarcopterygian.

divisions: The plant kingdom is partitioned into divisions as are several plant-like groups such as the Fungi Bacteria, and Protista. The animal kingdom is partitioned into phyla. See phylum and species.

DNA: Deoxyribonucleic acid is a long chain of deoxyribonucleotides. Each nucleotide is made of a deoxyribose sugar, a phosphate group, and one of four bases (adenine, thymine, guanine, or cytosine). DNA molecules are used for information storage in living cells. Genes and chromosomes are composed of DNA.

dorsal: [*dorsum*, back] The back or top side of an animal. See ventral which is the belly.

Dryopithecus: [*drus*, tree; *pithekos*, ape] A genus of apes that includes some of the earliest known fossil apes.

Echinodermata: [*echinos*, spiny; *derma*, skin] An animal phylum containing: sea stars, brittle stars, sea urchins, sand dollars, sea cucumbers, and crinoids. They share several common features with the chordates.

Ediacara fauna: [from the Ediacara Hills in South Australia] Name given to the soft-bodied multicellular animals that lived between about 635 and 541 mya (near the end of the Precambrian). The Ediacara fauna are the earliest known examples of multicellular animal life. Some of the worm-like creatures were almost half a meter long.

egg: See gamete.

embryology: [*embruon*, grow into; *logia*, study of] A study of the way organisms develop from a fertilized egg. Embryology can tell us if similar looking structures are formed in the same way. For example, the vertebrate eye and the eye of a squid look very similar, but they develop in quite different ways, indicating that they are analogous structures, not homologous structures. Embryology can also show us how very different looking animals can be derived from very similar embryos. Minor deviations at various stages of development can become accentuated and can produce great differences in the adult animals. Most of us would have trouble distinguishing an early salamander embryo from an early human embryo.

embryophyte: [*embruon*, grow into; *phyton*, plant] The group of plants in which the fertilized egg develops into a small plant or embryo while still in the female reproductive organ. Bryophytes and tracheophytes belong in this group.

endemic: [*endemos*, native] A plant or animal that is restricted to a given area is said to be endemic to that area. For example, lemurs are found only on the island of Madagascar so we say lemurs are endemic to Madagascar.

entropy: [*en*, within; *trop*, change] A measure of how chaotic a system is on the microscopic level. The second law of thermodynamics says the entropy of closed (isolated) systems must increase or stay the same. However, this does not prevent the entropy of an open system from decreasing. For example, a body of water can freeze and become a more ordered state (less chaotic at the microscopic level) only if energy is removed by the surroundings. This can only happen if the water is not insulated from the outside world. Any object that is giving off energy is losing entropy (becoming less chaotic at the microscopic level).

Eocene: [*eos*, dawn (of); *kainos*, recent (forms)] The second epoch of the Cenozoic Era. The Eocene lasted from about 56 to 34 mya and was a time when the mammals were experiencing rapid divergence. It was followed by the Oligocene and was preceded by the Paleocene Epoch.

eon: [*aion*, age] The largest divisions of geologic time are called eons or ages. The current geologic eon, the Phanerozoic, began with the Cambrian Period. The Phanerozoic was preceded by the Proterozoic, which was preceded by the Archean. The Hadean Eon is the time preceding the Archean. Often, the time before the Phanerozoic is simply called the Precambrian. See Phanerozoic, Proterozoic, Archean, Hadean, and epoch.

epoch: [*epoche*, pause] Geologic time is divided into eons, eons are divided into eras, eras are divided into periods, and periods are divided into epochs.

era; [*aera*, interval] A major division of geologic time. See epoch.

Euarchontoglires: [*eu*, true; *archon*, rulers; + *glires*, dormice] A superorder of mammals that evolved in Laurasia during the late Cretaceous. It includes the primates, tree shrews, colugos, and the Glires (rodents and rabbits).

eukaryote: [*eu*, true; *karyon*, kernel (i.e. true nucleus)] A member of the group of organisms with their DNA organized into chromosomes and held in a nucleus by a nuclear membrane. They generally reproduce sexually and have organelles such as mitochondria and plastids (in plants). Many are multicellular, and their cells are typically 10 to 100 microns long. Plants, animals, fungi, and protists are eukaryotes. See prokaryote.

eurypterid: [*eurys*, broad; *pteron*, fin] Large aquatic arthropods (10 cm to 2 meters long), also know as sea scorpions. They appeared in the Ordovician and disappeared during the great extinction at the end of the Permian.

Eusthenopteron: [*eustheno*, strong; *pteron*, wing or fin] A genus of rhipidistian fish that is believed to be very similar to the form that evolved into the amphibians. *Acanthostega* and *Ichthyostega*, two of the earliest known amphibians, share many common features with *Eusthenopteron*, such as similar skull bones, labyrinthodont teeth, and fin rays in their tails.

eutherian: [*eu*, true; *therion*, small animal] Mammals whose fetus is nourished with a placenta. All mammals except for marsupials and monotremes are placental mammals. See placental.

falsifiable: A word used with scientific theories to indicate that the potential for disproving scientific theories always exists. Falsifiable theories make definite predictions, and if a prediction is shown to be incorrect, the theory must be false.

family: Orders are subdivided into families. Species that belong to a given family are more closely related to each other than they are to members of other families. For example, a fox and a wolf belong to the dog family and are more closely related to each other than either is to a lion (a member of the cat family). See species.

fission, nuclear: [*fissio*, to split] A nuclear reaction that involves the splitting of a large nucleus into two smaller nuclei. Energy is generally released if the original nucleus is much larger than iron. Nuclear fission of either uranium-235 or plutonium-239 is the process that occurs in atom bombs.

fission, binary: A type of cell division found in bacteria. The DNA is duplicated and the cell divides into two approximately equal parts, but the process is not the same as mitosis.

flying lemur: see colugo.

foraminiferan: [*foramen*, opening; *ferre*, to carry] A single-celled protist with an easily fossilized calcite shell, usually called a test. They first appeared in the Cambrian and are the most common protist fossil.

fungus: A member of the kingdom Fungi, which includes the molds, yeast, and mushrooms. They lack chlorophyll and most are multicellular, but the partitions between the cells are often absent or do not completely divide the cells. This type of cell structure is referred to as multinucleate.

fusion, nuclear: [*fusus*, pour or melt] A nuclear reaction that involves the union of two nuclei to form one larger nucleus. Energy is generally given off if the fusion produces a nucleus smaller than iron. This is the type of nuclear reaction that occurs in a hydrogen bomb and in the deep interior of the Sun and other stars.

gamete: [*gamos*, marriage] A sex cell with a haploid chromosome number. In fertilization, two gametes unite to form a cell (called a zygote) with a diploid number of chromosomes. Eggs (or ova) and sperms are gametes.

gametophyte: [gamete plant] A plant or plant structure whose cells have a haploid number of chromosomes. It produces gametes by mitosis. In many plants, the gametophyte is a reasonably large plant, but in the flowering plants, the female gametophyte is the inner portion of the ovule, and the male gametophyte matures into a pollen grain. See sporophyte.

gamma radiation: Gamma radiation or gamma rays are the most energetic electromagnetic waves. They are generally emitted by the nuclei of certain radioactive elements.

gar: See Bowfin.

gastropod: [*gaster*, stomach; *pod*, foot] A class of mollusks whose members include snails and slugs. They appeared in the early Cambrian.

gene: [*genea*, generation] The recipe for a protein molecule that is coded on a strand of DNA.

gene frequency: The fraction of a particular allele of a given gene in a population. For example, the fraction of eye color genes in a population that are blue eyed genes.

gene pool: The total collection of genes held by all the breeding members of a population.

generation: In biology, the average time from the birth of one organism to the birth of that organism's offspring. In particle physics, a repeated pattern of fundamental particles that are the basic building blocks of matter. Members of the first generation are the up quark, the down quark, the electron, and the electron neutrino.

genetic drift: The frequency of a particular allele may fluctuate in a population due to random chance. This fluctuation is called genetic drift.

genotype: The genetic information carried by an individual organism, as opposed to the individual's visible features. See phenotype.

genus: Families are subdivided into genera. Species that belong to a given genus are more closely related to each other than they are to members of other genera. For example, a lion and a tiger belong to the same genus and are more closely related to each other than either is to a cheetah (a member of a different genus). See species.

gibbons: The smaller or lesser apes which, along with the great apes, are called hominoids. One gibbon is called the Siamang.

Glires: [*glires*, dormice] A clade of mammals that consists of the rodent order and the lagomorph order (rabbits, hares, and pikas). Lagomorphs and rodents split about 55 mya. They all have incisors that continually grow and must be worn down by gnawing or they will grow too long and the animal will starve. Rodents have two upper incisors and lagomorphs have four, but the second pair is smaller and located behind the first pair.

Gondwana: [from Gondwanaland, an Indian kingdom] The southern portion of Pangaea that was composed of South America, Antarctica, Australia, Africa, Madagascar, and India.

granite: The most common type of rock that forms the continents. Since it cooled relatively slowly, it is a course grained igneous rock. See basalt.

gymnosperm: [*gymnos*, naked; *sperma*, seed] A general term used to describe members of several groups of seed plants. The gymnosperms have naked seeds attached to the surface of cone scales. Flowering plants have seeds enclosed in an ovary, which develops into the fruit. Gymnosperms include the cycads, ginkgoes, and conifers.

Hadean: [from Hades, the Greek god of the underworld] The first of 4 eons that ended 4,000 mya and began with the formation of the planets about 4,600 mya. During the Hadean, the Earth's surface was too hot for large blocks of solid material to form.

hagfish: A living member of the order of jawless fish. See agnathan.

half-life: The time it takes for 50% (one half) of a given amount of radioactive material to be transformed.

haploid: [*haploos*, single] A haploid cell has only one set of chromosomes. Gametes (like eggs and sperms) are haploid cells, but normal body cells have 2 sets of chromosomes and are called diploid cells.

haplorhines: [*haploos*, simple; *rhinos*, nose] Members of the Haplorhini suborder of primates, which are characterized by dry noses. They are the tarsiers of Southeast Asia and the anthropoids (monkeys and apes). See strepsirhines.

heterochrony: [*heteros*, different; *chronos*, time] An evolutionary change caused by altering the timing of the development of different structures. For example, the larger human brain may, in part, be the result of mutations that provided an extended period of growth compared to the development of the rest of the body. A type of heterochrony known as neoteny results if the development of somatic features is slowed down relative to sexual maturation. Neoteny produces sexually mature individuals with juvenile features. See neoteny.

heterosporous: [*heteros*, different; *sporos*, seed] A plant that produces both male microspores and female megaspores. Megaspores grow into egg-producing female structures called gametophytes. Microspores grow into sperm-producing gametophytes. In higher vascular plants like angiosperms and gymnosperms, the female gametophyte is the inner portion of the ovule while the microspores become pollen grains.

heterozygous: [*heteros*, different; *zygotos*, joined] A word used to describe the condition of having two different genes for the same trait. For example, a person with the gene for brown eyes and the gene for blue eyes is said to be heterozygous for eye color. If a person has two identical eye-color

genes, they would be homozygous for eye color.

Holocene: [*holos*, whole; *kainos*, recent (kinds)] The last and current epoch of the 7 epochs of the Cenozoic Era, the Holocene began about 12 thousand years ago with the waning of the last glacial period. It was preceded by the Pleistocene Epoch.

holostean: [*holos*, whole; *osteo*, bone] A name used to describe the most primitive neopterygians (modern ray finned fishes). The only members are the gars and the Bowfin.

homeobox: [*homos*, same; box] All homeotic genes have a 180 base pair region which is called the homeobox. The homeobox codes for a 60 amino acid sequence called the homeodomain. The homeodomain is a helical structure that binds to a specific gene location on a DNA spiral and regulates the activity of that protein coding gene.

homeodomain: See homeobox.

homeotic complex: Many homeotic genes are linked in an array on a DNA strand. This array is called a homeotic complex. The order of the genes in the array corresponds to the order with which they are expressed in the developing embryo. The first gene controls the development of the front segment, and the last gene controls the last segment, etc. The homeotic genes in a homeotic complex are called Hox genes. See homeotic gene.

homeotic gene: One of several genes that are activated in the various segments of a developing embryo. Each homeotic gene produces proteins with a homeodomain helix that binds to a specific place on a DNA molecule and coordinates the production of the proteins that form that specific segment of the embryo. See homeobox.

hominid: [*homo*, man; *eidos*, form] Animals belonging to the great ape family Hominidae which includes orangutans, gorillas, chimpanzees, humans, and their closely related extinct relatives. The hominids probably evolved from an animal similar to *Proconsul* who lived about 20 million years ago.

hominoid: A member of the superfamily Hominoidea, whose living members include the gibbons and great apes.

homologous: [*homologos*, agreeing] Structures that share a common evolutionary history. The bones in the wing of a bat and the wing bones of a bird are homologous since they evolved from the front leg bones of a common reptile ancestor. However, as wings these structures are said to be analogous since each evolved independently as structures that function as wings. See analogous.

homoplastic: [*homos*, alike; *plastos*, formed] Structures that perform the same function but do not have a common origin like the wings of birds and butterflies. It is a synonym for analogous. See analogous.

homosporous: [*homos*, alike; *sporos*, seed] A plant that produces only one kind of spore. This spore produces a small plant, called a gametophyte, that in turn produces both eggs and sperms. Most living ferns are homosporous.

homozygous: See heterozygous.

honeycreeper: Neotropical birds in the tanager family or a group of Hawaiian birds that belong to the endemic subfamily Drepanidinae. This group is a good examples of adaptive radiation. Their bills range from thick seed-eating types to long thin sickle bills used for sipping nectar. When the Polynesians arrived there were about 60 species, but more than 40 have been exterminated because of the activities of humans.

honeyeater: A bird belonging to the honeyeater family of Australia and the islands of the Southwest Pacific. One member reached the Hawaiian Islands where a different species evolved on each of the four main islands. All of these unusual Hawaiian birds (of the genus *Moho*), have been exterminated by humans.

hormone: [*hormon*, set in motion] Molecules that are produced in one part of a plant or animal and initiate activities in other parts of the organism. Estradiol is a hormone that is produced in the ovaries and regulates the development of sex characteristics in women like enlarged breasts. Some hormones are proteins.

horsetail: A primitive vascular plant belonging to the class Sphenopsida. Horsetails are homosporous plants with jointed stems and scalelike leaves. They first appeared in the late Devonian and were very abundant in the Carboniferous coal forests. Many were large trees, but they declined after the Carboniferous, and all living forms are relatively small.

Hox gene: See homeotic complex.

hydrogen bonding: Molecules that have a positive side and a negative side are said to be polar molecules. If the negative portion of one molecule sticks to the positive portion of another molecule, the attraction is called a hydrogen bond.

Iapetus Ocean: [after Iapetus, the father of Atlas in Greek mythology] Name given the body of water that separated Laurentia and Baltica before the Devonian. It is sometimes called the Proto-Atlantic.

ichthyosaur: [*ichthys*, fish; *sauros*, lizard] A member of the order Ichthyosauria, dolphin-like marine reptiles that had a dorsal (back) fin and a large caudal (tail) fin that was partially supported by a downward bend in the vertebral column. Early forms from near the beginning of the Triassic still had five distinct fingers, but forms in the Jurassic and Cretaceous had many more than the usual number of bones in their fins. Unborn embryos in several German fossils indicate that this reptile gave birth to live young.

Ichthyostega: [*ichthys*, fish; *stega*, plate] One of the oldest known genera of amphibians. It had fishlike features such as fin-rays in its tail, and its skull bones match those of the rhipidistian fish. It had labyrinthodont teeth as did the rhipidistian fish, and its leg bones closely matched the bones in the fins of the rhipidistian fish. It was about one meter long, and lived in the late Devonian, about 370 mya.

igneous rock: [*ignis*, fire] Rock that is formed by the solidification of molten material. Igneous rocks are important in radiometric dating.

inductive reasoning: Reasoning that is used to formulate a general statement from specific observations. This type of reasoning may lead to incorrect conclusions, especially if the conclusions are based on a limited number of observations. In science, theories are formulated using inductive reasoning. Therefore, the potential always exists that a theory will be proven incorrect as more observations and experiments are carried out.

integument: [*integumentum*, cover] The covering of a seed. See seed.

ion: An atom that has lost (or captured) one or more electrons. If the atom has lost one or more electrons, it will have a net positive charge (a positive ion) and if the atom has captured one or more electrons it will have a net negative charge (a negative ion).

ionic bond: A chemical bond formed when an atom captures one or more electrons from another atom. The atom with the extra electron(s) has a net negative charge and the other atom has a net positive charge. Therefore the two atoms (called ions) attract each other.

isochron: [*isos*, equal; *chronos*, time] In radiometric dating, a mineral isochron is the straight line obtained by plotting the relative amount of daughter product versus the relative amount of radioactive parent for several minerals from a rock. The slope or steepness of the line will yield the age of the rock. If rocks from different parts of a formation are used, the line is called a whole-rock isochron.

isotope: [*isos*, equal; *topos*, place] Different forms of the same element are called isotopes. Isotopes of a specific element differ in neutron number. For example, uranium has 92 protons, but is a mixture of two isotopes; uranium-235 that has 143 neutrons and uranium-238 that has 146 neutrons.

jumping gene: see transposon.

Jurassic: [after the Jura Mountains in Switzerland and eastern France] The Jurassic was the second of the three periods of the Mesozoic Era and extended from about 201 to 145 mya. During this period, the dinosaurs took over the land and many large forms evolved. Birds first appeared in this period. It was followed by the Cretaceous Period and was preceded by the Triassic Period.

Kazakhstania: [after Kazakhstan, a republic in Central Asia] One of the continental fragments that existed before the formation of Pangaea. After uniting with Siberia, this unified land mass collided with Laurussia to form Laurasia during the Permian.

Kelvin temperature: [after Lord Kelvin (William Thomson)] The absolute temperature scale whose zero point is the coldest temperature possible. Zero Kelvin corresponds to 273 degrees below zero on the Celsius scale. A temperature change of one Kelvin is equivalent to a change of one degree Celsius, which means that the divisions on the two thermometers are the same size.

kingdom: Any of the major groupings of organisms that include the Bacteria (bacteria and cyanobacteria), Archaea, Protista, Fungi, Plantae, and Animalia. Closely related organisms in a kingdom are placed in a phylum (or division in the case of plants or certain plant-like organisms). See species and phylum.

labyrinthodont: [*labyrinthus*, maze; *odont*, tooth] A member of the subclass Labyrinthodontia, a group of amphibians, named because of the maze-like infoldings present in their teeth. This group includes two of the earliest amphibians, *Ichthyostega* and *Acanthostega*, that lived in the late

Devonian. The labyrinthodonts became extinct near the end of the Triassic.

lancelet: See amphioxus.

Latimeria: The genus name of the two species of coelacanths which are the only living crossopterygian fishes. Crossopterygians comprise one of the two orders of sarcopterygian or lobe-finned fishes. Amphibians evolved from an early crossopterygian fish. See coelacanth and sarcopterygian.

Laurasia: The northern portion of Pangaea that included North America, Europe, and Asia. Laurasia was formed when Laurussia united with Siberia and Kazakhstania during the Permian.

Laurasiatheria: [Laurasia, the land masses of North America and Eurasia; *theria*, wild animals] A superorder of mammals that includes the even-toed and odd-toed ungulates, carnivorans, bats, the whales and dolphins, the moles, shrews, hedgehogs, and solenodons. The group evolved in Laurasia during the late Cretaceous or early Cenozoic.

Laurentia: [after the Laurentian Mountains in southern Quebec] A land mass that included most of North America. Laurentia was an isolated land mass during the early Paleozoic, before the formation of Laurussia.

Laurussia: The land mass that was formed near the end of the Silurian and beginning of the Devonian when Laurentia (North America) collided with Baltica. The Caledonian Mountains in Britain and Scandinavia were formed along the suture line.

lemur: [*lemures*, ghosts] A group of primates restricted to the island of

Madagascar. Lemurs are prosimians. See prosimian.

lithosphere: [*lithos*, stone; *sphaera*, ball] The crust and upper mantle of the Earth form a relatively rigid layer that is called the lithosphere. It varies in thickness from about 40 to 280 kilometers. It floats on the asthenosphere, a more fluid layer. Beneath the deep oceans, it is thinner and more dense than the continental lithosphere. It is broken into about a half dozen large plates and several smaller ones, that move relative to one another and create various geologic features at their boundaries. For example, at a boundary where a continental plate and an ocean plate converge, the ocean plate will be subducted beneath the continental plate. This process results in the formation of a deep ocean trench and a line of volcanic mountains along the continental plate. The Andes Mountains were formed by this process. See asthenosphere.

Litopterna: [*litos*, smooth; *pteme*, heel bone] An order of South American ungulates whose members resembled horses and camels. The last members died out about the time humans arrived in South America.

lycopod: [*lukos*, wolf; *opsis*, appearance] A primitive vascular plant belonging to the division (phylum) Lycopodiophyta or Lycophyta. They first appeared in the Middle Devonian and were very common in the Carboniferous coal forests. Many grew to be large trees, but near the end of the Paleozoic they were replaced by the ferns and gymnosperms. Some lycopods were homosporous, but others were heterosporous. About 1000 small species are living today.

Macrauchenia: [*macra*, big; *auchenia*, llama] Genus name of an extinct 1000 kg camel-like notoungulate (South American ungulate).

mammal: [*mamma*, breast] A member of the class Mammalia, vertebrate animals whose living members have hair or fur and are characterized by the presence of milk-glands in the female. Fossil specimens have a jaw articulation formed by the squamosal and dentary bones. After this articulation was formed, the quadrate and articular bones became the ear bones we call the incus (anvil) and malleus (hammer) respectively. Mammals evolved close to the boundary between the Triassic and the Jurassic, about 200 million years ago. Today there are about 5400 species of living mammals grouped into about 29 orders and 146 families, of which about 1690 (40%) are rodents and about 850 (20%) are bats.

marsupial: [*marsupium*, pouch] A group of non placental mammals whose young are born in a very immature stage and are raised in a pouch by their mother. Marsupials may have as many as 5 upper and 4 lower incisors on each side, compared to a maximum of 3 in the placentals. Most of the common native Australian mammals are marsupials, as are the opossums of the Americas. Marsupials are also known as metatherians.

megaspore: A large spore that grows into a female gametophyte. This gametophyte produces the egg cells of the plant.

meiosis: [*meion*, smaller] The type of cell division that occurs during the formation of gametes (like eggs or sperms). It results in the production of four cells, each containing half as many chromosomes (called a haploid number) as normal body cells (a diploid number). When an egg and sperm unite, they form a cell with a diploid number of chromosomes. Meiosis appears to be a slightly modified form of the normal type of cell division that is called mitosis.

meridiungulates: [*meridies*, southern; *ungula*, hoof] A group of hoofed mammals that differentiated in South American. They split from the odd-toed ungulates about 70 mya and are members of the Laurasiatheria. Most died out after the Great American Interchange.

Mesozoic: [*meso*, middle; *zoic*, life] It is the middle era between the Paleozoic and Cenozoic of the Phanerozoic Eon. The Mesozoic extended from about 250 to 66 mya and is divided into three periods: the Triassic, Jurassic, and Cretaceous. The Mesozoic is often called as the age of reptiles. Birds, mammals, and flowering plants first appeared in the Mesozoic.

metamorphic rock: [*meta*, change of; *morphe*, form] Rocks that have been changed by heat and pressure. The pressure will squeeze the crystals together, sometimes giving the rock a layered appearance as in gneiss.

metatherian: [*meta*, beyond; *therion*, small animal] A marsupial mammal. See marsupial.

microspore: A small spore that develops into a male gametophyte plant that produces the sperm cells.

Mid-Continental Seaway: A large body of water that existed during the last half of the Cretaceous. It separated western North America (sometimes called Westamerica) from Eastern

North America and Europe, which were connected during that time. The eastern land mass is sometimes called Euramerica.

Miocene: [*meion*, less; *kainos*, recent (forms)] The fourth of the 7 epochs of the Cenozoic Era. It lasted from about 23 to 5.3 mya. Monkeys and apes diversified in the many tropical forests that flourished during this epoch. It was followed by the Pliocene and was preceded by the Oligocene Epoch.

Mississippian: [after the Mississippi River] The earlier subperiod of the Carboniferous Period that lasted from 359 to 323 mya.

mitochondria: [*mitos*, thread; *khondros*, granule] Organelles in the cells of eukaryotes that produce energy for the cell by aerobic respiration. Mitochondria contain genes and likely originated as independent bacteria that invaded the cell of an early organism.

mitosis: [*mitos*, thread] The normal type of cell division found in eukaryotes in which the chromosomes are duplicated, and two cells are formed that have the same number of chromosomes as the original cell.

moa: Any of the dozen or so species of large flightless birds that inhabited New Zealand. All became extinct soon after the Polynesians arrived.

Mollusca: [*molluscus*, soft] The animal phylum that contains the gastropods (snails), pelecypods (bivalves like the clams), and cephalopods (squids and octopuses). The ammonites were an important group of extinct mollusks. The body of a mollusk is composed of three parts, a muscular foot that is modified into tentacles in squids and octopuses, the digestive system and internal organs, and a protective covering called the mantle that often contains glands that secrete a shell. Mollusks appear to have evolved from either an early type of flatworm or annelid worm. See *Neopilina*.

moneran: [*monos*, single] A member of the group known as the prokaryotes. See prokaryote.

monophyletic: [*mono*, one; *phylon*, tribe] A group of organisms that evolved from the same common ancestor. A complete branch of an evolutionary family tree. See polyphyletic.

monoplacophoran: [*monos*, single; *plax*, plate] A member of a primitive class of limpet-shaped mollusks that appeared in the early Cambrian. See *Neopilina*.

monotreme: [*mono*, single; *trema*, hole] A member of the Monotremata order of egg laying mammals which includes the Platypus and the spiny anteaters or echidnas. In these mammals, eggs (in females), urine, and fecal matter are emptied into a chamber called the cloaca before they are passed to the outside through a single opening, hence the name monotreme. This condition is also found in reptiles and birds. Most female placental mammals have three separate openings. Monotremes are sometimes known as atherians.

Morganucodon: [after Glamorgan, South Wales; *odon*, tooth] Genus name of an early mammal with a body about the size of a mole or small rat, that possessed both the reptilian and the mammalian jaw articulations. These animals lived during early Jurassic times (about 200 mya).

mosaic evolution: Term that describes intermediate forms that display characteristics of two different groups. For example, *Archaeopteryx* had feathers and a furculum or wishbone like a bird, but also had the tail, head, and teeth of a reptile.

mosasaur: [*meuse*, river; *sauros*, lizard] A large fishlike marine lizard. Some were about 10 meters long. They appeared in the middle Cretaceous and became extinct at the end of the Cretaceous. Mosasaurs evolved from a group of aquatic lizards (aigialosaurs) that appeared in the late Jurassic.

multituberculates: [*multus*, many; *tuberculum*, lump] A group of rodent-like mammals distinguished by the many cusps (bumps) on their molar teeth. They were the most successful group of mammals known, lasting from the later Jurassic until the early Oligocene (a 130 million year span).

mutation: [*mutare*, to change] A base change in a gene is called a point mutation. These changes may produce a different amino acid sequence in the corresponding protein molecule, which in turn may show up as a modification in the organism. If the point mutation does not produce an amino acid change, it is called a synonymous or silent mutation. A missense mutation causes one amino acid to change to another, a nonsense mutation changes an amino acid to a termination code, and the insertion or deletion of a base is called a frame shift mutation. If a mutation causes one pyrimidine to be replaced by the other (adenine to guanine or visa versa) or one purine to be replaced by the other (cytosine to thymine or visa versa), it is called a transition. A transversion, which is about one-third as common, is where a pyrimidine replaces a purine or visa versa. Chromosomal rearrangements are also mutations as are any spontaneous changes in the genotype.

myoglobin: A protein molecule that stores oxygen in muscle cells.

natural selection: An idea put forward by Charles Darwin to explain why populations evolve. He realized that all organisms produce more offspring than could possibly survive to reproduce. The offspring exhibit various traits, and those individuals that are best suited to the environment will be more likely to survive and reproduce, passing their traits to the next generation.

nautiloid: [*nautilos*, sailor] A cephalopod mollusk that looks somewhat like a squid with short tentacles, except its body is protected by a large coiled shell like a snail. They first appeared in the late Cambrian.

Neanderthal: [from the Neander Valley in Germany] An archaic human that is considered a subspecies of *Homo sapiens*. Neanderthals lived between 250,000 and 40,000 years ago in Europe and Central Asia. They differed from modern humans by having shorter thicker bones, stronger muscles, a slightly larger brain, a prominent ridge across their brow, and a low sloping forehead. Some of their DNA is found in Europeans and Asians.

Neogene Period: [*neos*, young; *genos*, kind] The middle of the three periods of the Cenozoic Era. The Neogene began about 24 mya and is divided into two epochs: the Miocene and Pliocene. The Miocene accounts for about 85% of the period. The Neogene was preceded by the Paleogene Period and followed by the Quaternary, the last period of the era.

Neopilina: The genus name of a living mollusk (class Monoplacophora) with a single shell and other primitive features. Its internal structure may exhibit patterns of segmentation which has been interpreted as evidence that early mollusks evolved from an annelid (segmented worm) type ancestor. Mollusks and annelids share similar patterns of early development and are closely related. Monoplacophorans first appeared in the Cambrian.

neopterygian: [*neos*, new; *pterygion*, fin] A member of the modern group of ray-finned fishes. The neopterygians arose near the end of the Paleozoic and have radiated extensively. They have an advanced jaw structure that allows the mouth to push forward and be opened very wide. Except for the Bowfin (*Amia*), and the gars, all other neopterygians belong to an even more advanced group known as the teleosts. See Bowfin and actinopterygian.

neoteny: [*neos*, youth; *teinein*, to extend] A phenomenon in which an ancestral youthful character is retained in a sexually mature adult descendant. A possible example is the downy plumage of the ratites (large flightless birds like the ostrich and rhea). These adult birds have retained the plumage of a downy chick. Also, the skull of a human more closely resembles the skull of a young chimpanzee than it does the skull of an adult chimp. The flower may represent a modified leaf bud that never fully opened. See heterochrony.

neutral evolution: The idea that certain mutations are selectively neutral (they confer no advantage or disadvantage) and, therefore, can spread throughout a population by pure chance.

niche: [*nicher*, nest] A term used to describe an environment that supplies the needs of a specific plant or animal.

nothosaur: [*nothos*, false; *sauros*, lizard] A group of marine reptiles that grew up to 4 meters long. They are known from the middle Triassic and may be the ancestors of the plesiosaurs.

notochord: [*noton*, back; *chorde*, cord] A stiff rod located just below the dorsal nerve cord that runs down the back of animals belonging to the phylum Chordata. All vertebrates possess a notochord during their embryonic development and are, therefore, classified as chordates. Amphioxus is a chordate, but not a vertebrate.

Notoungulata: [*notos*, south; *ungulatus*, hoof] An order of South American hoofed mammals some of which resembled rhinoceroses and hippopotamuses.

nucleon: [*nucleus*, kernel or core] A particle in the nucleus of an atom, either a proton or a neutron.

nucleotide: A molecule composed of a five-carbon sugar, a phosphate group, and one of four different bases. Nucleic acids are long chains composed of the four different nucleotides. See DNA and RNA.

nudibranch: [*nudus*, naked; *brankhia*, gills] A group of marine gastropod mollusks, which are often brightly colored. Their name is derived from the fact that some breathe with gills protruding from their back. There are over 3000 species.

Ockham's (or Occam's) Razor: A principle attributed to William of Ockham, a fourteenth century philosopher. In science, the principle states that one must pick the simplest theory that is consistent with the observations.

Oldowan tool: [after the Olduvai Gorge in Tanzania] Name of the earliest type of tools, made between 2 and 1.5 mya. They consist of sharp flakes used as crude cutting tools and scrapers and were probably produced by *Homo habilis*.

Oligocene: [*oligo*, few; *kainos*, (of) recent (forms)] The third epoch of the Cenozoic Era, the Oligocene lasted from about 34 to 23 mya. The mammals were well established by the Oligocene, and many large forms developed during this epoch. It was followed by the Miocene and was preceded by the Eocene Epoch.

Onychophora: [*onych*, claw; *phoros*, carrying] The phylum to which the velvet worms belong. Velvet worms, like *Peripatus*, have features found in both the annelid worms and the arthropods. They are believed to resemble the common ancestor of these two closely related phyla. A 530 million year old Burgess Shale fossil of an animal called *Aysheaia* closely resembles the velvet worms of today.

order: Classes are subdivided into orders. Species that belong to a given order are more closely related to each other than they are to members of other orders. For example, lions and wolves belong to the carnivore order and are more closely related to each other than either is to a beaver (a member of the rodent order). See species.

Ordovician: [from Ordovices, the Latin name for an ancient tribe of people in Wales] The second period of the Paleozoic Era, the Ordovician extended from about 485 to 443 mya.

Fragmentary evidence believed to be from jawless fish comes from this period, but fish were certainly a very minor part of the fauna if they were present. Trilobites were still fairly common and mollusks became very abundant. It was followed by the Silurian and was preceded by the Cambrian Period.

ornithischian: [*ornis*, bird; *ischion*, hip] A member of the order Ornithischia, one of the two orders of reptiles commonly referred to as dinosaurs. This order developed strong birdlike hip bones, but were not the ancestors of birds. All members of this order were plant eaters, and its members included: *Stegosaurus*, *Ankylosaurus*, *Iguanodon*, and *Triceratops*. The ornithischians lived during Jurassic and Cretaceous times, 200 to 65 mya.

orogeny: [*oros*, mountain; *genesis*, creation or origin] A word referring to the mountain building events that occur when two tectonic plates converge, either two continental plates or an oceanic with a continental plate.

Osteichthyes: [*osteon*, bone; *ichthys*, fish] The group of fishes commonly known as the bony fish, the most advanced fish. They first appeared in the late Silurian (about 420 mya). This group is divided into two classes, the lobe-finned fishes (Sarcopterygii) and the ray-finned fishes (Actinopterygii).

ostracoderm: [*ostrakon*, shell; *derma*, skin] A general term used to describe the early jawless fish (agnathans) that were covered with bony plates. Plate fragments suggest that they first appeared in the Ordovician, but more complete fossils are first found in the Silurian. They became extinct near the end of the Devonian.

ovary: [*ovum*, egg] A structure that produces eggs or ova. In plants it encloses the ovules. In flowering plants, the ovary develops into the fruit.

ovule: [*ovum*, egg] A structure in higher plants that produces haploid megaspores by meiosis. Megaspores grow into gametophytes that produce egg cells. After fertilization, the diploid zygote grows into an embryo plant. See megaspore.

ozone: An oxygen molecule composed of three oxygen atoms instead of two as with ordinary molecular oxygen. There is a concentration of ozone between the altitudes of 10 and 50 kilometers that is formed by the action of sunlight. Ozone is a very efficient absorber of short wavelength ultraviolet light, and the ozone layer protects us from this dangerous form of radiation. Without the ozone layer, life would not be possible on the land. See chlorofluorocarbons.

paddlefish: A living chondrostean (primitive ray-finned fish). Paddlefish are found in the Mississippi Valley (genus *Polyodon*) and China (genus *Psephurus*). The Chinese Paddlefish is most likely extinct.

Paleocene: [*paleo*, ancient; *kainos*, recent (forms)] The first of the seven epochs of the Cenozoic Era. It lasted from about 66 to 56 mya and was a time of great mammal diversification after the extinction of the dinosaurs at the end of the Mesozoic. It was followed by the Eocene Epoch.

Paleogene Period: [*palaios*, ancient; *genes*, specific kind] The first of the three periods of the Cenozoic Era, the Paleogene extended from about 65 to 24 mya. Since it began just after the dinosaurs became extinct, the Paleogene was a time when the mammals experienced their greatest divergence. Many new forms arose to fill the niches left vacant by the dinosaurs. The Paleogene is composed of three epochs, the Paleocene, Eocene, and Oligocene. It was followed by the Neogene Period and was preceded by the Cretaceous Period.

paleomagnetism: [*palaios*, ancient; magnetism] The study of rock magnetization which is used to determine the location of a land mass. The latitude, but not the longitude, can be determined from the inclination of the Earth's magnetic field that was frozen into igneous rocks when they solidified from magma.

Paleozoic: [*paleo*, ancient; *zoic*, life] The first of the three eras of the Phanerozoic Eon, the Paleozoic began with the start of the Cambrian about 541 mya and ended with a great extinction at the close of the Permian approximately 250 mya. The Paleozoic is divided into six periods: the Cambrian, Ordovician, Silurian, Devonian, Carboniferous, and Permian. Life in the sea diversified greatly during the Paleozoic. Although no life existed on the land at the beginning of the era, by the end of the era many plants and animals had adapted to life on the land.

Panderichthys: [after Christian H. Pander; *ichthys*, fish] Genus name of a 380 million year old sarcopterygian fish fossil considered to be a transitional form because its pectoral girdle is more developed, allowing the fish to support itself in shallow water.

Pangaea: [*pan*, whole; *gaia*, Earth] A supercontinent formed when the major continental blocks were assembled near the beginning of the Mesozoic Era, about 250 mya. The portion north of the Equator is called Laurasia and the southern portion is called Gondwana.

pelecypod: [*pelecys*, axe; *pod*, foot] Any of the bivalve mollusks that includes the clams, mussels, oysters, and scallops. These animals have a two-part shell and first appeared in the early part of the Cambrian.

pelycosaur: [*pelyc*, bowl; *sauros*, lizard] A member of the group Pelycosauria, and one of the two groups of the Synapsida. These lizard-like reptiles with differentiated teeth were close ancestors of the mammals. They lived from middle Carboniferous to middle Permian times, about 330 to 270 mya. A well-known example of the group is *Dimetrodon*, but the earliest known member is *Archaeothyris*, whose fossils have been found in the hollow stumps of lycopod trees in Nova Scotia. See synapsid.

Pennsylvanian: [after Pennsylvania] Named because of the state's extensive coal seams. This later subperiod of the Carboniferous lasted from 323 to 299 mya.

period: A unit of geologic time generally tens of millions of years long. See epoch.

perissodactyl: [*perissos*, odd; *dactylos*, fingers] A mammal belonging to the order Perissodactyla, commonly known as the odd-toed ungulates. They were once more common, but their numbers have declined in the later Cenozoic. Living representatives of this group are the horses, rhinoceroses, and tapirs.

Permian: [after the town of Perm in central Russia] The sixth and final period of the Paleozoic Era, the

Permian extended from about 299 to 252 mya. The Permian was a time when the mammal-like reptiles rose to be the dominate land animals. Perhaps the greatest extinction of all times marked the end of this period. It was followed by the Triassic and was preceded by the Carboniferous Period.

permineralized: [*per-*, by means of] The process of fossilization where minerals are deposited between and within the cells of an organism. Also known as petrification, it often preserves microscopic details of the internal parts of the organism.

Phanerozoic: [*phaneros*, visible; *zoic*, life] The fourth and last eon, it followed the Proterozoic and extended from the beginning of the Cambrian Period to the present day (a 541 million year span). The Phanerozoic is divided into three eras: Paleozoic, Mesozoic, and Cenozoic.

phenotype: [*phainein*, to show; *typos*, type] The visible characteristics of an organism such as their blue eyes and brown hair. See genotype.

phylum: [*phyton*, plant] Kingdoms are subdivided into phyla. Since the word phylum comes from the word meaning plant, the corresponding groupings in the plant kingdom are called divisions. See species.

phytoplankton: [*phyton*, plant; *plankton*, wandering] See plankton.

placental: [*placenta*, flat cake] A mammal whose young develop in their mother's uterus before birth. While in the uterus, they are attached by an umbilical cord to a structure called the placenta that is attached to the uterine wall. The placenta facilitates the flow of nutrients to the embryo and wastes to the mother. The only non placental mammals are the marsupials and the egg laying monotremes. Placentals are also called eutherians.

placoderm: [*plax*, plate; *derma*, skin] A member of the class Placodermi, one of the four classes of fish. Placoderms were heavily armored with bony plates and were among the first fish to have jaws and paired fins. They appeared in the early Devonian and diversified greatly, but became extinct near the end of the same period.

placodont: [*plax,* plate; *odont,* tooth] A short legged marine reptile, some of which were covered with bony armor and looked somewhat like turtles, although they were not closely related. They apparently ate shelled mollusks like clams. They are known only from the later half of the Triassic.

plankton: [*plangktos*, wandering] A general term for the masses of small organisms that float near the surface of the ocean. The animal component is called zooplankton, and the plant component is called phytoplankton.

plant: A member of the kingdom Plantae. The plants include green algae, the bryophytes, and the vascular plants.

plastid: [*plastos*, moulded] Organelles in the cells of certain eukaryotes that carry out the process of photosynthesis. Plastids (called chloroplasts in the higher plants and algae) contain genes and represent the descendants of prokaryotes that invaded the cells of early organisms. This invasion may have happened more than once since the plastids in red algae (rhodophytes) are different from the plastids of the green algae (chlorophytes) and higher plants. See *Prochloron*.

platyrrhine: [*platys*, wide; *rhinos*, nose] A member of the New World monkeys, many of which have a prehensile tail that functions as a hand. Their noses have widely separated round nostrils that point to the sides.

Pleistocene: [*pleistos*, most; *kainos*, recent (forms)] The sixth of seven epochs that comprise the Cenozoic Era. The Pleistocene, that began about 2.6 million years ago and ended about 12 thousand years ago, is best known for the great ice sheets that waxed and waned during this epoch. The Woolly Mammoth and Woolly Rhinoceros roamed the North, and early humans evolved to become *Homo sapiens*. It was followed by the Holocene and was preceded by the Pliocene Epoch.

plesiosaur: [*plesios*, near; *sauros*. lizard] A long necked sea serpent-like marine reptile belonging to the order Plesiosauria. They had large paddle-like flippers and relatively short tails, but some reached lengths of over 12 meters. Plesiosaurs lived from early Jurassic times until the great extinction at the end of the Cretaceous.

Pliocene: [*pleion*, more; *kainos*, recent (forms)] The fifth of the 7 epochs of the Cenozoic Era. The Pliocene lasted from about 5.3 to 2.6 mya. It was followed by the Pleistocene and was preceded by the Miocene Epoch.

polyphyletic: [*polys*, many; *phylon*, tribe] A group of organisms whose last common ancestor was not a member of the group. For example, monotremes may have evolved from a different reptile ancestor than other mammals. If it is true, then the mammals form a polyphyletic group. All the organisms that evolved from a single ancestor are said to form a monophyletic group. See clade.

polyploidy: A condition where an organism has more than two sets (2n) of chromosomes. The organism is called a polyploid. It is a method of speciation that commonly occurs in plants, where a diploid plant (2n) produces a tetraploid offspring (4n). The tetraploid is a new species since it cannot pollinate with the original plants. 30% to 80% of all plant species arc polyploid. Common examples are the potato, peanut, some apples, and some wheat. In animals, polyploidy is more common in invertebrates, but several fish and amphibian species are polyploids. About 15% of human miscarriages are triploid fetuses (3n), usually due to the egg being fertilized by two sperms.

Precambrian: The interval of time before the Cambrian. Precambrian is not an official time interval, but is often used when referring to times before the Cambrian which began about 541 million years ago.

primitive: [*primitivus*, first of its kind] A word used in evolution to describe an organism or part of an organism that is more like its ancestral form. A feature that has undergone a great deal of change is said to be advanced. A more accurate way of describing a primitive feature would be to say it is not as specialized or as highly evolved.

Prochloron: [*pro*, before; chloroplasts] Genus of a prokaryote that has both chlorophyll a and b as do the chloroplasts of higher plants. *Prochloron* lives in close association with a group of animals called tunicates.

Proconsul: [*pro*, before; Consul, a chimpanzee who lived in the 1900s] The oldest fossil ape, *Proconsul africanus*, is a possible ancestor of the great apes. Some scientists place this 20 million year old fossil in the genus *Dryopithecus*.

progymnosperm: [*pro*, before; gymnosperm] A member of the class of vascular plants that were the ancestors of the gymnosperms. They had the leaves of a free-sporing fern and secondary wood with annual growth rings characteristic of modern conifers. An example is *Archaeopteris*.

prokaryote: [*pro*, before; *karyon*, kernel (i.e. before the nucleus)] Members of the kingdoms Bacteria (the bacteria and blue-green algae or cyanobacteria) and the Archaea. These organisms do not have a membrane bounded nucleus as eukaryotes do. Prokaryotic cells divide by binary fission, not by mitosis as do the cells of eukaryotes. They are generally single-celled organisms or colonies of organisms, and their cells are 1 to 10 microns across, much smaller than a typical eukaryotic cell. See eukaryote.

prosimian: [*pro*, before; *simia*, apes] A member of the so called lower primate group. They include the tarsiers of Southeast Asia, the lemurs of Madagascar, galagos or bushbabies, and the Bosman's Potto of Africa and the lorises of India, Sri Lanka, and southeast Asia.

protein: [*proteion*, first] A long chain of amino acids, often folded in complicated ways. Protein molecules give structure to living organisms and help carry out many of their important bodily functions.

Proterozoic: [*proteros*, first; *zoon*, life] The third eon, it followed the Archean Eon and preceded the Phanerozoic Eon. The Proterozoic began about 2,500 mya and came to a close at the beginning of the Cambrian.

protist: [*protistos*, first of all] A member of the kingdom Protista, single-celled or simple multicellular organisms. Some members are animal-like, some are plant-like, and some are fungal-like.

protostome: [*protos*, first; *stoma*, mouth] See deuterostome.

Psilotopsida: The class of primitive vascular plants that have many algae-like characteristics. *Psilotum nudum*, a living member of this group, is believed to be a close relative of members of the class Rhyniopsida that are the most primitive vascular plants known only from fossils. Examples of this extinct group include *Cooksonia* and *Rhynia*.

pterosaur: [*pteron*, wing; *sauros*, lizard] Any of the group of winged reptiles that first appeared in the late Triassic and lived throughout the rest of the Mesozoic Era. They were not dinosaurs, but they had a diapsid skull as the dinosaurs did. They had four fingers. The first three were short with claws, but the fourth was greatly elongated and supported the wing membrane. In bats, the wing membrane is supported by four fingers. Well-known pterosaurs are *Pteranodon* and *Pterodactylus* [meaning *wing finger*].

punctuated equilibrium: The idea that most species remain relatively stable with little change over relatively long periods of time and changes occur quickly during relatively short periods of time. Therefore, the fossil record is punctuated with sudden changes.

quark: One of the fundamental particles out of which composite particles like the proton and neutron are formed. While single quarks have never been detected, their postulated existence has explained all the exotic particles that have been made in particle accelerators. There appear to be six kinds (called flavors) of quarks: up, down, strange, charmed, top, and bottom. All the observed composite particles can be explained as combinations of three bound quarks or as a bound quark and antiquark.

Quaternary Period: [*quaterni*, fourth] Originally it was the name used for the fourth period of the history of Earth. Now, it is the name of the third and last period of the Cenozoic (the last 2.6 million years). It is divided into two epochs, the Pleistocene and Holocene and it followed the Neogene Period.

radiolarian: [*radiolus*, ray, spike] A single-celled organism that has a shell or test made of silica. They first appeared in the Cambrian Period and are still found today.

radius: Of the two bones in the lower part of the arm, it is the one on the thumb side. The ulna is on the little finger or elbow side.

rank: As it is applied to taxonomic classification, a rank is the name given to a level of taxonomic classification. For example, the highest level of classification is the rank of kingdom, the next highest is the rank of phylum or division, next is class, next is order, next is family, next is genus, and the lowest level is the rank of species.

Recent Epoch: See Holocene Epoch.

Reedfish: See bichir.

reptile: [*repere*, to crawl] A member of the class Reptilia. This class of vertebrates includes every amniote not closely resembling living birds or mammals. It is a rather disjointed group. For example, crocodiles are more closely related to birds than they are to the turtles. There are about 10,800 living species divided into 4 orders: crocodilians (25 species), turtles (350 species), Tuatara (1 species), snakes (3,700 species) and lizards (6,700 species). Snakes and lizards belong to the same order (Squamata), and about 12 other orders are known only from fossils.

rhipidistian: [*rhipis*, fan] A type of sarcopterygian or lobe-finned fish. See sarcopterygian.

Rhynchocephalia: [*rhyngchos*, snout; *kephale*, head] An ancient order of true diapsid reptiles whose only living member is the Tuatara. See Tuatara.

ribosome: [*ribo*, RNA; *soma*, body] A structure composed of protein and RNA molecules, that uses an RNA template (the recipe) to synthesize protein molecules.

RNA: Ribonucleic acid, a long chain of molecules called ribonucleotides. A ribonucleotide is made of a ribose sugar, a phosphate group, and one of four different bases (adenine, uracil, guanine, or cytosine).

Rodinia: [from the Russian meaning *homeland*] A supercontinent that is believed to have formed about 1100 mya. It began fragmenting about 750 mya.

sarcopterygian: [*sarkodes*, fleshy; *pterygion*, fin] A member of the Sarcopterygii (the lobe-finned fish). Members of this class had paired pectoral (front) and paired pelvic (rear) fins that had fleshy bases. Muscles connected to the bones of the fin could manipulate the fin. This class is divided into two groups, an extinct group called the rhipidistians and the coelacanths, which contains two living species of the genus *Latimeria*. *Eusthenopteron*, or a close relative, was a rhipidistian from which the amphibians evolved. See dipnoan.

saurischian: [*sauros*, lizard; *ischion*, hip] A member of the order Saurischia, one of the two ancient orders of reptiles commonly referred to as dinosaurs. Saurischians had hip bones similar to their more primitive ancestors. This order, which is usually divided into the sauropods and theropods, contained the biggest dinosaurs as well as the most powerful carnivores. Members include: *Diplodocus*, *Brachiosaurus*, *Allosaurus*, and *Tyrannosaurus*. Saurischians lived during the Jurassic and Cretaceous Periods, 200 to 65 mya.

sauropod: [*sauros*, lizard; *pod*, feet] Sauropods are one of the two groups of saurischian dinosaurs. They were herbivorous dinosaurs who walked on all four legs. Examples are *Diplodocus* and *Brachiosaurus*.

sauropterygian: [*sauros*, lizard; *pterygion*, fin] A member of the order Sauropterygia, sea serpent-like reptiles that lived during most of the Mesozoic Era. Most of the prominent members belonged to a group called the plesiosaurs. An early sauropterygian that lived during the Triassic Period was *Nothosaurus*.

scale: A generic term used to describe the plate-like coverings of fish, reptiles, and butterflies. Fish scales are plates of bone and are classified into four types. Placoid scales are toothlike structures that cover the skin of sharks, rays, and skates. Shark teeth are modified scales.

Ganoid scales are non-overlapping, diamond-shaped structures composed of bone covered with a shiny substance called ganoin. They are present on primitive fish like the gars. Cycloid and ctenoid scales are thin, light and flexible structures arranged in overlapping rows on modern bony fish. Ctenoid scales have a comb-like edge and cycloid scales have rounded edges. Reptile scales are localized thickenings of the protein keratin. They are connected by regions of thinner material and are completely different structurally from fish scales. Reptile-type scales are found on the legs of birds and the tails of some mammals.

sea scorpion: See eurypterid.

sedimentary rock: [*sedimentum*, settling] Layered rock formed by the accumulation of small pieces of other rocks and organic material.

seed: A structure in higher plants that forms after the egg is fertilized by the sperm. It contains the embryo plant, stored food, and a covering called the seed coat or integument. Seeds have a diploid number of chromosomes and they grow into a sporophyte plant.

Silurian: [from the Silures, an ancient Celtic tribe] The third period of the Paleozoic Era, the Silurian extended from about 443 to 419 mya. The first clear evidence of fish, including the first fish with jaws, comes from this period. The first vascular land plants are from the end of this period. It was followed by the Devonian and was preceded by the Ordovician Period.

Solenodon: [*solen*, pipe; *odon*, tooth] Genus of an insectivorous mammal that looks like a long nosed rat and is one of the few venomous mammals.

One species is found on the island of Cuba and another on Hispaniola.

somatic cells: [*soma*, body] The body cells of an organism as opposed to the reproductive or germ cells like the egg and sperm. Somatic cells have a diploid number of chromosomes, but germ cells have a haploid number (half as many).

species: [*specere*, to look] An isolated group of organisms whose members are capable of interbreeding and producing fertile offspring. See rank.

sperm: See gamete.

Sphenodon: [*sphen*, wedge; *odon*, teeth] Genus of Tuatara. See Tuatara.

sphenophyte or sphenopsid: [*sphen*, wedge; *phyton*, plant; *opsis*, appearance] See horsetail.

spore: [*sporos*, seed] A reproductive cell that is produced by meiosis, so they have a haploid number of chromosomes. Spores grow into a gametophyte plant.

sporophyte: [*sporos*, spore or seed; *phyton*, plant] A plant with a diploid number of chromosomes in its cells. The plant produces haploid spores by meiosis, and the spores grow into the gametophyte plant. In most well known plants, the sporophyte is the plant we see. See gametophyte.

Squamata: [*squama*, scale] The reptile order of snakes and lizards.

stratigraphy: [*stratus*, spread out] The study of rock layers or strata.

strepsirhines: [*strepsis*, curved; *rhinos*, nose] Members of the Strepsirhini suborder of primates, characterized by wet noses. The group includes all the prosimians except the tarsiers. All members except the Aye-aye have toothcombs made of tightly clustered lower incisors and canine teeth that are used for grooming. See prosimian.

stromatolite: [*stroma*, bed covering; *lithos*, rock] A layered, mound or column shaped structure, formed by a colony of cyanobacteria. Limestone, dolomite, chert and other minerals are trapped or precipitated within the colony. Fossilized stromatolites are most common in Precambrian deposits, and some contain algal filaments. Living stromatolites can be found in Shark Bay, Western Australia.

sturgeon: A chondrostean or primitive ray-finned fish. Sturgeons have a few rows of thick bony plates (called ganoid scales) on their sides and an asymmetrical tail which are indications of their primitive nature. See scale.

subfossil: A term used to describe the remains of recently extinct animals like the moas of New Zealand and elephant birds of Madagascar. These animals have become extinct in historical times, due largely to human actions.

symbiotic: [*sun*, with; *bios*, live] Organisms that live together for the benefit of both are said to have a symbiotic relationship.

synapsid: [*sun*, with; *apsid*, arch] A member of the reptile subclass Synapsida, but also a general term used to describe this group of reptiles and their ancestors, the mammals. Synapsid reptiles, commonly known as mammal-like reptiles, have a single opening on each side of the skull below the postorbital and squamosal bones. This opening allowed for the efficient attachment of the jaw muscles and led to the development of very strong jaws. The reptiles belonging to this group are the pelycosaurs and the therapsids. The synapsid reptiles became extinct in the Jurassic after giving rise to the mammals.

taxon: [*taxis*, arrangement] A group of organisms that form a specific unit. Usually the taxon is given a rank such as phylum, family, species, etc., and a name. For example, the taxon of animals that is given the name birds (Aves), is given the taxonomic rank of class.

taxonomy: [*taxis*, arrangement; *nomos*, law or knowledge] The branch of science that describes and classifies organisms.

teleost: [*telos*, end; *osteon*, bone] The most advanced group of neopterygians (modern ray-finned fish). Teleosts have small overlapping scales, and their tail is very symmetrical as the backbone stops at the base of the tail. All but a very few of the living bony fish belong to this group.

tenrec: Insectivorous mammals that are endemic to Madagascar. Tenrecs look like long nosed rats although they are not closely related. They are members of the superorder Afrotheria.

terrane: A geologic entity whose history is distinct from the adjoining land. For example, if an oceanic plate is subducted beneath a continental plate an island on an oceanic plate will eventually run into a continent. The island will be scraped off and will become a terrane as it is sutured to the continent.

Tertiary: [*tertius*, third] Originally, the history of the Earth was divided into four periods and this was the third.

Later, Tertiary was used for the first period of the Cenozoic Era. It contained the first five epochs. Currently the Cenozoic is divided into three periods: the Paleogene, Neogene, and Quaternary. Tertiary is no longer used.

Tethys Sea: [in Greek mythology Tethys was the daughter of Uranus and Gaia. She and her brother Oceanus, they gave birth to all the waters of Earth] The body of water that partly separated Laurasia from Gondwana (they were connected at the west end). A widening of the Tethys Sea in the Triassic and Jurassic initiated the breakup of Pangaea.

tetrapod: [*tetra*, four; *pod*, feet] A general term used to describe any of the four-legged vertebrates which are the amphibians, reptiles, birds, and mammals.

thallophyte: [*thallos*, young shoot; *phyton*, plant] A general term used to describe those plants (green, red, and brown algae) that are not differentiated into leaves, stems, and roots. The entire plant is called a thallus.

thecodont: [*theca*, sheath; *odon*, teeth] A member of the order Thecodontia, a primitive group of Triassic reptiles that included the ancestors of the dinosaurs. Some thecodonts developed bipedal locomotion.

theory: A word commonly used by scientists to describe anything from a wild speculation to an idea that provides our most accurate description of nature. Although not precisely defined, in this book we have reserved the word theory to describe a broad general statement that explains and unifies many observations and experimental results. For example, the theory of quantum mechanics. Because of their broad application, theories cannot be proven. Words like law, principle, and theory are often applied somewhat arbitrarily.

therapsid: [*ther*, beast; *apsid*, arch] A member of the order Therapsida, which is one of the two groups of the subclass Synapsida (the mammal-like reptiles). These reptiles had differentiated teeth and limbs similar to those of mammals. They lived from the middle Permian through the Triassic, about 270 to 200 mya. The largest and most important group of therapsids where the cynodonts, the group from which the mammals evolved. See pelycosaur.

therian: [*theria*, wild animals] A member of the subclass Theria, mammals that give birth to live young. They are the marsupials (sometimes called the metatherians) and the placental mammals (the eutherians). The monotremes are not included.

theropod: [*ther*, beast; *pod*, foot] A bipedal carnivorous saurischian dinosaur. The saurischian dinosaurs are usually divided into two groups, the sauropods (which were herbivorous and walked on all four legs) and the theropods. Well-known theropods are *Allosaurus* and *Tyrannosaurus*.

Tiktaalik: [Inuktitut word for large freshwater fish] Genus name of a sarcopterygian fish that is a transitional form. They had front limbs and shoulders that were like the first tetrapods, a skull that was flattened more like a crocodile, and eyes on top of its head.

tonne or metric ton: A unit of mass that is equivalent to 1000 kg. In the US it is often called a metric ton, and its weight is about 2200 pounds.

Toxodon: [*tox*, bow; *odont*, tooth] Genus name of an extinct rhinoceros sized notoungulate (South American ungulate) and close relative of the *Mixotoxodon.*

tracheophyte: [*trachia*, windpipe; *phyton*, plant] A vascular plant whose members form one of the two major groupings of the plant kingdom (the other is the bryophyte group). Some examples include the club mosses, horsetails, ferns, gymnosperms, and angiosperms (flowering plants). They have chlorophyll a and b, store true starch, and have cellulose in their cell walls. The oldest known vascular plant fossil (*Cooksonia*) appeared in middle Silurian times, about 420 million years ago. See bryophyte and Psilotopsida.

transposon: Sometimes called a jumping gene, it is a piece of DNA that can make a duplicate of itself and then that duplicate can insert somewhere else in the genome of the organism. This insertion can change a gene or the way a gene is expressed.

Triassic: [named for the three distinct rock layers in southern Germany] The first of three periods of the Mesozoic Era, the Triassic extended from about 252 to 201 million years ago. Reptiles became common sea creatures and the land remained in control of the reptiles, but no really large dinosaurs lived during this period. The earliest mammal fossils are from the Triassic. It was followed by the Jurassic Period and preceded by the Permian Period.

triconodonts: [*tri*, three; *conus*, peak; *odont*, teeth] A group of mammals that lived from Middle Jurassic through the Cretaceous in Laurasia and northern Africa. Their lower molars had three conical cusps lined up in a row. The central cusp was larger with smaller ones in front and back.

Tuatara: [Maori for peaks on the back] A reptile of the genus *Sphenodon* and order Sphenodontida, an ancient order of diapsid reptiles. The Tuatara is the only living member of the order. It is endemic to New Zealand.

tubercle: [*tuberculum*, lump] A point or bump on the chewing surface of a tooth. See cusp.

Turgai Sea: [after the Turgi River in Kazakhstan] The name given to the shallow body of water that separated Europe and Asia from Middle Jurassic to Oligocene times.

ulna: [*ulna*. elbow] the bone on the little finger side of the forearm that joins the elbow.

unconformity: [*un*, not; conform] A surface that has been worn down by erosion. Unconformities occur where sediment was not deposited, and they are discontinuities in the geologic column at that location.

ungulate: [*ungula*, hoof] A general term used to describe any of the hoofed mammals. Ungulates walk on their hoofs (which are modified claws) and have lost their clavicle or collarbone which enables them to run very fast.

Urodela: [*oura*, tail; *delos*, visible] An order of amphibians, the salamanders.

ventral: [*venter*, belly] The belly side of an animal. See dorsal.

vestigial: [*vestigium*, trace or footprint] A structure or organ that no longer performs its original function or its original function is greatly reduced. For example, a few cave dwelling fish and salamanders have vestigial eyes that do not function.

virus: [*virus*, poisonous liquid] A strand of nucleic acid, either DNA or RNA, that is enclosed in a protein coat. Many scientists consider them to be nonliving. Viruses are able to take over the protein replication mechanism of a cell and use it to reproduce themselves. Viruses may be either very primitive life forms or highly modified cellular life forms that have adapted to a parasitic lifestyle.

Xenarthra: [*xenos*, strange; *arthron*, joint] A superorder of mammals that diverged very early from other placentals and evolved in South America. Xenarthrans have an additional pair of joints on their vertebrae, hence the name. The 30 living members are anteaters, armadillos, and sloths.

zygote: [*zygotos*, joined] A cell formed by the union of two gametes. In higher animals the gametes are the egg and sperm. The zygote has a diploid number of chromosomes. See gamete.

414

Figure Credits

2.9 Harris, W.E. (1996). A Catalog of Parameters for Globular Clusters in the Milky Way. *Astron.J.* **112**: 1487-1488.

2.24 Anders, Edward and Nicolas Grevesse. (1989). Abundances of the elements: Meteoritic and solar. *Geochim. Cosmochim. Acta* **53**: 197-214.

3.5 Gale, N.H., R.D. Beckinsale, and A.J. Wadge. (1979). A Rb-Sr whole rock isochron for the Stockdale Rhyolite of the English Lake District and a revised mid-Palaeozoic time-scale. *J. geol Soc London* **136**: 235-242.

3.6 Reimer, P.J. et al. (2004). IntCal04 Terrestrial radiocarbon age calibration. *Radiocarbon* **46**: 1029-1058.

Table 3.2 Baadsgard, H. (1973). U-TH-Pb dates of zircons from the early Precambrian Amitsoq gneisses, Godthaad district, West Greenland. *Earth Planet. Sci. Lett.* **19**: 22-28.
Pettingill, H.S. and P.J. Patchett. (1981). Lu-Hf total-rock age for the Amitsoq gneisses, west Greenland. *Earth Planet. Sci. Lett.* **55**: 150-156.
Moorbath, S., R.K. O'Nions, R.J. Pankhurst. (1975). The evolution of early Precambrian crustal rocks at Isua, West Greenland—geochemical and isotopic evidence. *Earth Planet. Sci. Lett.* **7**: 229-239.

4.3 Manhes, G., J.F. Minster, and C.J. Allegre. (1978). Comparative uranium-thorium-lead and rubidium strontium study of the Saint Severin amphoterite: consequences for early solar system chronology. *Earth Planet. Sci. Lett.* **39**: 14-24.
Murthy, V.R. and C.C. Patterson. (1962). Primary isochron of zero age for meteorites and the Earth. *Geophys. Res. J.* **67**: 1161.
York, D. and R.M. Farquhar. (1972). *The Earth's age and geochronology.* Pergamon Press, Oxford.

4.5 Papanastassiou, D.A. and G.J. Wasserburg. (1975). Rb-Sr study of a lunar dunite and evidence for early lunar differentiation. *Proceedings of the Sixth Lunar Science Conference.* 1467-1489.

4.19 McElhinny, M.W. (1973). *Paleomagnetism and Plate Tectonics.* Cambridge University Press, p186. (Figure 102). (from Martin J. Aitken 1970a)
Aitken, Martin J., (1970a). Dating by archaeomagnetic and thermoluminescent methods. *Phil. Trans. Roy. Soc. London.* **A269**, 77-88.

Table 5.2   Miller, S.L. (1987). Which Organic Compounds Could Have Occurred on the Prebiotic Earth?. *Cold Spring Harbor Symposia on Quantitative Biology.* **52**: 17-27. Cold Spring Harbor Laboratory Press.

7.16   Watson, D.M.S. (1937). Acanthodian fishes. *Phil. Trans. R. Soc.* (B) **228**: 49-146.

7.22   Bendix-Almgreen, S.E. (1975). The paired fins and shoulder girdle in *Cladoselache*, their morphology and phyletic significance. *Colloques int. Cent. natn. Rech. scient.* **218**: 111-123.
         Zangerl, R. (1981). Chondrichthyes I. Paleozoic elasmobranchs. in H.P.Schultze (ed.), *Handbook of Paleoichthyology, Vol. 3. Elasmobranchi.* Gustav Fisher Verlag, Stuttgart.

7.29   Morgan, J. (1959). The morphology and anatomy of American species of genus *Psaronius*. *Illinois Biological Monograms.* **27**: 1-108.

8.3   Jarvik, Erik. (1980) *Basic Structure and Evolution of Vertebrates Volume 1.* Academic Press Inc., London. (Figure 102. on page 141)

8.16   Kermack, K.A., F. Mussett, and H.W. Rigney. (1981). The skull of *Morganucodon*. *Zool. J. Linn. Soc.* **71**: 1-158.
         Romer, A.S. (1970). The Chanares (Argentina) Triassic reptile fauna. VI. A chiniquodontid cynodont with an incipient squamosal-dentary jaw articulation. *Breviora.* **344**: 1-18.

8.37   Franzen, Jens L., Philip D. Gingerich, Jörg Habersetzer, Jørn H. Hurum, Wighart von Koenigswald, B. Holly Smith (2009). Complete Primate Skeleton from the Middle Eocene of Messel in Germany: Morphology and Paleobiology. PLoS ONE 4(7): e5723. doi:10.1371/journal.pone.0005723.

10.2   Kimura, Motoo. (1983). *The Neutral Theory of Molecular Evolution.* Cambridge University Press, Cambridge. (Figure 4.4 on page 70)

Table 10.3 Dickerson, Richard E., and Irving Geis (1983). *Hemoglobin: Structure, Function, Evolution, and Pathology.* Benjamin/Cummings Publishing Co., Inc., Menlo Park, CA. (Table 3.3)

Table 10.4 Efstratiadis, A., J.W. Posakony, T. Maniatis, R.M. Lawn, C. O'Connell, R. Spritz, J.K. DeReil, B.G. Forget, S.M. Weissman, J.L. Slightom, A.E. Blechl, O. Smithies, F.E. Baralle, C.C. Shoulders, and N.J. Proudfoot (1980). The structure and evolution of the human b-globin gene family. *Cell* **21**: 653-668.

Index

www.ingramcontent.com/pod-product-compliance
Lightning Source LLC
Chambersburg PA
CBHW082348230326
41599CB00058BA/7144